INTEGRATION AND MEASURE

VOLUME TWO

GENERAL INTEGRATION AND MEASURE

CONTENTS

CONTENTS

PREFACE

This volume is considerably more abstract than the first, and is pitched at the level of an elementary graduate course. The construction of the Daniell integral in Chapter 8 and the consequent notions of measurable function and measure follow almost exactly the same pattern as our earlier construction of the Lebesgue integral and measure. If the first volume were given as a second year undergraduate course then the present volume might well be given as a special option for honours students in their third year. In this sense Volume 1 is a concrete introduction to an abstract course in measure theory where the order of generality is roughly that of the classical book of Halmos [11]. The early emphasis on integration gives an economic and highly motivated approach to a subject that is often regarded as rather difficult.†

For ease of reading we have tried to make the present volume as 'self-contained' as possible. In doing this we have occasionally had to repeat arguments that are direct and rather obvious generalisations of passages in the earlier book, but we are prepared to face criticism on this point as the repetition may in fact be a valuable part of the learning process. (Incidentally, we shall now presume an elementary knowledge of topological spaces including the notion of a compact space and the Heine–Borel Theorem which states that the compact subsets of Euclidean space $\mathbf{R}^k$ are the bounded closed subsets of $\mathbf{R}^k$. This material is covered in the Appendix to Volume 1.)

In Chapter 8 the Daniell integral is introduced and the fundamental Monotone and Dominated Convergence Theorems are established. Measurable functions are defined in terms of integrable functions using the idea of truncation (as in §6.1), and the notion of measure follows naturally. To draw out the parallel with the classical case of Lebesgue measure we assume the truth of Stone's Condition (1) on p. 22 which ensures that the constant functions are measurable, and enables us to prove Stone's famous theorem (Theorem 8.5.1) relating measure and integration.

The Lebesgue–Stieltjes integrals of Chapter 9 are precisely the Daniell integrals on spaces $L^1(\mathbf{R}^k)$ which contain all the step functions

† Added in proof: the excellent book by G. E. Shilov and B. L. Gurevich [20] has recently come to our attention. It is very close in spirit to our two volumes but goes much more deeply into the question of differentiation.

(or equivalently, contain all continuous functions on $\mathbf{R}^k$ which vanish outside a compact subset).

A particular feature of the Daniell integral is the ease with which it may be used to prove Riesz's Representation Theorem for the bounded linear functionals on the space $C(X)$ of continuous functions on a compact topological space X. In its various forms this theorem provides the subject matter of Chapter 10. (A subsequent generalisation to locally compact Hausdorff spaces is given in §15.6.)

In Chapter 11 the general notion of measure is introduced. We begin with a non-empty collection $\mathscr{C}$ of subsets of X which satisfies the following *Disjointness Condition*:

if $A, B \in \mathscr{C}$ then $A \cap B$, $A \setminus B$ are finite disjoint unions of sets in $\mathscr{C}$.

We would claim, on the strength of §11.2 and §11.3, that a collection $\mathscr{C}$ which satisfies the Disjointness Condition is a natural domain on which to define an elementary measure. We show how to extend a measure on $\mathscr{C}$ to a complete measure on a σ-algebra: this is done by means of the Daniell integral in §11.4, but the classical construction of Carathéodory [5] is given in §13.5 and shown to be equivalent in all respects to that of Daniell.

The construction of product measures in Chapter 14 follows the same pattern of extension from a measure on elementary product sets $A \times B$ to a complete measure on a σ-algebra. In §14.3 we define a product measure in the non-σ-finite case where the measure theorists' standard convention

$$\infty . 0 = 0 . \infty = 0$$

seems to break down. Our approach may be compared with the treatment in Berberian [3] where a 'smallest possible' product measure is constructed.

The Borel sets of Chapter 15 are essentially topological in nature and it is possible to reconstruct some results on the geometric properties of Lebesgue measure (§6.3, 4) first by proving them for Borel sets, for which they are quite natural, and then by applying a simple completion argument to establish them for Lebesgue measurable sets. In §15.6 we pay a brief visit to the more general realm of locally compact Hausdorff space where the distinction has to be made between Baire and Borel sets and their corresponding measures.

In the last chapter we prove the famous Radon–Nikodym Theorem which is of considerable importance as an abstract theorem and has many applications to the general theory of measure, not least in establishing the duality properties of the spaces $\mathscr{L}^p, \mathscr{L}^q$ $(1/p+1/q = 1)$. But perhaps its greatest interest for us is that it illuminates the whole approach to the transformation of integrals given in §6.4 where the 'measure derivative' $\Delta_{T'}$ (i.e. the absolute value of the determinant of the Jacobian matrix of T') is seen to be the Radon–Nikodym derivative defined (to within a null function) by means of the 'density' property

$$\mu(TI) = \int_I \Delta_{T'} d\mu.$$

Once again we stress the importance of the exercises and have provided solutions to almost all of them.

In a brief Appendix we have listed the definitions of some closely related notions such as semi-rings, rings, algebras, σ-rings, σ-algebras, together with references to some appropriate extension theorems.

Dr John Haigh read both volumes and made meticulous and extremely helpful comments on the manuscript. The illuminating Exx. 11.3.2, 11.4.13 and the construction in Ex. 11.4.19 are due to Dr Dan Newton who has come to the rescue on several occasions. Grateful thanks to both of them and to all the other colleagues who have contributed to the writing of this volume.

As always the Cambridge University Press have been helpful and understanding at every stage in the production of the book.

A. J. W.

8

GENERAL INTEGRATION

The integral which we discuss in this chapter was first presented around 1918 by P. J. Daniell [7]; his contribution was to develop a theory of integration which applies to real valued functions on a *completely arbitrary set* X. Our exposition of the Daniell integral is in the slightly modified form due to F. Riesz [17] and is virtually the same as the treatment of the Lebesgue integral and measure given in Chapters 3, 4, 5, and 6 of Volume 1. We have repeated the argument here, in a slightly briefer style, so that the second volume may be read independently of the first, though it goes without saying that much of the motivation for the abstract Daniell integral will be lost if the reader is not acquainted with the more concrete case of the Lebesgue integral.

In four important papers around 1948 [21] M. H. Stone gave further insight in linking the Daniell integral to the concepts of measurable function and measure. These ideas are the substance of § 8.4 and § 8.5.

8.1 The Daniell integral

Let X be a non-empty, but otherwise arbitrary, set and let L be a linear space of functions $\phi\colon X \to \mathbf{R}$; assume also that L is closed with respect to the operation of taking absolute values, i.e.

$$\phi \in L \Rightarrow |\phi| \in L.$$

In view of the relations

$$\max\{\phi, \psi\} = \tfrac{1}{2}(\phi+\psi)+\tfrac{1}{2}|\phi-\psi|,$$

$$\min\{\phi, \psi\} = \tfrac{1}{2}(\phi+\psi)-\tfrac{1}{2}|\phi-\psi|,$$

L is closed with respect to the operations max, min. The notation

$$\max\{\phi, \psi\} = \phi \vee \psi,$$

$$\min\{\phi, \psi\} = \phi \wedge \psi,$$

is often used and $\vee$, $\wedge$ referred to as the 'lattice operations' on L. A linear space of real valued functions which is closed with respect to these lattice operations will also be closed with respect to absolute values:

$$\phi \in L \Rightarrow |\phi| = \phi \vee 0 - \phi \wedge 0 \in L.$$

Here $\phi \vee 0 = \phi^+$, $\phi \wedge 0 = -\phi^-$ in a familiar notation. We shall therefore say that L is a *linear lattice.* (In the current literature, a linear space is often called a *vector space* but a linear lattice is almost always called a *vector lattice.*) We refer to the members of L as *elementary functions*; they generalise the ideas of step functions (§ 4.1) and simple functions (§ 6.2).

An *elementary integral* $\int$ on the linear lattice L is a positive linear operator $\int : L \to \mathbf{R}$ which satisfies the following condition

$$\phi_n \downarrow 0 \Rightarrow \int \phi_n \to 0. \tag{1}$$

We shall refer to (1) as the *Daniell condition*; it plays a fundamental role in all that follows.

When our elementary integral is related to some other integral it is often convenient to use another symbol, such as P (for positive), in place of $\int$. See Example 5 and Theorem 8.5.1.

To say that $\int : L \to \mathbf{R}$ is a *linear operator* means that

$$\int (c_1\phi_1 + c_2\phi_2) = c_1 \int \phi_1 + c_2 \int \phi_2 \tag{2}$$

for any c_1, c_2 in $\mathbf{R}$, ϕ_1, ϕ_2 in L. In order to emphasise the fact that the values of $\int$ lie in the field of real numbers (and not in a general linear space over $\mathbf{R}$) we shall follow the majority of texts and refer to $\int$ as a positive linear *functional* on L. To say that $\int$ is *positive*† means that

$$\phi \geqslant 0 \Rightarrow \int \phi \geqslant 0 \quad (\phi \in L). \tag{3}$$

As usual, (2) and (3) together give

$$\phi \geqslant \psi \Rightarrow \int \phi \geqslant \int \psi \quad (\phi, \psi \in L). \tag{4}$$

The Daniell condition (1) takes the place of Lemma 3.2.1 although we have weakened its statement by omitting the phrase 'almost everywhere' which is as yet undefined in our present abstract setting. As there are no subsets of X which obviously fulfil the role of intervals, we shall use the analogue of Theorem 3.2.1 and Converse to define a null set in X: a subset A of X is said to be *null* if there exists an increasing sequence $\{\psi_n\}$ of elementary functions for which $\int \psi_n$ is bounded and $\psi_n(x)$ diverges for every x in A. By subtracting ψ_1, if necessary, we may arrange that all these ψ_n are positive.

† As in Volume 1, we say that a real number x is *positive* if and only if $x \geqslant 0$.

Note that two of our hardest results in § 3.2 are now *assumptions* in the abstract setting: for this reason abstract integration theory often appears easier and more elegant, but of course the reckoning comes when the abstract theory has to be applied to a concrete situation.

Our first and motivating example of an elementary integral is the Lebesgue integral on the linear lattice of step functions. But the following examples already show something of the wider scope of the abstract definition.

Example 1. Let X be the set $\{1, 2, 3, \ldots\}$ of natural numbers and L the set of functions $\phi: X \to \mathbf{R}$ each of which vanishes outside a finite set (depending on ϕ). In other words, ϕ is a sequence whose terms, from a certain point onwards, are all zero. It is easy to check that L is a linear lattice and that

$$\int \phi = \phi(1) + \phi(2) + \ldots$$

(which is effectively a finite sum) defines an elementary integral on L (Ex. 1). The only null set is the empty set: for if $\{\psi_n\}$ is an increasing sequence of positive elementary functions, then $\int \psi_n \geqslant \psi_n(x)$ for any x in X.

Example 2. Let X be an uncountable set (e.g. the interval $[0, 1]$), let L be the set of functions $\phi: X \to \mathbf{R}$ each of which vanishes outside a finite subset of X (depending on ϕ), and let $\int \phi = 0$ for all ϕ in L. The Daniell condition is quite obvious in this case. The null sets are just the *countable* subsets of X. Suppose that $\psi_n \uparrow \infty$ on A; the sets $S_n = \{x \in X: \psi_n(x) \neq 0\}$ are finite, $\bigcup S_n$ is countable (Proposition 2.1.1) and so A is countable. On the other hand, if $A = \{x_1, x_2, \ldots\}$, define $\psi_n(x) = n$ for $x = x_1, x_2, \ldots, x_n$, $\psi_n(x) = 0$ otherwise. Then $\psi_n \uparrow \infty$ on A and $\int \psi_n = 0$.

Example 3. Let L consist of all continuous functions $\phi: \mathbf{R}^k \to \mathbf{R}$ and define $\int \phi = \phi(0)$. Again the Daniell condition is obvious as

$$\phi_n \downarrow 0 \Rightarrow \phi_n(0) \downarrow 0.$$

A set A is null if and only if A does not contain the origin 0 (Ex. 3). This elementary integral is often called the *Dirac δ-function* at 0.

Example 4. Let L be the space of integrable simple functions defined in § 6.2 and $\int$ the Lebesgue integral (on $\mathbf{R}^k$). Then the null sets are exactly the sets of Lebesgue measure zero. This follows from the Monotone Convergence Theorem.

Example 5. Let L be the space $C[a,b]$ of all real valued continuous functions ϕ on the compact interval $[a,b]$ and P *any* positive linear functional on $C[a,b]$. The Daniell condition is automatically satisfied (by Dini's Theorem 10.1.1). The null sets depend on P: for example, if

$$P(\phi) = \int_a^b \phi(x)\,dx,$$

the 'integral from a to b' in the sense of (Riemann or) Lebesgue, then, as we shall see in § 8.3 the null sets are exactly the sets of Lebesgue measure zero; or, if $[a,b] = [-1,1]$ and $P(\phi) = \phi(0)$, as in Example 3, the null sets are all the subsets which do not contain 0.

Example 6. Let L consist of the step functions ϕ on the interval $[0,2]$ and define

$$\int \phi = \phi(1-0)$$

(where $\phi(1-0)$ is the limit of $\phi(1-t)$ as t tends to 0 through positive values). It may be verified that $\int$ is a positive linear functional on L but is *not* an elementary integral, as the sequence $\{\phi_n\}$ defined by

$$\phi_n = \chi_{(1-\frac{1}{n},1)}$$

decreases to the limit zero and $\int \phi_n = 1$ for all n.

In view of our new definition of null set it is necessary to give a new proof of the following fact (cf. Proposition 2.2.3).

Proposition 1. *The union of a sequence of null sets in X is again a null set.*

Proof. Let $A = \bigcup A_n$ where each A_n is null. For each $m \geqslant 1$ there is an increasing sequence $\{\psi_{mn}\}$ of positive elementary functions which diverges on A_m and such that $\int \psi_{mn} \leqslant 2^{-m}$. (To achieve this inequality, multiply ψ_{mn} by a suitably small constant, if necessary.) Let

$$\psi_n = \psi_{1n} + \psi_{2n} + \dots + \psi_{nn};$$

then the increasing sequence $\{\psi_n\}$ diverges on A and $\int \psi_n < 1$. Thus A is null.

Note that single points may not be null in the general theory (cf. Examples 1 and 3) and so we must certainly not assume that countable sets are null!

A property $\mathfrak{P}$ which holds for all points of X outside some null set is said to hold *almost everywhere* (*a.e.*) in X; or the property $\mathfrak{P}(x)$ holds for *almost all* x (*a.a.* x) in X. Of course, in a case like Example 1 above, where the only null set is the empty set, 'almost everywhere' and 'everywhere' are synonymous.

Suppose that $\{\phi_n\}$ is an increasing sequence of elementary functions for which $\int \phi_n$ is bounded and let S be the set of all points x for which $\{\phi_n(x)\}$ diverges. According to the definition on p. 2, S is *null*. Define the function $f_0 : X \to \mathbf{R}$ as follows.

$$\begin{aligned} f_0(x) &= \lim \phi_n(x) \quad \text{if} \quad x \notin S, \\ &= 0 \qquad\qquad\quad \text{if} \quad x \in S. \end{aligned}$$

Then $\{\phi_n\}$ converges to f_0 almost everywhere. Note that f_0 takes values in $\mathbf{R}$ and that the 'value' ∞ is not permitted at this stage.

Quite generally, suppose that a function $f: X \to \mathbf{R}$ is the limit almost everywhere of an increasing sequence $\{\phi_n\}$ of elementary functions whose integrals are bounded. We denote the set of all such f by L^{inc} and define

$$\int f = \lim \int \phi_n.$$

In general, any function f in L^{inc} may be expressed in infinitely many ways as the limit almost everywhere of an increasing sequence of elementary functions, and so the consistency of this definition must be checked. It depends on two lemmas the first of which is a slightly stronger version of the Daniell condition (cf. Lemma 3.2.1).

Lemma 1. *Let $\{\phi_n\}$ be a decreasing sequence of positive elementary functions converging almost everywhere to* 0. *Then*

$$\int \phi_n \to 0.$$

Proof. Suppose that $\phi_n \downarrow 0$ outside the null set A. Given any $\epsilon > 0$, there is an increasing sequence $\{\psi_n\}$ of positive elementary functions diverging on A such that $\int \psi_n \leqslant \epsilon$. (Multiply ψ_n by a suitably small constant, if necessary.) Now $(\phi_n - \psi_n) \downarrow -\infty$ on A so that $(\phi_n - \psi_n)^+ \downarrow 0$ *everywhere*. Thus

$$\int (\phi_n - \psi_n)^+ \to 0$$

by the Daniell condition. We may therefore find N so that

$$\int (\phi_n - \psi_n)^+ < \epsilon$$

for $n \geqslant N$. Thus $$\int \phi_n - \int \psi_n = \int (\phi_n - \psi_n) < \epsilon$$

for $n \geqslant N$ and so $$0 \leqslant \int \phi_n < 2\epsilon$$

for $n \geqslant N$.

Lemma 2. *Let $\{\phi_n\}$, $\{\psi_n\}$ be increasing sequences of elementary functions whose integrals are bounded and let $\{\phi_n\}$, $\{\psi_n\}$ converge almost everywhere to f, g, respectively. If $f \geqslant g$ almost everywhere, then*

$$\lim \int \phi_n \geqslant \lim \int \psi_n.$$

Proof. For any $m, n \geqslant 1$,

$$\int \psi_m - \int \phi_n = \int (\psi_m - \phi_n) \leqslant \int (\psi_m - \phi_n)^+.$$

For any fixed $m \geqslant 1$, the sequence $(\psi_m - \phi_1)^+, (\psi_m - \phi_2)^+, (\psi_m - \phi_3)^+, \ldots$ satisfies the conditions of Lemma 1, and so

$$\int (\psi_m - \phi_n)^+ \to 0 \quad \text{as} \quad n \to \infty.$$

Hence $$\int \psi_m \leqslant \lim \int \phi_n.$$

As this is true for all $m \geqslant 1$ the result follows.

To prove the consistency of our definition of $\int f$ let $\{\phi_n\}$, $\{\psi_n\}$ be two increasing sequences of elementary functions whose integrals are bounded and suppose that both sequences converge to f almost everywhere. We may take $g = f$ in Lemma 2 and deduce

$$\lim \int \phi_n \geqslant \lim \int \psi_n$$

and also $$\lim \int \phi_n \leqslant \lim \int \psi_n$$

from which $$\lim \int \phi_n = \lim \int \psi_n.$$

Thus the value of $\int f$ does not depend on the particular sequence of elementary functions chosen.

If $f_1, f_2 \in L^{\text{inc}}$ and if c_1, c_2 are *positive* constants, then $c_1 f_1 + c_2 f_2$ clearly belongs to L^{inc} and the most elementary theorem on limits

(Proposition 1.1.1) applied to equation (2) gives

$$\int (c_1 f_1 + c_2 f_2) = c_1 \int f_1 + c_2 \int f_2 \tag{5}$$

($f_1, f_2 \in L^{\text{inc}}$, $c_1, c_2 \geqslant 0$).

The last step in the construction of the Daniell integral is to denote by L^1 the set of all $f = g - h$, where g, $h \in L^{\text{inc}}$ and to define

$$\int f = \int g - \int h.$$

In this case the consistency of the definition is much more easily verified. Suppose that $f = g_1 - h_1 = g_2 - h_2$, where $g_1, g_2, h_1, h_2 \in L^{\text{inc}}$. As $g_1 + h_2 = g_2 + h_1$ it follows from (5), with $c_1 = c_2 = 1$, that

$$\int g_1 + \int h_2 = \int g_2 + \int h_1,$$

whence

$$\int g_1 - \int h_1 = \int g_2 - \int h_2,$$

as required for consistency.

It now follows that L^1 is a linear space over $\mathbf{R}$ and the extended integral is a positive linear functional on L^1; we call this functional the *Daniell integral* on L^1 (extending the given elementary integral on L) and refer to L^1 as the space of *integrable functions*.

Observe that all functions in L^{inc} and L^1 take values in $\mathbf{R}$. We have not allowed the values $\pm\infty$ in this context; if we were to do so then there would be difficulty in defining sums and differences of functions, and L^1 would not be a linear space. (See Exx. A–G below and also § 17.1 where the space $\mathscr{L}^1$ of equivalence classes is defined.)

Theorem 1. *The space L^1 is a linear lattice and the Daniell integral is a positive linear functional on L^1.*

Proof. (i) Let $f_1 = g_1 - h_1$, $f_2 = g_2 - h_2$ where g_1, g_2, h_1, $h_2 \in L^{\text{inc}}$ and consider

$$c_1 f_1 + c_2 f_2 = c_1 g_1 - c_1 h_1 + c_2 g_2 - c_2 h_2,$$

where $c_1, c_2 \in \mathbf{R}$. First of all, if $c_1, c_2 \geqslant 0$,

$$c_1 f_1 + c_2 f_2 = (c_1 g_1 + c_2 g_2) - (c_1 h_1 + c_2 h_2),$$

where the terms in parentheses belong to L^{inc} so that $c_1 f_1 + c_2 f_2 \in L^1$ and

$$\begin{aligned}\int (c_1 f_1 + c_2 f_2) &= \int (c_1 g_1 + c_2 g_2) - \int (c_1 h_1 + c_2 h_2) \\ &= c_1 \int g_1 + c_2 \int g_2 - c_1 \int h_1 - c_2 \int h_2 \quad \text{(by (5))} \\ &= c_1 \int f_1 + c_2 \int f_2.\end{aligned}$$

If $c_1 \leqslant 0$ then we may write $c_1 f_1 = (-c_1)(-f_1)$ where $-c_1 \geqslant 0$, $-f_1 = h_1 - g_1$, and note that

$$c_1 \int f_1 = (-c_1) \int (-f_1).$$

This observation, and a similar one for c_2, allow us to prove that $c_1 f_1 + c_1 f_2 \in L^1$ and

$$\int (c_1 f_1 + c_2 f_2) = c_1 \int f_1 + c_2 \int f_2$$

for *any* c_1, c_2 in $\mathbf{R}$.

(ii) Let $f = g - h$, where $g, h \in L^{\text{inc}}$. If $f \geqslant 0$ then $g \geqslant h$ and Lemma 2 gives

$$\int g \geqslant \int h.$$

In view of (i) this is equivalent to

$$\int f \geqslant 0.$$

(iii) Let $f = g - h$ where g, h are the limits almost everywhere of increasing sequences $\{\phi_n\}$, $\{\psi_n\}$ of elementary functions whose integrals are bounded. By subtracting $\min\{\phi_1, \psi_1\}$ from ϕ_n, ψ_n, g, h we may suppose that ϕ_n, ψ_n are positive. It follows at once from the definitions that

$$|f| = \max\{g, h\} - \min\{g, h\}$$

and the increasing sequences $\{\max\{\phi_n, \psi_n\}\}$, $\{\min\{\phi_n, \psi_n\}\}$ converge almost everywhere to the functions $\max\{g, h\}$, $\min\{g, h\}$ (see Ex. 1.1.2). Moreover, as ϕ_n, ψ_n are positive,

$$\int \min\{\phi_n, \psi_n\} \leqslant \int \max\{\phi_n, \psi_n\} \leqslant \int \phi_n + \int \psi_n$$

are all bounded above and so $\max\{g, h\}, \min\{g, h\} \in L^{\text{inc}}$ and $|f| \in L^1$.

According to the above definition, if $f_1 \in L^{\text{inc}}$ and $f_2 = f_1$ almost everywhere, then $f_2 \in L^{\text{inc}}$ and

$$\int f_2 = \int f_1.$$

This extends to L^1.

Theorem 2. *If $f_1 \in L^1$ and $f_2 = f_1$ almost everywhere, then $f_2 \in L^1$ and*

$$\int f_2 = \int f_1.$$

Proof. Let $f_1 = g_1 - h_1$, where $g_1, h_1 \in L^{\text{inc}}$ and define

$$g_2 = g_1, \quad h_2 = h_1 + (f_1 - f_2)$$

so that $g_2, h_2 \in L^{\text{inc}}$ (the latter since $h_2 = h_1$ almost everywhere). But now $f_2 = g_2 - h_2$ is in L^1 and

$$\int f_2 = \int g_2 - \int h_2 = \int g_1 - \int h_1 = \int f_1.$$

A function $f: X \to \mathbf{R}$ which equals zero almost everywhere is called a *null function*. In rough terms, Theorem 2 says that null functions are 'negligible' in the general theory of integration.

Exercises

1. In Example 1,

(i) check the details mentioned in the text;

(ii) describe the elements of L^{inc};

(iii) show that L^1 is the set of sequences f for which $\Sigma f(x)$ is absolutely convergent, and

$$\int f = \Sigma f(x).$$

2. In Example 2 show that $L^{\text{inc}} = L^1$ consists of all functions which vanish outside a countable subset of X, and that all integrals are zero.

3. In Example 3,

(i) verify that the null sets A are those which do not contain the origin;

(ii) show that $L^{\text{inc}} = L^1$ consists of all functions $f: \mathbf{R}^k \to \mathbf{R}$, and

$$\int f = f(0).$$

4. In Example 4,

(i) identify the null sets by means of the Monotone Convergence Theorem;

(ii) show that L^{inc} consists of the Lebesgue integrable functions that are essentially bounded below;

(iii) show that $L^1 = L^1(\mathbf{R}^k)$.

(A function f is *essentially bounded below* if there is a null set A for which $f\chi_{A^c}$ is bounded below.)

5. Show that the Daniell condition is equivalent to each of the following.

(i) If $\{\phi_n\}$ is an increasing sequence of functions in L and if ϕ is a function in L such that $\phi \leqslant \lim \phi_n$, then

$$\int \phi \leqslant \lim \int \phi_n.$$

(ii) If $\{\psi_n\}$ is a sequence of positive functions in L and if ϕ is a function in L such that $\phi \leqslant \Sigma\psi_n$, then

$$\int \phi \leqslant \Sigma \int \psi_n.$$

(These limits and sums may take infinite values, in which case we agree that the corresponding inequalities are satisfied.)

6. (i) Let L denote the linear lattice of step functions on $\mathbf{R}$. For any ϕ in L let

$$P(\phi) = \int x^2 \phi(x)\, dx.$$

Show that P is an elementary integral on L.

(ii) Extend P to a Daniell integral P on $L^1(P)$. Show that $f \in L^1(P)$ if and only if the function $g\colon \mathbf{R} \to \mathbf{R}$ defined by

$$g(x^3) = f(x)$$

for all x in $\mathbf{R}$, is Lebesgue integrable, and that in this case

$$3P(f) = \int g.$$

(Cf. the rule for integration by substitution given in Proposition 5.1.3.)

7. If f is integrable show that there is a sequence $\{\phi_n\}$ of elementary functions which converges almost everywhere to f and such that

$$\int |\phi_n - f| \to 0.$$

If f is positive show that the ϕ_n's may be chosen to be positive.

Upper and Lower Daniell Integrals

For brevity denote $\mathbf{R} \cup \{\infty, -\infty\}$ by $\bar{\mathbf{R}}$. Daniell's original definition allowed for the integration of functions f on X whose values are in $\bar{\mathbf{R}}$. In the context of these exercises we shall introduce a convention for linear combinations of such functions, viz. we shall write $h = af + bg$ $(a, b \in \mathbf{R})$ for *any* function $h\colon X \to \bar{\mathbf{R}}$ which satisfies $h(x) = af(x) + bg(x)$ whenever $f(x),\ g(x) \in \mathbf{R}$. Roughly, this means that we ignore any infinite values that arise.

Daniell begins with a linear lattice L of real valued functions and a positive linear functional P on L satisfying the condition

$$\phi_n \downarrow 0 \;\Rightarrow\; P(\phi_n) \to 0 \quad (\phi_n \in L).$$

A. For any increasing sequence $\{\phi_n\}$ of functions in L, $\phi_n \to u$ where $u\colon X \to \mathbf{R} \cup \{\infty\}$. Denote the set of all such functions u by L^+ and show that P extends to L^+ by the definition

$$P(u) = \lim P(\phi_n)$$

where this limit may be infinite. (Adapt Lemma 2 to check the consistency.)

B. For any $f: X \to \bar{\mathbf{R}}$ let $S = \{P(u): u \in L^+, u \geqslant f\}$ and define

$$P^*(f) = \inf S,$$

where this infimum is $-\infty$ if S is unbounded below, and ∞ if $S \cap \mathbf{R}$ is empty. Also define

$$P_*(f) = -P^*(-f)$$

for any $f: X \to \bar{\mathbf{R}}$. P^* and P_* are the *upper* and *lower* Daniell integrals corresponding to the given elementary integral P on L.

Denote by L_1 the set of all $f: X \to \bar{\mathbf{R}}$ for which $P^*(f)$, $P_*(f)$ are finite and equal. If $f \in L_1$ write

$$Q(f) = P^*(f) = P_*(f).$$

Show that L_1 is a linear space over $\mathbf{R}$, i.e. if $f, g \in L_1$ and $h = af+bg$ (in the sense of the above convention) then $h \in L_1$. Show also that Q is a positive linear functional on L_1. (Note that there is no question of Q taking infinite values.)

C. Let $h: X \to \bar{\mathbf{R}}$ be a null function, i.e. $h(x) = 0$ for all x outside a null set (as defined on p. 2). Show that $h \in L_1$ and $Q(h) = 0$.

D. Extend the definitions of L^{inc} and L^1 to allow functions $f: X \to \bar{\mathbf{R}}$. Show that any function in L^{inc} is equal almost everywhere to a function of $L^+ \cap L_1$. Hence show that $L^{\text{inc}} \subset L_1$ and $L^1 \subset L_1$.

E. Let $f \in L_1$. Find u_n, $-l_n$ in L^{inc} such that

$$l_n \leqslant f \leqslant u_n \quad \text{and} \quad P(u_n) - P(l_n) < 1/n.$$

Arrange so that $\{u_n\}$ is decreasing and $\{l_n\}$ is increasing. Appeal to the Monotone Convergence Theorem of the next section to prove that $f \in L^1$.

F. Combining D and E we see that $L_1 = L^1$. Show that Q coincides with the Daniell integral on L^1 as previously constructed (but allowing for the extension to functions $f: X \to \bar{\mathbf{R}}$).

G. In this question interpret addition of infinite values according to the natural rules $\infty + \infty = \infty$, $-\infty + (-\infty) = -\infty$ but leave $\infty + (-\infty)$ and $-\infty + \infty$ undefined.

Let A and B be elementary integrals on L. Show that $C = A + B$ is an elementary integral on L and

$$C^*(f) = A^*(f) + B^*(f),$$
$$C_*(f) = A_*(f) + B_*(f)$$

for any $f: X \to \bar{R}$, provided the right hand sides are defined.

Deduce that

$$L_1(C) = L_1(A) \cap L_1(B).$$

As a postscript, let us note that to any function f in L_1 there is a function u in L^+ with $u \geqslant f$ and $P(u)$ finite. By our definition of null set on p. 2 this implies that u, and hence f, can only take the value ∞ on a null set. In the same way, f can only take the value $-\infty$ on a null set. Our convention about linear combinations $af+bg$ was really framed with this in mind.

8.2 The convergence theorems

As L^{inc} was defined in terms of increasing sequences of elementary functions, it is natural to consider the same process applied to increasing sequences of functions in L^{inc}. We shall see that this process gives nothing new.

Theorem 1. *Let $\{f_n\}$ be an increasing sequence of functions in L^{inc} whose integrals are bounded. Then $\{f_n\}$ converges almost everywhere to a function f, where f lies in L^{inc} and*

$$\int f = \lim \int f_n.$$

Proof. For each $m \geqslant 1$ let $\{\phi_{mn}\}$ be an increasing sequence of elementary functions which converges to f_m outside a null set S_m, so that

$$\int f_m = \lim_{n\to\infty} \int \phi_{mn}.$$

Let
$$\phi_n = \max\{\phi_{ij}: 1 \leqslant i,j \leqslant n\};$$

the sequence $\{\phi_n\}$ of elementary functions so defined is clearly increasing. For each $n \geqslant 1$,

$$\phi_n \leqslant f_n$$

outside the null set $S_1 \cup S_2 \cup \ldots \cup S_n$, and so the integrals $\int\phi_n$ are bounded. Thus $\{\phi_n\}$ converges outside a null set S to a function f of L^{inc} and

$$\int \phi_n \to \int f.$$

By our construction,
$$\phi_{mn} \leqslant \phi_n$$

for $m \leqslant n$; let n tend to infinity and deduce that

$$f_m \leqslant f$$

outside $S_m \cup S$. But now
$$\phi_n \leqslant f_n \leqslant f$$

and
$$\phi_n \to f$$

outside the null set $S \cup S_1 \cup S_2 \cup \ldots$ (Proposition 8.1.1). Thus $\{f_n\}$ is 'squeezed' and converges to f almost everywhere.

In the same way

$$\int \phi_n \leqslant \int f_n \leqslant \int f$$

and
$$\int \phi_n \to \int f$$

so that
$$\int f_n \to \int f.$$
This completes the proof.

This result may be extended to L^1 by means of the following simple lemma.

Lemma 1. *Let $f \in L^1$. Given $\epsilon > 0$; there exist g, h in L^{inc} such that $f = g - h$, where h is positive and*
$$\int h < \epsilon.$$

Proof. Let $f = g_1 - h_1$, where $g_1, h_1 \in L^{\text{inc}}$. Then, by the definition of L^{inc}, there is an elementary function ψ such that $h_1 - \psi$ is positive almost everywhere and
$$\int (h_1 - \psi) < \epsilon.$$
To satisfy the conditions of the lemma, set $h = (h_1 - \psi)^+$ and $g = f + h$.

Theorem 2 (The Monotone Convergence Theorem). *Let $\{f_n\}$ be a monotone sequence of functions in L^1 whose integrals are bounded. Then $\{f_n\}$ converges almost everywhere to a function f, where f lies in L^1, and*
$$\int f = \lim \int f_n.$$

Proof. Without loss of generality we may assume that $\{f_n\}$ is an *increasing* sequence of *positive* functions in L^1. Let
$$a_1 = f_1, \quad a_n = f_n - f_{n-1}$$
for $n \geqslant 2$ so that
$$f_n = a_1 + a_2 + \ldots + a_n$$
and the terms a_n are all positive. Apply Lemma 1, with $\epsilon = 2^{-n}$, to find positive functions b_n, c_n in L^{inc} such that $a_n = b_n - c_n$ and
$$0 \leqslant \int c_n < 2^{-n}.$$
Let
$$g_n = b_1 + b_2 + \ldots + b_n,$$
$$h_n = c_1 + c_2 + \ldots + c_n,$$
so that
$$f_n = g_n - h_n,$$
where $g_n, h_n \in L^{\text{inc}}$ and the sequences $\{g_n\}, \{h_n\}$ are increasing. Moreover,
$$\int h_n < 2^{-1} + 2^{-2} + \ldots + 2^{-n} < 1$$
and
$$\int g_n = \int f_n + \int h_n$$

are bounded. By Theorem 1, $\{g_n\}$, $\{h_n\}$ converge almost everywhere to functions g, h of L^{inc} and hence $\{f_n\}$ converges almost everywhere to $f = g - h$ which is an element of L^1. Finally,

$$\int g_n \to \int g, \quad \int h_n \to \int h$$

and so

$$\int f_n \to \int f.$$

Corollary. *If $f \in L^1$ is positive and $\int f = 0$, then $f = 0$ almost everywhere.*

Proof. Apply the Monotone Convergence Theorem to the increasing sequence $\{nf\}$.

In the context of the Daniell construction, the Monotone Convergence Theorem is of profound importance. It shows first of all that the Daniell integral on L^1 itself satisfies the Daniell condition:

$$f_n \downarrow 0 \Rightarrow \int f_n \to 0 \quad (f_n \in L^1).$$

In other words, the Daniell integral is an elementary integral on the linear lattice L^1. As such it is subject to the same extension procedure, but the Monotone Convergence Theorem shows that this procedure gives us nothing new. In this sense the Daniell integral is 'complete'. We must be careful not to misinterpret this property. For example, suppose that L consists of these functions $\phi: [0, 1] \to \mathbf{R}$ which take a constant value c outside a finite subset of $[0, 1]$ and define $\int \phi = c$. It is easy to verify that L is a linear lattice and that the elementary integral so defined coincides with the Lebesgue integral on L. This elementary integral extends to a Daniell integral P on a space $L^1(P)$ which is complete in the sense just described. But $L^1(P)$ is much smaller than the space $L^1[0, 1]$ of Lebesgue integrable functions on $[0, 1]$! In fact, $L^1(P)$ consists of functions $f: [0, 1] \to \mathbf{R}$ which assume a constant value c outside a countable subset, and $\int f = c$ (Ex. 1). Thus the fact that $L^1(P)$ is complete does not rule out the possibility that the integral on $L^1(P)$ may be extended in some way – other than the Daniell construction – to a larger space.

There is another formulation of the Monotone Convergence Theorem in terms of absolutely convergent series.

Theorem 3 (The Absolute Convergence Theorem). *Let Σa_n be a series of functions in L^1 for which the series $\Sigma \int |a_n|$ is convergent. Then*

Σa_n is absolutely convergent pointwise almost everywhere to a function f in L^1 and

$$\int f = \Sigma \int a_n.$$

Proof. Apply the Monotone Convergence Theorem to the series Σa_n^+, Σa_n^- (which are dominated by $\Sigma |a_n|$) and subtract.

For any f in L^1, write

$$\|f\| = \int |f|.$$

The condition in Theorem 3 is that $\Sigma \|a_n\|$ should be convergent: if this is satisfied we say that Σa_n is *absolutely convergent* in L^1. Thus Theorem 3 relates absolute convergence in this sense and pointwise absolute convergence almost everywhere. This may be compared with Theorem 7.2.1 which was used to prove the completeness of the normed linear space L^1 in the case of the Lebesgue integral; we shall return to this question in the last chapter when the general spaces L^p and $\mathscr{L}^p$ are defined. The Daniell construction and the whole theory of the Daniell integral may be worked out in terms of absolute convergence. This approach is beautifully expounded in the book by Asplund and Bungart [1].

There are two other famous results which follow from the Monotone Convergence Theorem.

***Theorem 4* (*The Dominated Convergence Theorem*).** *Suppose that $\{f_n\}$ is a sequence of integrable functions which converges almost everywhere to a function f and that there is a positive integrable function g satisfying*

$$|f_n| \leqslant g$$

for all n. Then f is integrable and

$$\int f_n \to \int f.$$

Proof. If $\{f_n(x)\}$ is convergent (and therefore bounded), let

$$l_n(x) = \inf\{f_n(x), f_{n+1}(x), \ldots\},$$
$$u_n(x) = \sup\{f_n(x), f_{n+1}(x), \ldots\},$$

and otherwise let $l_n(x) = u_n(x) = 0$. The *monotone* sequences $\{l_n\}$, $\{u_n\}$ so defined converge almost everywhere to f and satisfy

$$l_n \leqslant f_n \leqslant u_n$$

almost everywhere. Let us assume for the moment that l_n and u_n are integrable. As $\{l_n\}$ is increasing and $\{u_n\}$ is decreasing, the Monotone Convergence Theorem shows that f is integrable and

$$\int l_n \to \int f, \quad \int u_n \to \int f.$$

But
$$\int l_n \leqslant \int f_n \leqslant \int u_n,$$

and the familiar squeezing argument gives

$$\int f_n \to \int f.$$

To show that l_n and u_n are integrable note first of all that

$$l_{nk} = \min\{f_n, f_{n+1}, \ldots, f_{n+k}\},$$
$$u_{nk} = \max\{f_n, f_{n+1}, \ldots, f_{n+k}\}$$

are integrable. For each $n \geqslant 1$, as k varies, we obtain two monotone sequences whose limits almost everywhere are l_n, u_n and the domination by the integrable function g ensures that

$$\left|\int l_{nk}\right| \leqslant \int g, \quad \left|\int u_{nk}\right| \leqslant \int g.$$

It is now clear by the Monotone Convergence Theorem that l_n and u_n are integrable, and this completes the proof.

***Theorem 5* (*Fatou's Lemma*).** *Suppose that* $\{f_n\}$ *is a sequence of positive integrable functions which converges almost everywhere to a function* f *and that the integrals* $\int f_n$ *are bounded. Then* f *is integrable and*

$$\int f \leqslant \liminf \int f_n.$$

Proof. Let
$$l_n = \inf\{f_n, f_{n+1}, \ldots\},$$
$$l_{nk} = \min\{f_n, f_{n+1}, \ldots, f_{n+k}\}.$$

For each $n \geqslant 1$ the decreasing sequence $l_{n1}, l_{n2}, \ldots$ of positive integrable functions converges to l_n and so, by the Monotone Convergence Theorem, l_n is integrable. Moreover

$$\int l_n \leqslant \int f_{n+k}$$

for all $k \geqslant 0$ and so
$$\int l_n \leqslant \inf\left\{\int f_n, \int f_{n+1}, \ldots\right\}.$$

As the increasing sequence $\{l_n\}$ converges almost everywhere to f and the integrals $\int f_n$ are bounded, the Monotone Convergence Theorem finally shows that $f \in L^1$ and

$$\int f \leqslant \liminf \int f_n.$$

Exercises

1. Let L consist of functions $\phi: [0,1] \to \mathbf{R}$ which take a constant value c outside a finite subset of $[0,1]$ (the finite subset and the constant c both depending on ϕ) and let $P(\phi) = c$. Show that P is an elementary integral on L and identify the corresponding spaces $L^{\text{inc}}(P)$ and $L^1(P)$.

2. Let $\{a_n\}$ be a sequence of functions in L^1 for which the series $\Sigma \int |a_n|$ is convergent. Show that $\{a_n\}$ converges to zero almost everywhere.

3. Notation: let A be a subset of X; if $f\chi_A \in L^1$ then we write

$$\int f\chi_A = \int_A f;$$

the set of all such functions f is denoted by $L^1(A)$.

Let $\{S_n\}$ be an increasing sequence of sets whose union is X. If $f \in L^1(S_n)$ for $n \geqslant 1$ and if the integrals

$$\int_{S_n} |f|$$

are bounded above, show that $f \in L^1(X)$ and

$$\int_X f = \lim \int_{S_n} f.$$

(Cf. Proposition 5.1.1.)

4. Prove ***The Bounded Convergence Theorem:*** *Let $\{f_n\}$ be a sequence of integrable functions which converges almost everywhere to a function f. If the constant function* 1 *is integrable and if there is a real number K such that*

$$|f_n(x)| \leqslant K$$

for all n and all x, then f is integrable and

$$\int f_n \to \int f.$$

5. Prove the following generalisation of ***Fatou's Lemma.*** *Suppose that $\{f_n\}$ is a sequence of positive integrable functions and that the integrals $\int f_n$ are bounded. Then* $\liminf f_n$ *is integrable and*

$$\int \liminf f_n \leqslant \liminf \int f_n.$$

6. Let E be a subset of $[0,1]$ for which χ_E is Lebesgue integrable and consider the sequence $\{f_n\}$ defined by

$$f_n = \chi_E \quad \text{if } n \text{ is odd,}$$
$$= 1-\chi_E \quad \text{if } n \text{ is even.}$$

What is the relevance of this sequence to Ex. 5 on Fatou's Lemma?

8.3 The Comparison Theorem

In Example 8.1.5 we considered the elementary integral P on $C[a,b]$ defined by

$$P(\phi) = \int_a^b \phi(x)\,dx,$$

where the right hand side is interpreted as the (Riemann or) Lebesgue 'integral from a to b' of the continuous function ϕ. Now P extends by the standard construction to a Daniell integral, and it is natural to compare this Daniell integral with the Lebesgue integral on $[a,b]$, derived as it was from a similar elementary integral on the space of step functions. The following fundamental theorem allows such a comparison to be made.

Theorem 1. *Let P, Q be Daniell integrals on $L^1(P)$, $L^1(Q)$ obtained by the standard construction from elementary integrals on linear lattices $L(P)$, $L(Q)$, respectively.*

(i) *If $L(P) \subset L^1(Q)$ and*

$$P(\phi) = Q(\phi)$$

for any ϕ in $L(P)$, then $L^1(P) \subset L^1(Q)$ and

$$P(f) = Q(f)$$

for any f in $L^1(P)$.

(ii) *Suppose that* (i) *holds. If, also, $L(Q) \subset L^1(P)$ and*

$$P(\psi) = Q(\psi)$$

for any ψ in $L(Q)$, then $L^1(P) = L^1(Q)$ and

$$P(f) = Q(f)$$

for any f in $L^1(P) = L^1(Q)$.

Proof. (i) Since P and Q agree on $L(P)$, any P-null set A is automatically a Q-null set. For, if $\{\phi_n\}$ is an increasing sequence of elements of $L(P)$ which diverges on A and for which $P(\phi_n)$ is bounded, then

$Q(\phi_n) = P(\phi_n)$ is bounded and the Monotone Convergence Theorem (applied to Q) shows that A is Q-null. It follows at once by the Monotone Convergence Theorem that

$$L^{\text{inc}}(P) \subset L^1(Q)$$

and P, Q agree on $L^{\text{inc}}(P)$. Finally, by taking differences,

$$L^1(P) \subset L^1(Q)$$

and P, Q agree on $L^1(P)$.

(ii) This is obvious by interchanging the roles of P and Q.

Returning to the example quoted above, let P also denote the Daniell integral obtained from the elementary integral P on $C[a,b]$, and let Q denote the Lebesgue integral on $[a,b]$. Then condition (i) of the Comparison Theorem is satisfied by the definition of P on $C[a,b]$. If I is any closed interval in $[a,b]$ then we may express χ_I as the limit of a decreasing sequence of continuous functions from $C[a,b]$ (see Fig. 1

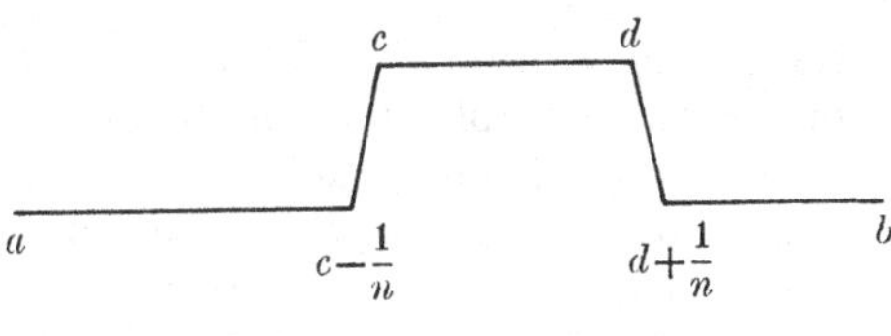

Fig. 1

and Ex. 1). Thus $\chi_I \in L^1(P)$ by the Monotone Convergence Theorem. Any step function on $[a, b]$ may be expressed as a linear combination of such characteristic functions χ_I (cf. Ex. 4.1.6) and so condition (ii) is also satisfied. Theorem 1 now identifies P and Q. In other words the Lebesgue integral on $[a,b]$ may equally well be constructed starting with the linear lattice $C[a,b]$ of continuous functions on $[a,b]$, provided, of course, that we have some other means (such as the Riemann integral) of defining

$$\int_a^b \phi(x)\,dx.$$

In this example we have used the Comparison Theorem to identify two integrals and so it may be considered as a uniqueness theorem: we shall meet it again in §13.4 in the role of uniqueness theorem.

Exercises

1. (i) See Fig. 1. Suppose that $a \leqslant c \leqslant d \leqslant b$. For $n \geqslant 1$ let

$$\begin{aligned} \phi_n(x) &= 0 && \text{if } a \leqslant x \leqslant c-\frac{1}{n} \\ &= n(x-c)+1 && \text{if } c-\frac{1}{n} \leqslant x \leqslant c \\ &= 1 && \text{if } c \leqslant x \leqslant d \\ &= -n(x-d)+1 && \text{if } d \leqslant x \leqslant d+\frac{1}{n} \\ &= 0 && \text{if } d+\frac{1}{n} \leqslant x \leqslant b; \end{aligned}$$

to take care of the possibility that $c-1/n < a$ or $b < d+1/n$ we also assume that $a \leqslant x \leqslant b$. Show that $\{\phi_n\}$ is a decreasing sequence of functions in $C[a, b]$ which converges to $\chi_{[c,d]}$.

(ii) Show that any step function on $\mathbf{R}$ may be expressed as a linear combination of characteristic functions of (bounded) closed intervals.

2. Let Q denote the Lebesgue integral on $\mathbf{R}^k$ and let P be the restriction of Q to the space L of integrable simple functions defined in §6.2 (L consists of the functions in $L^1(Q)$ that attain only a finite set of values). Show that P is an elementary integral on L and that P extends to Q by the Daniell construction. (In other words, the Lebesgue integral may be constructed starting with the simple functions rather than the step functions of Chapter 4.)

3. A *piecewise linear* function ϕ on $\mathbf{R}$ is a continuous function on $\mathbf{R}$ whose graph consists of finitely many straight line segments. Let L denote the collection of piecewise linear functions ϕ on $\mathbf{R}$ which vanish outside a bounded interval (depending on ϕ). Show that L is a linear lattice and that the restriction P of the (Riemann or) Lebesgue integral to L is an elementary integral. Show that P extends by the Daniell construction to the Lebesgue integral on $\mathbf{R}$.

8.4 Measurable functions and measure

Following Stone [21] we shall say that a function $f: X \to \mathbf{R}$ is *measurable* if the truncated function

$$\text{mid}\{-g, f, g\}\dagger$$

belongs to L^1 for all positive functions g in L^1.

Suppose that

$$\text{mid}\{-\phi, f, \phi\}$$

† Recall that $\text{mid}\{f, g, h\} = \max\{\min\{g, h\}, \min\{h, f\}, \min\{f, g\}\}$.

is integrable for all positive *elementary* functions ϕ and that g is an arbitrary positive integrable function. Then we may find a sequence $\{\phi_n\}$ of positive elementary functions converging almost everywhere to g (Ex. 8.1.7). The truncated function

$$h_n = \text{mid}\{-\phi_n, f, \phi_n\}$$

is integrable for all n and

$$h_n \to \text{mid}\{-g, f, g\}$$

almost everywhere. If we truncate again and apply the Dominated Convergence Theorem to the sequence $\{\text{mid}\{-g, h_n, g\}\}$ it follows that

$$\text{mid}\{-g, f, g\}$$

is integrable. This means that we may replace the positive integrable function g in the above definition by an arbitrary positive function ϕ in L.

The good behaviour of the class of measurable functions is guaranteed by the following result.

Theorem 1. (i) *All integrable functions are measurable.*

(ii) *If f, g are measurable, then so are $|f|$, f^+, f^-, $af+bg (a, b \in \mathbf{R})$, $\max\{f, g\}$ and $\min\{f, g\}$.*

(iii) *If $f_n \to f$ almost everywhere and f_n is measurable for $n = 1, 2, \ldots$ then f is measurable.*

More briefly: the measurable functions form a linear lattice which contains all the integrable functions and is closed with respect to almost everywhere convergence of sequences.

Proof. Throughout the proof let ϕ be an arbitrary positive elementary function.

(i) If f is integrable then $\text{mid}\{-\phi, f, \phi\}$ is integrable.

(ii) The truncated function

$$\text{mid}\{-\phi, |f|, \phi\} = |\text{mid}\{-\phi, f, \phi\}|$$

is integrable and so $|f|$ is measurable.

Let f_n, g_n be the integrable functions defined by

$$f_n = \text{mid}\{-n\phi, f, n\phi\},$$

$$g_n = \text{mid}\{-n\phi, g, n\phi\}.$$

If $\phi(x) = 0, f_n(x) = g_n(x) = 0$, and if $\phi(x) > 0, f_n(x) = f(x), g_n(x) = g(x)$ for $n \geqslant N$ (where N depends on x). Thus

$$\text{mid}\{-\phi, af_n + bg_n, \phi\} \to \text{mid}\{-\phi, af + bg, \phi\}.$$

The functions on the left hand side are integrable and are dominated by ϕ. The Dominated Convergence Theorem therefore shows that

$$\operatorname{mid}\{-\phi, af+bg, \phi\}$$

is integrable and so $af+bg$ is measurable.

The measurability of f^+, f^-, $\max\{f,g\}$, $\min\{f,g\}$ now follows from the simple relations

$$f^+ = \tfrac{1}{2}(|f|+f),$$
$$f^- = \tfrac{1}{2}(|f|-f),$$
$$\max\{f,g\} = \tfrac{1}{2}(f+g)+\tfrac{1}{2}|f-g|,$$
$$\min\{f,g\} = \tfrac{1}{2}(f+g)-\tfrac{1}{2}|f-g|.$$

(iii) The truncated functions

$$\operatorname{mid}\{-\phi, f_n, \phi\}$$

are integrable for $n = 1, 2, \ldots$, are dominated by ϕ, and

$$\operatorname{mid}\{-\phi, f_n, \phi\} \to \operatorname{mid}\{-\phi, f, \phi\}$$

almost everywhere. Thus $\operatorname{mid}\{-\phi, f, \phi\}$ is integrable and so f is measurable.

In the light of Theorem 1 it is often possible to recognise that a function f is measurable, and then the following result is a practical criterion for the integrability of f.

Proposition 1. *If f is measurable and $|f| \leqslant g$, where g is integrable, then f is integrable.*

Proof. $\operatorname{mid}\{-g, f, g\} = f$.

Corollary. *If f is measurable and $|f|$ is integrable, then f is integrable.*

With the Lebesgue integral in mind we may well ask whether or not the continuous functions on X are measurable. But *what* continuous functions? In general X has no topological structure and continuity is not even defined! We cannot even be sure that the constant functions are measurable. It was Stone who saw clearly how important it is to satisfy this latter condition and his name is usually attached to it.

The Stone Condition:

$$\min\{1, \phi\} \in L^1 \quad \text{for any positive } \phi \text{ in } L. \tag{1}$$

In view of the above definition of a measurable function, (1) is equivalent to demanding that the constant function 1 (i.e. χ_X) should be measurable and hence that the constant functions $c = c1$ should all be measurable.

Condition (1) involves the space L^1, and for some purposes it is convenient to have a stronger condition only involving the space L of elementary functions:

$$\min\{1, \phi\} \in L \quad \text{for any positive } \phi \text{ in } L. \tag{2}$$

A still stronger condition is often satisfied, viz.

$$1 \in L. \tag{3}$$

We shall say that a subset S of X is *measurable* if the characteristic function χ_S is measurable; moreover, we define a *measure* μ on the collection of all measurable sets S as follows:

$$\begin{aligned} \mu(S) &= \int \chi_S \quad \text{if} \quad \chi_S \in L^1, \\ &= \infty \qquad \text{otherwise.} \end{aligned}$$

In terms of this definition, the Stone Condition (1) is equivalent to demanding that the whole space X should be measurable. If condition (3) is satisfied then certainly $1 \in L^1$ and so the whole space X has finite measure.

Let A be a subset of X. If $f\colon X \to \mathbf{R}$ and $f\chi_A$ is integrable, then we write

$$\int f\chi_A = \int_A f,$$

and say that f is *integrable on* A. The set of all such functions f is denoted by $L^1(A)$. In practice we shall only be interested in $L^1(A)$ when A is a measurable subset of X, but it is convenient to have the notation for an arbitrary subset A of X (cf. Ex. 8.2.3).

Example 1. Let $X = (0, 1]$ and let L consist of all the linear functions ϕ on X defined by $\phi(x) = ax$ for some a in $\mathbf{R}$. Define

$$\int \phi = a/2,$$

viz. the area under the graph (Fig. 2). The only null set is the empty set (because the divergence of an increasing sequence $\{\phi_n\}$ at any point of $(0, 1]$ would imply the divergence of $\{\int \phi_n\}$). It is then easy to see that L^{inc} and L^1 are both equal to L, and in fact, the measurable functions are also just the elements of L. The only element of L that is of the form χ_A is the zero function and so the only measurable set is $\varnothing$. Thus the Stone Condition (1) is not satisfied in this case.

Example 2. Let $X = [0, 1]$ and let L consist of all continuous functions $\phi: X \to \mathbf{R}$ which have a 'tail', however short, of the form $\phi(x) = ax$ – rather like a pointer dog whose quarry is at the origin! More formally $\phi: X \to \mathbf{R}$ is continuous and there is a real number k in $[0, 1)$ and a real number a (k and a both depending on ϕ) such that $\phi(x) = ax$ for all x in $[k, 1]$. For any given ϕ in L, a suitable horizontal truncation would remove part of the tail (Fig. 3): thus the stronger Stone Condition (2)

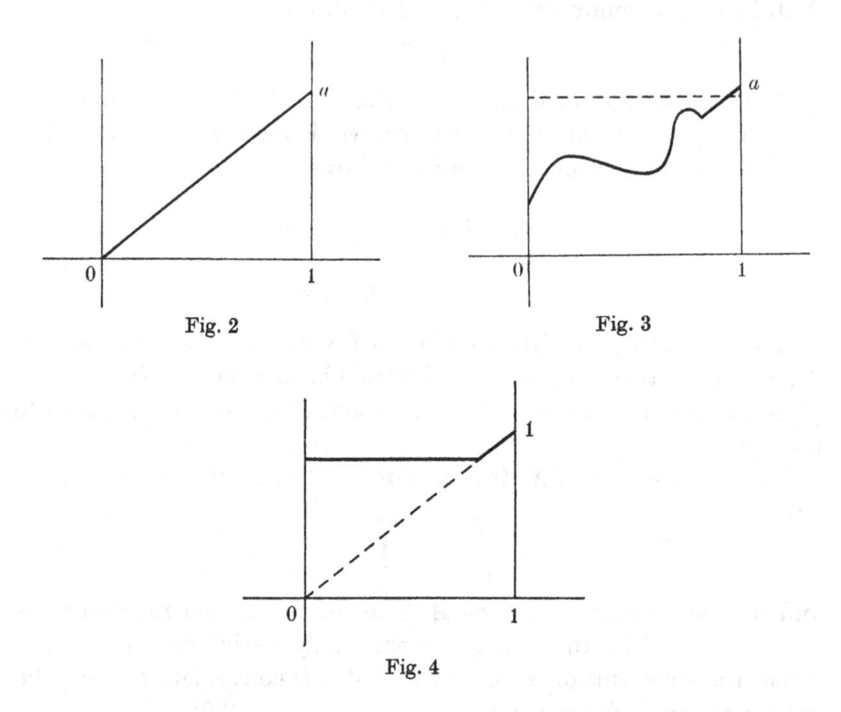

Fig. 2

Fig. 3

Fig. 4

is not satisfied. On the other hand, (no matter what elementary integral we define on L) there is an increasing sequence $\{\phi_n\}$ of functions in L whose limit is 1 (everywhere), e.g.

$$\phi_n(x) = 1 - \frac{1}{n} \quad \text{for} \quad 0 \leqslant x \leqslant 1 - \frac{1}{n},$$

$$= x \quad \text{for} \quad 1 - \frac{1}{n} \leqslant x \leqslant 1$$

(Fig. 4). Thus 1 is measurable, i.e. the weaker Stone Condition (1) is satisfied.

Some of the most important properties of the measure μ are summarised in

Theorem 2. (i) *Let S, T be measurable sets; then so are $S \cup T$, $S \cap T$, $S \setminus T$ and $S \,\Delta\, T$, and*

(*a*) $\mu(S) = 0$ *if and only if S is null;*

(*b*) $\mu(S) \geqslant \mu(T)$ *if* $S \supset T$;

(*c*) $\mu(S \cup T) + \mu(S \cap T) = \mu(S) + \mu(T)$.

(ii) *If $\{S_n\}$ is an increasing sequence of measurable sets and*

$$S = \bigcup_{n=1}^{\infty} S_n$$

then S is measurable and $\mu(S) = \lim \mu(S_n)$.

(iii) *If $\{S_n\}$ is a sequence of measurable sets and*

$$S = \bigcup_{n=1}^{\infty} S_n$$

then S is measurable and $\quad \mu(S) \leqslant \sum_{n=1}^{\infty} \mu(S_n)$;

moreover, if the sets S_n are disjoint, then

$$\mu(S) = \sum_{n=1}^{\infty} \mu(S_n).$$

Theorem 2 has a natural interpretation for sets of infinite measure. For example, in (ii), if any $\mu(S_n) = \infty$, or if $\{\mu(S_n)\}$ diverges, then $\mu(S) = \infty$. With this in mind we agree to write

$$a < \infty$$

for any real number a and interpret the inequalities of (i) (*b*) and (iii) accordingly: thus in (i) (*b*), if $S \supset T$ and $\mu(T) = \infty$, then $\mu(S) = \infty$. It is important to note that the spurious 'difference' $\infty - \infty$ must be avoided, and care may be required in calculating the measure of a difference set $S \setminus T$.

Proof. (i) The first part follows from Theorem 1 (ii) in view of the relations

$$\chi_{S \cup T} = \max\{\chi_S, \chi_T\},$$

$$\chi_{S \cap T} = \min\{\chi_S, \chi_T\},$$

$$\chi_{S \setminus T} = (\chi_S - \chi_T)^+,$$

$$\chi_{S \Delta T} = |\chi_S - \chi_T|.$$

(*a*) If S is null then $\chi_S = 0$ almost everywhere and so $\int \chi_S = 0$. If $\int \chi_S = 0$ then $\chi_S = 0$ almost everywhere by the Corollary to the Monotone Convergence Theorem. In other words, S is null.

(*b*) If we assume that S has finite measure, i.e. χ_S is integrable, then the inequality $\chi_S \geqslant \chi_T$ shows that χ_T is integrable and

$$\int \chi_S \geqslant \int \chi_T.$$

By the same token, if T has infinite measure then S must also have infinite measure.

(*c*) We have $$\chi_{S\cup T} + \chi_{S\cap T} = \chi_S + \chi_T.$$

The various interpretations of this equation, in the case of infinite measures, are quite obvious using (*b*).

(ii) Apply the Monotone Convergence Theorem to the increasing sequence $\{\chi_{S_n}\}$ which converges to χ_S.

(iii) For each $n \geqslant 1$ let

$$T_n = S_1 \cup S_2 \cup \ldots \cup S_n.$$

By (i) (*c*) we deduce that

$$\mu(T_2) \leqslant \mu(S_1) + \mu(S_2),$$

and by a simple induction

$$\mu(T_n) \leqslant \mu(S_1) + \mu(S_2) + \ldots + \mu(S_n).$$

The result now follows if we apply part (ii) to the increasing sequence $\{T_n\}$.

In all that follows we shall assume that Stone's Condition (1) holds.

This is vital in linking Stone's definition of a measurable function with the classical definition. We approach this by Lebesgue's method of 'horizontal approximation'.

In the first place assume that $f: X \to \mathbf{R}$ is bounded. Thus there exist real numbers p, q such that the values of f all lie in the half open interval $[p, q)$. Let us divide $[p, q)$ into r disjoint intervals $[c_i, c_i + \epsilon)$ $(i = 1, 2, \ldots, r)$ each of length $\epsilon = (q-p)/r$ and define the sets

$$S_i = \{x \in X : c_i \leqslant f(x) < c_i + \epsilon\}. \tag{4}$$

Then the function $$\phi = c_1\chi_{S_1} + \ldots + c_r\chi_{S_r} \tag{5}$$

satisfies $$0 \leqslant f(x) - \phi(x) < \epsilon,$$
for all x in X.

We shall say that the function ϕ defined by (5) is a *simple function* if each of the sets S_i is measurable. Fig. 5 illustrates the case where X is a subset of $\mathbf{R}$, but the construction is valid quite generally.

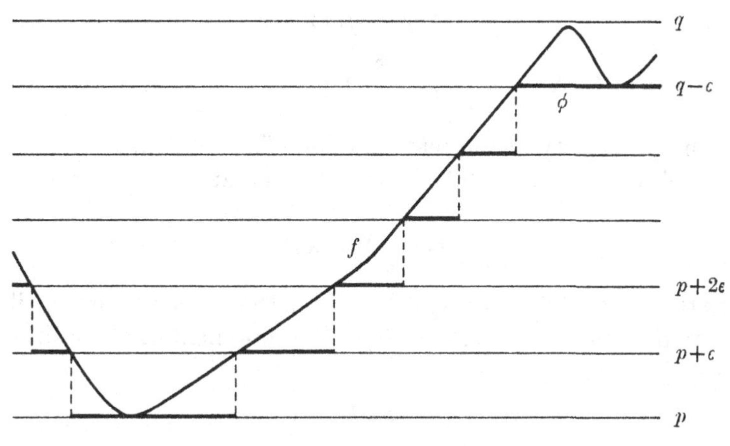

Fig. 5

Proposition 2. *The function $f: X \to \mathbf{R}$ is measurable if and only if the sets*

$$A_c = \{x \in X : f(x) \geqslant c\}$$

are measurable for all c in **R**.

Proof. Assume that f is measurable in the sense of Stone; then so are the functions f_n defined by

$$f_n(x) = n\left(\min\{f(x), c\} - \min\left\{f(x), c - \frac{1}{n}\right\}\right).$$

(This is where we use the fact that the constant functions are measurable.) If $f(x) \geqslant c$, $f_n(x) = 1$, and if $f(x) < c$, $f_n(x) = 0$ for $n \geqslant N$ (depending on x). Thus

$$f_n \to \chi_{A_c}$$

which is therefore measurable by Theorem 1 (iii).

On the other hand, if A_c is measurable for every c in **R** then we may apply the method of horizontal approximation to the bounded function

$$f_n = \operatorname{mid}\left\{-n, f, n - \frac{1}{n}\right\}$$

to obtain a simple function ϕ_n which satisfies

$$0 \leqslant f_n(x) - \phi_n(x) < \frac{1}{n}$$

for all x in X. (In this case we take $p = -n$, $q = n$, $\epsilon = 1/n$, $r = 2n^2$, and note that $S_i = A_{c_i} \setminus A_{c_i+\epsilon}$.) Now the sequence $\{\phi_n\}$ of simple functions, each of which is obviously measurable, converges to f, and so f is measurable. This completes the proof.

If we write $$B_c = \{x \in X : f(x) > c\},$$

then it is clear that $$B_c = \bigcup_{n=1}^{\infty} A_{c+1/n},$$

so the measurability of the sets B_c follows from the measurability of the sets A_c by Theorem 2 (ii). On the other hand

$$A_c = \bigcap_{n=1}^{\infty} B_{c-1/n}$$

and so the measurability of A_c follows from the measurability of all B_c. We may therefore replace A_c by B_c in the statement of Proposition 2. It also follows that the set

$$A_c \setminus B_c = \{x \in X : f(x) = c\}$$

is measurable, from which we deduce that the simple functions are precisely the measurable functions which assume only a finite number of values.

In the case of the Lebesgue integral on $\mathbf{R}^k$ the simple functions incorporated all the step functions, but were considerably more sophisticated. In the general theory an elementary function may take more than a finite set of values and so we cannot expect all elementary functions to be contained in the space of simple functions (cf. Examples 3, 5 of § 8.1). Nevertheless, we may still regard the simple functions as more sophisticated in the sense of the following proposition.

Proposition 3. *A measurable function $f: X \to \mathbf{R}$ may be expressed as the limit, everywhere, of a sequence of simple functions; if f is also positive, then f may be expressed as the limit, everywhere, of an increasing sequence of positive simple functions.*

Proof. The simple functions ϕ_n constructed in the proof of Proposition 2 satisfy the conditions for the first part. When $f \geqslant 0$ we may adapt the construction as follows: for each integer $n \geqslant 1$, take $p = 0$, $q = n$, $\epsilon = 2^{-n}$, $r = n2^n$ and let

$$f_n = \min\{f, n - 2^{-n}\}.$$

The standard horizontal approximation now gives a positive simple function ϕ_n which satisfies

$$0 \leqslant f_n - \phi_n < 2^{-n}$$

for all n and the choice of $\epsilon = 2^{-n}$ (rather than $\epsilon = 1/n$ as above) ensures that

$$\phi_n \leqslant \phi_{n+1}$$

for all n.

Any simple function ϕ may be expressed uniquely in the form

$$\phi = c_1\chi_{S_1} + \ldots + c_r\chi_{S_r},$$

where $c_1 < \ldots < c_r$ are the distinct values assumed by ϕ. The measurable sets $S_1, \ldots, S_r$ are disjoint and their union is the whole set X. If a second simple function ψ has expression

$$\psi = d_1\chi_{T_1} + \ldots + d_s\chi_{T_s},$$

where $d_1 < \ldots < d_s$ are the distinct values assumed by ψ, then the measurable sets

$$S_i \cap T_j \quad (i = 1, \ldots, r; j = 1, \ldots, s)$$

(some of which may be empty) are disjoint and their union is X. For *any* real valued function h whose domain of definition contains the ordered pairs (c_i, d_j) we now define a simple function θ where

$$\theta(x) = h(\phi(x), \psi(x)).$$

Using this idea we may apply Proposition 3 to prove another result which enhances Theorem 1 by showing that the class of measurable functions is closed with respect to certain operations of taking composites (but see Ex. 15).

Proposition 4. *If $h: \mathbf{R}^2 \to \mathbf{R}$ is continuous and if $f, g: X \to \mathbf{R}$ are measurable, then the composite function $F: X \to \mathbf{R}$ defined by*

$$F(x) = h(f(x), g(x))$$

for x in X, is again measurable.

Proof. By Proposition 3, f, g are limits (everywhere) of sequences $\{\phi_n\}$, $\{\psi_n\}$ of simple functions. The functions θ_n defined by

$$\theta_n(x) = h(\phi_n(x), \psi_n(x)),$$

are also simple. By the continuity of h,

$$\theta_n(x) \to h(f(x), g(x))$$

for all x, and so F is measurable.

Corollary. *If $f, g: X \to \mathbf{R}$ are measurable, then their product fg is also measurable.*

Exercises

1. In Example 8.1.1 show that any sequence is a measurable function and that $\mu(S)$ is the number of elements in S (possibly infinite). This measure μ is often called the *counting measure* on $\{1, 2, \ldots\}$.

2. In Example 8.1.2 show that any function $f: X \to \mathbf{R}$ is measurable, and that $\mu(S) = 0$ if S is countable and $\mu(S) = \infty$ if S is uncountable.

3. In Example 8.1.3 show that any function $f: \mathbf{R}^k \to \mathbf{R}$ is measurable, and that $\mu(S) = 1$ if $0 \in S$, $\mu(S) = 0$ if $0 \notin S$. This measure μ may be called the *unit mass concentrated at* 0.

4. In Example 8.1.4 show that the measurable functions are the Lebesgue measurable functions and that μ coincides with Lebesgue measure.

5. What becomes of Example 1 on p. 23 if X is replaced by $[0, 1]$?

6. Let L consist of the linear functions on $[0, 1]$ defined by $\phi(x) = ax$ for some a in $\mathbf{R}$ and let $\int \phi = 0$ for all ϕ in L. Show that the weaker Stone Condition is satisfied but not the stronger. Describe the corresponding Daniell integral and measure. (Cf. Ex. 5. above.)

7. Let L be the space of continuous functions on $[0, 1]$ with tails pointing at the origin as in Example 2 (Fig. 3) and interpret $\int \phi$ for any ϕ in L as a (Riemann or) Lebesgue integral. Show that the elementary integral so defined extends by the Daniell construction to the Lebesgue integral on $[0, 1]$.

8. Let f, g be measurable functions whose squares are integrable. Show that fg is integrable and that

$$\left(\int fg\right)^2 \leqslant \int f^2 \int g^2.$$

Under what circumstances can we have equality?

9. If f is integrable and g is measurable and bounded, show that fg is integrable.

10. If $1 \in L^1$ show that $f/(1+|f|) \in L^1$ for any measurable function f.

11. If f is measurable and satisfies $p \leqslant f(x) \leqslant q$ for almost all x, and if g is integrable, show that

$$\int f|g| = r\int |g|$$

for some real number r in $[p, q]$.

Can we replace $|g|$ by g in this equation?

12. Show that the integrable simple functions are precisely the integrable functions which assume only a finite number of values.

13. (i) Let h be a real valued function on $\mathbf{R}^k$ and $f_1, \ldots, f_k$ simple functions on X; define the composite function F on X by

$$F(x) = h(f_1(x), \ldots, f_k(x))$$

for all x in X. Show that F is simple.

(ii) If the simple functions $f_1, \ldots, f_k$ are also integrable, show that F is the sum of an integrable simple function and a constant.

14. Extend Proposition 4 to the case where h is a continuous real valued function on $\mathbf{R}^k$.

15. In Exx. 6.3.6, 8 we constructed a continuous strictly increasing function T of $[0,1]$ onto $[0,2]$ which carries a measurable subset of $[0,1]$ onto a non-measurable subset of $[0,2]$. Assuming the existence of such a function, find a measurable function $h: \mathbf{R} \to \mathbf{R}$ and a continuous function $f: \mathbf{R} \to \mathbf{R}$ such that the composite function $h \circ f$ is not measurable. (Cf. Ex. 14 above, in the case where $k = 1$.)

16. Let $\mathscr{M}$ denote the collection of all measurable sets for a given Daniell integral P on L^1. For any S in $\mathscr{M}$, let

$$\begin{aligned} \nu(S) &= P(\chi_S) \quad \text{if} \quad \chi_S \in L^1, \\ &= \sup\{\nu(A): A \subset S, \chi_A \in L^1\} \quad \text{otherwise.} \end{aligned}$$

(The supremum here is allowed to take the value ∞.) Show that ν satisfies the conditions of Theorem 2 (iii), i.e. ν is a measure on $\mathscr{M}$.

17. Suppose that X has finite measure (i.e. $1 \in L^1$).

(i) Let f, f_n $(n \geqslant 1)$ be measurable functions on X such that $f_n \to f$. For $m, n \geqslant 1$ let

$$A_{mn} = \{x \in X: |f_r(x) - f(x)| \geqslant 1/m \quad \text{for some} \quad r \geqslant n\}.$$

Show that, for each $m \geqslant 1$,

$$A_{mn} \downarrow \varnothing \quad \text{as} \quad n \to \infty.$$

(ii) Let $\epsilon > 0$ be given. To each integer m, there is an integer $n = n(m)$ (depending also on ϵ) such that $\mu(A_{mn}) < \epsilon/2^m$. Let

$$A = \bigcup_{m=1}^{\infty} A_{m\,n(m)}.$$

Show that $\mu(A) < \epsilon$ and also that $|f_r(x) - f(x)| < 1/m$ for all x in $X \setminus A$ provided $r \geqslant n(m)$.

(iii) Prove ***Egoroff's Theorem.*** *Suppose that X has finite measure (i.e. $1 \in L^1$). Let f, f_n $(n \geqslant 1)$ be measurable functions such that $f_n \to f$ almost everywhere on X. Then, to any $\epsilon > 0$, there is a subset A of X such that $\mu(A) < \epsilon$ and $f_n \to f$ uniformly on $X \setminus A$.*

The conclusion of this theorem is often expressed by saying that

$$f_n \to f \quad \textit{almost uniformly} \text{ on } X.$$

18. Let μ be the counting measure of Ex. 1. Let $S_n = \{1, 2, \ldots, n\}$ and let $f_n = \chi_{S_n}$. Show that $f_n(x) \to 1$ for all x but that f_n does not converge almost uniformly to 1. What does this example tell us about Egoroff's Theorem (Ex. 17)?

8.5 Stone's Theorem

To round off this chapter it is convenient to anticipate some of the ideas of Chapters 11 and 12. In § 8.4 we constructed a measure μ in terms of a Daniell integral. Is it possible to follow the classical pattern and deduce the Daniell integral starting from the measure μ?

For ease of notation we denote the Daniell integral by P. The most obvious functions to consider are the simple functions

$$\phi = a_1\chi_{A_1} + \ldots + a_m\chi_{A_m},$$

where $a_1, \ldots, a_m \in \mathbf{R}$ and $A_1, \ldots, A_m$ have *finite* measure with respect to μ. We may define

$$\int \phi\, d\mu = a_1\mu(A_1) + \ldots + a_m\mu(A_m).$$

In § 11.3 we shall see that this definition is unambiguous for any given measure μ, but in our present context we may use the fact that μ is derived from the Daniell integral P to note that

$$\int \phi\, d\mu = P(\phi)$$

is independent of the way in which ϕ is expressed as a simple function.

In view of Proposition 8.4.2, a measurable function f may be described in terms of μ, viz. f is measurable if and only if

$$\{x \in X : f(x) \geqslant c\}$$

is measurable with respect to μ for every c in $\mathbf{R}$. If f is a positive measurable function we may express f as the limit (everywhere) of an increasing sequence $\{\phi_n\}$ of positive simple functions (Proposition 8.4.3) and define

$$\int f\, d\mu = \lim \int \phi_n\, d\mu$$

provided this limit is finite. For a general measure μ we shall prove in § 12.2 that this definition is again unambiguous and we shall say that the *positive function f is integrable with respect to μ*. In our present context μ is derived from P and we may appeal to the Monotone Convergence Theorem to see that (our given positive measurable function) f lies in $L^1(P)$ if and only if $\lim P(\phi_n)$ is finite; moreover $P(f) = \lim P(\phi_n)$ in this case. Thus

$$\int f\, d\mu = P(f)$$

for all positive f in $L^1(P)$.

Finally, if the positive and negative parts f^+, f^- of f are integrable with respect to μ we say that *f is integrable with respect to μ* and define

$$\int f d\mu = \int f^+ d\mu - \int f^- d\mu.$$

Once again it is clear that

$$\int f d\mu = P(f).$$

We summarise this in the form of a theorem.

***Theorem 1* (*Stone*).** *Let P be a Daniell integral on L^1 for which the constant functions are measurable, and μ the corresponding measure. Then f is integrable with respect to μ if and only if $f \in L^1$, and*

$$P(f) = \int f d\mu$$

for all f in L^1.

Exercises

For the purpose of these exercises, a measure may be described as a function μ which satisfies the conditions of Theorem 8.4.2 (iii).

1. Let $X = \mathbf{R} \cup \{\omega\}$, where ω is any element not in $\mathbf{R}$, and let L consist of all functions $\phi: X \to \mathbf{R}$ which are Lebesgue integrable on $\mathbf{R}$ and vanish at ω. Let P be defined on L by

$$P(\phi) = \int_{\mathbf{R}} \phi(x)\, dx.$$

Show that P is an elementary integral on L. What is the corresponding measure μ? Let $\mathscr{M}$ be the collection of all measurable subsets of X. Show that there is another measure ν on $\mathscr{M}$ which satisfies the conclusion of Stone's Theorem.

2. Show that the measure ν of Ex. 8.4.16 satisfies the conclusion of Stone's Theorem, viz.

$$P(f) = \int f d\nu$$

for all f in L^1. If λ is any other measure on $\mathscr{M}$ satisfying this condition, and if μ is the measure of Stone's Theorem, show that $\nu \leqslant \lambda \leqslant \mu$, i.e.

$$\nu(E) \leqslant \lambda(E) \leqslant \mu(E)$$

for all E in $\mathscr{M}$.

9

LEBESGUE–STIELTJES INTEGRALS AND MEASURES

In view of the important role played by the step functions in the construction of the Lebesgue integral it is natural to ask what is the most general elementary integral defined on the linear lattice L of step functions on $\mathbf{R}^k$. According to the construction described in § 8.1, such an elementary integral extends to a Daniell integral $\int$ on a space L^1 and provides a measure μ for which

$$\mu(I) = \int \chi_I$$

is finite for each bounded interval I. This Daniell integral is a *Lebesgue–Stieltjes integral* and the corresponding measure is a *Lebesgue–Stieltjes measure*. In § 9.1 we consider the case $k = 1$ and show how this integral and measure may be described in terms of an increasing function $F\colon \mathbf{R} \to \mathbf{R}$. In § 9.2 this idea is extended to give the Lebesgue–Stieltjes integral and measure on $\mathbf{R}^k$ in terms of a rather more complicated Stieltjes function $F\colon \mathbf{R}^k \to \mathbf{R}$. This Stieltjes function depends heavily on the 'rectangular' nature of the coordinate axes in $\mathbf{R}^k$, but in § 9.3 we break free from the bondage of the axes in $\mathbf{R}^k$ and show that the Lebesgue–Stieltjes integrals are the ones which integrate all 'good' functions like the continuous functions on $\mathbf{R}^k$ vanishing outside a compact subset, and the Lebesgue–Stieltjes measures are the ones which give a measure, finite or infinite, to all 'good' sets like the open sets or the closed sets in $\mathbf{R}^k$.

9.1 Integrals on R

[A student who has not read Chapter 8 on the abstract Daniell integral but who is familiar with the construction of the Lebesgue integral in § 3.1 and § 3.2 may understand the main theorem of this section if he is prepared to begin at the paragraph containing equations (5) for μ_F. To follow the earlier paragraphs it is only necessary to think of a Daniell integral as a generalisation of the Lebesgue integral which satisfies the conditions of Theorem 3.2.2 (or 8.1.1) together with the Monotone Convergence Theorem 5.1.2 (or 8.2.2), and the corresponding measure as a generalisation of Lebesgue measure which satisfies the conditions of Theorem 6.2.1 (or 8.4.2).]

Let $\int$ be a Daniell integral on $\mathbf{R}$ whose space L^1 of integrable functions contains all the step functions $\phi: \mathbf{R} \to \mathbf{R}$, and let μ be the corresponding measure. For the moment let us also *assume that $\mu(\mathbf{R})$ is finite* and define

$$F(x) = \int \chi_{(-\infty, x]} = \mu(-\infty, x] \tag{1}$$

for all x in $\mathbf{R}$. It follows (either from the linearity of $\int$ or the additivity of μ) that

$$F(b) - F(a) = \int \chi_{(a,b]} = \mu(a, b] \tag{2}$$

for any $a < b$. It is clear that *F is an increasing function* so that the discontinuities of F, if any, are simple jumps, and they are countable (cf. § 3.3). If $b_n \downarrow b$ then $\chi_{(a, b_n]} \downarrow \chi_{(a,b]}$ and so $F(b_n) \downarrow F(b)$ (either by the Monotone Convergence Theorem for $\int$ or the σ-additivity of μ). In other words, the increasing function *F is continuous on the right* (cf. § 3.3 again). If $b_n \uparrow b$ ($b_n \neq b$) then $\chi_{(b_n, b]} \downarrow \chi_{[b, b]}$ and so

$$F(b) - F(b_n) \downarrow \int \chi_{[b, b]} = \mu\{b\}.$$

But $F(b_n) \uparrow F(b-0)$, so

$$\mu\{b\} = F(b) - F(b-0). \tag{3}$$

In other words *the measure of the single point b is the jump at b*. In particular this means that F is continuous at b if and only if $\mu\{b\} = 0$. From equations (2), (3) it is now clear that

$$\left.\begin{aligned} \mu(a,b] &= F(b) - F(a), \\ \mu(a,b) &= F(b-0) - F(a), \\ \mu[a,b] &= F(b) - F(a-0), \\ \mu[a,b) &= F(b-0) - F(a-0), \end{aligned}\right\} \tag{4}$$

for $a < b$. Thus $\mu(I)$ measures the 'variation of F on the interval I', sensible account being taken of whether or not the end points are included.

The fact that F is continuous on the right is convenient, but not important; if we had chosen to use intervals of type $(-\infty, x)$ in the definition, F would have been continuous on the left and we should have found ourselves discussing bounded intervals of type $[a, b)$.

We have so far assumed that $\mu(\mathbf{R})$ is finite. In this case, and in particular when μ is a *probability measure* on $\mathbf{R}$, i.e. when $\mu(\mathbf{R}) = 1$, F is usually called the (*cumulative*) *distribution function* of μ. Even if

$\mu(\mathbf{R}) = \infty$, we can always define an increasing function F_0 as follows:

$$F_0(x) = \int \chi_{(0,x]} = \mu(0,x] \quad \text{for} \quad x \geqslant 0,$$

$$= -\int \chi_{(x,0]} = -\mu(x,0] \quad \text{for} \quad x < 0.$$

This function $F_0: \mathbf{R} \to \mathbf{R}$ is not quite so natural as the distribution function F, especially as it satisfies the artificial condition

$$F_0(0) = 0,$$

and so gives the impression that the origin is specially favoured. But it is easy to verify (2) for the new F_0 by discussing the three cases $0 \leqslant a$, $a < 0 \leqslant b$, $b < 0$, separately, and hence to check as before that F_0 *is increasing and continuous on the right*. (Of course, in the finite case, the functions F, F_0 differ only by the constant $\mu(-\infty, 0]$, and so we should expect them to have these properties in common.)

Suppose now that an arbitrary increasing function $F: \mathbf{R} \to \mathbf{R}$ is given. If $a < b$ we may define

$$\left.\begin{aligned} \mu_F(a,b] &= F(b+0) - F(a+0), \\ \mu_F(a,b) &= F(b-0) - F(a+0), \\ \mu_F[a,b] &= F(b+0) - F(a-0), \\ \mu_F[a,b) &= F(b-0) - F(a-0). \end{aligned}\right\} \tag{5}$$

In particular, $\qquad \mu_F\{b\} = F(b+0) - F(b-0),$

the jump at b. From this it is clear that the definition of μ_F on these bounded intervals is independent of the value of F at any point of discontinuity. It is therefore permissible to adjust F so that

$$F(x) = F(x+0)$$

for all x, i.e. so that F is continuous on the right. This is convenient as we may now concentrate on intervals of type (,] and use the simplified formula

$$\mu_F(a,b] = F(b) - F(a) \tag{6}$$

for $a < b$.

From this starting point we may trace through the steps of § 3.1 replacing the length of an interval $(a, b]$ by the 'variation' $F(b) - F(a)$. Briefly the argument is as follows.

Denote by L' the linear space of step functions

$$\phi = c_1\chi_{I_1} + \ldots + c_r\chi_{I_r},$$

where $c_1, \ldots, c_r \in \mathbf{R}$ and $I_1, \ldots, I_r$ are intervals of type (,], and define the integral

$$\int \phi \, dF = c_1\mu_F(I_1) + \ldots + c_r\mu_F(I_r).$$

To check the consistency of this definition note that if $p \in (a,b]$, then

$$F(b)-F(a) = \{F(p)-F(a)\}+\{F(b)-F(p)\}$$

and so
$$\mu_F(a,b] = \mu_F(a,p]+\mu_F(p,b]. \tag{7}$$

Also, it is quite obvious that

$$(c_1+c_2)\,\mu_F(I) = c_1\mu_F(I)+c_2\mu_F(I) \tag{8}$$

for any c_1, c_2 in $\mathbf{R}$ and any interval I of type (,].

Let ϕ have another expression

$$\phi = d_1\chi_{J_1}+\dots+d_s\chi_{J_s},$$

where $d_1, \dots, d_s \in \mathbf{R}$ and $J_1, \dots, J_s$ are intervals of type (,]. Without loss of generality we may assume that $I_1, \dots, I_r, J_1, \dots, J_s$ are non-empty. If these intervals have n distinct end points we may express ϕ in terms of the $(n-1)$ *disjoint* intervals of type (,] which lie between them. The two expressions for ϕ therefore have a common refinement in terms of disjoint intervals. In each case this refining process may be carried out taking one interval and one point at a time. Using (7) each time we subdivide, and then using (8) each time we gather terms involving the same interval, we deduce that

$$c_1\mu_F(I_1)+\dots+c_r\mu_F(I_r), \quad d_1\mu_F(J_1)+\dots+d_s\mu_F(J_s)$$

are equal because they are individually equal to the corresponding expression for the common refinement. Once we have checked the consistency it follows immediately that $\int dF$ is a positive linear functional on the linear space L'.

There are two possible ways to continue. We may either modify the argument of § 3.2 or we may follow the pattern of the Daniell construction in § 8.1.

In the former approach a *null* set is redefined as a set which can be covered by a sequence $\{(a_n, b_n]\}$ of intervals for which the total variation $\Sigma\{F(b_n)-F(a_n)\}$ is arbitrarily small, and Theorem 3.2.1 extends immediately with minor changes in wording. The only snag arises in the proof of Lemma 3.2.1 (which we now recognise as a slightly strengthened form of the Daniell condition). The trouble is that any point b at which F is discontinuous has non-zero measure

$$\mu_F\{b\} = F(b)-F(b-0)$$

and so the end points of the intervals of the various step functions ϕ_n in Lemma 3.2.1 can no longer be dismissed as forming a null set. Let us sketch one way of meeting the difficulty. (For details see Exx. 1, 2, and for an alternative method see Ex. 6.)

Express F as the sum of two functions G, H where G is continuous and H picks out the discontinuities of F, i.e. H has a jump discontinuity whenever F has a jump discontinuity and H is constant between discontinuities of F. Then

$$\int \phi dF = \int \phi dG + \int \phi dH$$

and it suffices to prove Lemma 3.2.1 for G, H separately. The original proof is easily adapted to take care of G and the 'pure jump' function H can be treated as follows.

Suppose that ϕ_1 vanishes outside $(a,b]$ and that $\phi_1 \leqslant K$. The discontinuities of H are countable: let H have jumps j_m at the points p_m in $(a,b]$ $(m = 1, 2, \ldots)$. (We consider the case where there are infinitely many discontinuities: the finite case is a little simpler.) Then

$$\int \phi_n dH = \sum_{m=1}^{\infty} \phi_n(p_m) j_m,$$

where
$$\sum_{m=1}^{\infty} j_m = H(b) - H(a).$$

For any $\epsilon > 0$, find M so that

$$\sum_{M+1}^{\infty} j_m < \epsilon.$$

As $\phi_n(x) \downarrow 0$ for almost all x, we certainly must have $\phi_n(p) \downarrow 0$ at a point of discontinuity p (because $\{p\}$ is not null); now find N so that

$$\phi_n(p_m) < \epsilon$$

for $m = 1, 2, \ldots, M$ and for all $n \geqslant N$. Then

$$\int \phi_n dH \leqslant \epsilon\{H(b) - H(a)\} + K\epsilon$$

for $n \geqslant N$, from which it follows that

$$\int \phi_n dH \downarrow 0.$$

The analogue of Lemma 3.2.2 now follows by the same argument as before and we may extend $\int dF$ *unambiguously* by means of increasing sequences to a set L^{inc} and then to a linear space L^1 by taking differences. It is obvious that L^1 contains all the step functions on **R** (and not just the ones whose intervals are of type (,]). The Monotone Convergence Theorem 5.1.2 (or 8.2.2) now follows for $\int dF$ and the corresponding measure μ_F satisfies the conditions of Theorem 6.2.1 (or 8.4.2).

We may summarise our findings in the following theorem.

Theorem 1. *The existence of any one of the following implies the existence of the other two satisfying the equations* (9):

(i) *a Daniell integral $\int$ for which L^1 contains all the step functions on* $\mathbf{R}$,

(ii) *a measure μ which is finite on the bounded intervals in* $\mathbf{R}$,

(iii) *an increasing function $F: \mathbf{R} \to \mathbf{R}$ which is continuous on the right,*

where

$$\int \chi_{(a,b]} = \mu(a,b] = F(b) - F(a) \quad \textit{for} \quad a < b. \tag{9}$$

(In the context of this theorem a measure may be regarded as a function μ which satisfies the conditions of Theorem 8.4.2 (iii).)

The integral in (i) is usually written $\int dF$ and called the *Lebesgue–Stieltjes integral* with respect to F; likewise the measure in (ii) is written μ_F and called the *Lebesgue–Stieltjes measure* with respect to F.

Let us reconsider the final part of the proof of Theorem 1. Instead of following § 3.2 we may refer to the Daniell construction of § 8.1. It is convenient to define $\int dF$ on the linear space L consisting of all step functions

$$\phi = c_1 \chi_{I_1} + \ldots + c_r \chi_{I_r}$$

by the same formula as before, viz.

$$\int \phi \, dF = c_1 \mu_F(I_1) + \ldots + c_r \mu_F(I_r),$$

but where $I_1, \ldots, I_r$ are arbitrary bounded intervals on $\mathbf{R}$ and μ_F is defined by equations (5) (which simplify to look more like equations (4) as $F(x+0) = F(x)$ for all x). To establish the consistency of this definition we use a slightly different method of expressing ϕ in terms of disjoint intervals. If n distinct points are given on $\mathbf{R}$ then they may be used to express $\mathbf{R}$ as the union of $(2n+1)$ disjoint intervals, n of which consist of the points themselves, two are unbounded and $(n-1)$ are the open intervals between the points. Without loss of generality the intervals $I_1, \ldots, I_r$ are non-empty: if they have n distinct end points we may express each interval I_i as the union of certain of the $(2n-1)$ disjoint bounded intervals determined by them. Thus ϕ may be expressed in terms of disjoint intervals. (Incidentally this shows at once that $|\phi| \in L$ and so L is a *linear lattice*.) If p is a point of the bounded interval I then I may be expressed as the disjoint union

$$I = I' \cup \{p\} \cup I'',$$

where I', I'' consist of the points x of I which satisfy $x < p$, $x > p$, respectively. In place of equation (7) we now have

$$\mu_F(I) = \mu_F(I') + \mu_F\{p\} + \mu_F(I'') \tag{10}$$

which follows at once from the definition (5) of μ_F. If two expressions for ϕ are given, we may use the end points of all the intervals involved to find a common refinement just as on p. 37, and the same argument as before proves the consistency of the definition of $\int dF$ on L. It is now clear that $\int dF$ is a positive linear functional on the linear lattice L. All that remains to check is that $\int dF$ satisfies the Daniell condition

$$\phi_n \downarrow 0 \Rightarrow \int \phi_n dF \to 0$$

for ϕ_n in L, and hence is an elementary integral on L. To this end it is convenient to introduce a new idea.

A function $f: \mathbf{R} \to \mathbf{R}$ is said to be *upper semicontinuous* if

$$\{x \in \mathbf{R}: f(x) < c\}$$

is an open subset of $\mathbf{R}$ for every real number c. (See Exx. 3, 4 and Ex. 10.1.2 for further details.) For example, if $A_1, \ldots, A_m$ are disjoint *closed* intervals and $a_1, \ldots, a_m$ are positive, then the step function

$$\phi = a_1 \chi_{A_1} + \ldots + a_m \chi_{A_m}$$

is upper semicontinuous. We may now prove that the Daniell condition holds for upper semicontinuous step functions.

Lemma 1. *Let $\{\theta_n\}$ be a decreasing sequence of upper semicontinuous step functions which converges (everywhere) to zero. Then*

$$\int \theta_n dF \to 0.$$

Proof. Let θ_1 vanish outside the compact interval $[a_1, b_1]$. Let $\epsilon > 0$ be given and define

$$G_n = \{x \in \mathbf{R}: \theta_n(x) < \epsilon\}.$$

According to the above definition G_n is open, and clearly $G_n \subset G_{n+1}$ for all $n \geqslant 1$. As $\theta_n(x) \downarrow 0$ for all x, it follows that $\mathbf{R}$ is the union of the open sets G_n for $n \geqslant 1$.

Now we may apply the Heine–Borel Theorem which shows that

$$[a_1, b_1] \subset G_N$$

for a suitably large integer N (depending on ϵ). Thus

$$\theta_n(x) < \epsilon$$

for all x in $[a_1, b_1]$ and all $n \geqslant N$. Hence

$$\int \theta_n dF \leqslant \epsilon \mu_F[a_1, b_1]$$

for $n \geqslant N$ and this proves the lemma.

Now for the Daniell condition. Let $\{\phi_n\}$ be a decreasing sequence of step functions in L converging (everywhere) to zero and let $\epsilon > 0$ be given. Express ϕ_n in terms of disjoint intervals and replace each non-closed interval by a slightly smaller closed subinterval, obtaining in this way an upper semicontinuous step function ψ_n satisfying

$$0 \leqslant \psi_n \leqslant \phi_n$$

and

$$\int (\phi_n - \psi_n) dF < \epsilon/2^n.$$

The sequence $\{\psi_n\}$ may not be decreasing, so define

$$\theta_n = \min\{\psi_1, \ldots, \psi_n\}.$$

It follows at once from the definition that θ_n is upper semicontinuous, $0 \leqslant \theta_n \leqslant \psi_n \leqslant \phi_n$ and $\theta_n \downarrow 0$. By Lemma 1 we may find N such that

$$\int \theta_n dF < \epsilon$$

for $n \geqslant N$. But

$$\begin{aligned}\int (\phi_n - \theta_n) dF &= \int \max\{\phi_n - \psi_1, \phi_n - \psi_2, \ldots, \phi_n - \psi_n\} dF \\ &\leqslant \int \max\{\phi_1 - \psi_1, \phi_2 - \psi_2, \ldots, \phi_n - \psi_n\} dF \\ &\leqslant \int \{(\phi_1 - \psi_1) + (\phi_2 - \psi_2) + \ldots + (\phi_n - \psi_n)\} dF \\ &< \epsilon/2 + \epsilon/2^2 + \ldots + \epsilon/2^n < \epsilon\end{aligned}$$

and so

$$\int \phi_n dF < 2\epsilon$$

for $n \geqslant N$, which establishes the Daniell condition.

As on p. 23 we shall denote

$$\int f\chi_A dF \quad \text{by} \quad \int_A f dF,$$

provided the former integral exists. In particular, if A is one of the intervals $(a, b]$, (a, b), $[a, b]$, $[a, b)$, we may be tempted to use the notation

$$\int_a^b f dF,$$

but this will be ambiguous unless we are sure that F is continuous at both points a, b. (Recall that $\mu_F\{p\}$ is the jump at p.)

If the functions f, F are given in terms of formulae it is very convenient to use the classical notation

$$\int f(x)\, dF(x).$$

For example,
$$\int_1^2 x\, dx^3$$
is a natural way of writing
$$\int_A f dF,$$
where $f(x) = x$, $F(x) = x^3$ for all x in $\mathbf{R}$ and where A is one of the four intervals with end points 1, 2. From experience with calculus we may be confident in evaluating this integral as

$$\int_1^2 x\, 3x^2 dx = \tfrac{3}{4}[x^4]_1^2 = \tfrac{45}{4}.$$

The justification for this manipulation may be given at different levels of difficulty. From the practical point of view the simplest result is often the easiest to use.

Proposition 1. *Let G have continuous positive derivative G' at every point of $\mathbf{R}$. If $f \in L^1(\mu_G)$, then*

$$\int f dG = \int f G',$$

or, in the classical notation,

$$\int_{-\infty}^{\infty} f(x) dG(x) = \int_{-\infty}^{\infty} f(x) G'(x) dx.$$

(Cf. Proposition 3.4.2 on Integration by Substitution.)

Proof. (i) By the simplest form of the Fundamental Theorem of the Calculus (Theorem 3.4.2)

$$G(b) - G(a) = \int_a^b G'(x) dx.$$

In other words, our result is true when $f = \chi_{(a,b]}$.

(ii) The result extends by linearity to the case where f is a step function in L' (with intervals of type (,]).

(iii) Suppose that $\mu_G(A) = 0$. Then there is an increasing sequence $\{\psi_n\}$ of step functions in L' such that $\{\psi_n\}$ diverges on A and $\int \psi_n dG$ is bounded. Let $S = \{x \in \mathbf{R}: G'(x) \neq 0\}$, so that $\{\psi_n G'\}$ diverges on $A \cap S$, and $\int \psi_n G' = \int \psi_n dG$ is bounded. Thus $A \cap S$ has Lebesgue measure zero.

Now suppose that $\{\phi_n\}$ is an increasing sequence of step functions of L' which converges to f outside the set A with $\mu_G(A) = 0$. Then $\{\phi_n G'\}$ converges to fG' outside the Lebesgue null set $A \cap S$ so that $fG' \in L^{\text{inc}}$ and $\int f dG = \int fG'$.

(iv) The extension to $L^1(\mu_G)$ is immediate by linearity.

This proof may be compared profitably with the proofs of Proposition 5.1.3 and Theorem 6.4.1. Almost the same proof applies if we assume only that there is a positive function g which satisfies

$$G(x) = \int_c^x g + C$$

for all x in $\mathbf{R}$, and replace the second integral by $\int fg$ (see Ex. 11). As a matter of fact, we shall prove in Theorem 17.3.1 that the equation

$$\int f dG = \int fg$$

may be interpreted in the most generous possible way, viz. the existence of either integral implies the existence of the other, and they are then equal.

These results assume a good deal about the Stieltjes function G. A general method of expressing a Lebesgue–Stieltjes integral as a Lebesgue integral is given in Ex. 8.

Exercises

1. Let $G: \mathbf{R} \to \mathbf{R}$ be increasing and continuous. Extend Lemma 3.2.1 to the case of $\int dG$ (see pp. 37, 8).

2. Suppose that $F: \mathbf{R} \to \mathbf{R}$ is increasing, continuous on the right and bounded below. Show that $F = G + H$, where G is increasing and continuous and where

$$H(x) = \sum_{p \leqslant x} \{F(p) - F(p-0)\} \quad (x \in \mathbf{R})$$

(the sum being taken over points of discontinuity p of F).

Extend this construction to the case where F is unbounded below.

3. A function $f: \mathbf{R} \to \mathbf{R}$ is *lower semicontinuous* if

$$\{x \in \mathbf{R} : f(x) > c\}$$

is an open subset of $\mathbf{R}$ for all c in $\mathbf{R}$; similarly, f is *upper semicontinuous* if

$$\{x \in \mathbf{R} : f(x) < c\}$$

is an open subset of $\mathbf{R}$ for all c in $\mathbf{R}$.

Show that f is continuous if and only if f is both lower and upper semicontinuous.

Show that the characteristic function χ_A is lower semicontinuous if and only if A is open, and upper semicontinuous if and only if A is closed.

4. Show that the supremum of any bounded collection of lower semicontinuous functions is lower semicontinuous, and that the infimum of any bounded collection of upper semicontinuous functions is upper semicontinuous.

5. Let L denote the lattice of step functions on $\mathbf{R}$. For what increasing functions F is

$$\int \phi\, dF = \phi(0)$$

for all ϕ in L? For what increasing functions F is

$$\int \phi\, dF = \int x^2 \phi(x)\, dx$$

for all ϕ in L?

6. Let $F\colon \mathbf{R} \to \mathbf{R}$ be increasing and continuous on the right.

(i) The set $\{x \in \mathbf{R}\colon F(x) \geqslant c\}$ is either empty, or an interval $[d, \infty)$, or the whole line $\mathbf{R}$: in the second case define

$$G(c) = d.$$

If we dismiss the trivial case where F is constant, show that G is defined on the interval J which is the union of all the intervals $(F(a), F(b)]$ $(a, b \in \mathbf{R})$ and that

$$G(c) \in (a, b] \Leftrightarrow c \in (F(a), F(b)].$$

(ii) For any step function ϕ in L' (using intervals of type (,] as on p. 36) define the (composite) step function $\psi = \phi \circ G$ on J by the rule

$$\psi(c) = \phi(G(c))$$

and show that the Lebesgue integral

$$\int_J \psi$$

is equal to the Lebesgue–Stieltjes integral

$$\int \phi\, dF.$$

(iii) Deduce the Daniell condition for the Lebesgue–Stieltjes integral.

7. Show that the function G of Ex. 6 is increasing and continuous on the left. What are the composite functions $F \circ G$, $G \circ F$?

If F is strictly increasing and continuous, show that G is the inverse of F.

8. We use the notation of Ex. 6.

(i) Let $\{\phi_n\}$ be an increasing sequence of step functions in L' which diverges on A and satisfies $\int \phi_n dF \leqslant K$. If $\psi_n = \phi_n \circ G$, show that $\{\psi_n\}$ diverges on $G^{-1}(A)$ and $\int_J \psi_n \leqslant K$.

(ii) If the Lebesgue–Stieltjes integral $\int f dF$ exists and $g = f \circ G$, show that the Lebesgue integral $\int_J g$ exists and these two integrals are equal.

(iii) Let $g = f \circ G$. As G is constant on certain intervals I_r of type (,], the same is true of g. Let $\{\psi_n\}$ be an increasing sequence of step functions on J which converges almost everywhere to g and such that $\int \psi_n \leqslant K$. Adapt ψ_n by replacing ψ_n by its mean value on each interval of constancy I_r which contains a discontinuity of ψ_n. Show that the adapted step function ψ_n converges to g on each I_r and satisfies $\int \psi_n \leqslant K$ and $\psi_n \circ F \circ G = \psi_n$.

(iv) Let $g = f \circ G$. If $g \in L^1(J)$ show that $\int f dF$ exists and equals $\int_J g$.

Combining (ii), (iv). Let $g = f \circ G$; then

$$\int f dF = \int_J g$$

in the sense that if one side exists, so does the other, and they are equal.

9. If $F(x) = x^3$ $(x \in \mathbf{R})$ find the corresponding function G of Ex. 6 and evaluate

$$\int_{[1,2]} x^2 dx^3$$

by means of Ex. 8. Use Proposition 1 to check the result.

10. For x in $\mathbf{R}$ let $F(x)$ be the smallest integer n satisfying $n > x$. Show that F is increasing and continuous on the right. Find the corresponding function G of Ex. 6 and the composite functions $F \circ G$, $G \circ F$.

Use Ex. 8 to evaluate

$$\int_I x^{-2} dF(x)$$

where I is the open interval $(0, \infty)$. (The answer may be given as the sum of a series.) Is there an easier way of evaluating this integral?

11. Prove the following generalisation of Proposition 1. *Suppose that there is a positive function g such that*

$$G(x) = \int_c^x g + C$$

for all x in $\mathbf{R}$. If $f \in L^1(\mu_G)$, then

$$\int f dG = \int fg.$$

The Riemann–Stieltjes Integral

Let f, G be real valued functions on the compact interval $[a, b]$ and consider the sum

$$S = \sum_{i=1}^{r} f(t_i)\{G(x_i) - G(x_{i-1})\},$$

where $a = x_0 < x_1 < \ldots < x_r = b$ and $t_i \in [x_{i-1}, x_i]$ for $i = 1, \ldots, r$. If this sum tends to a limit l as $\max(x_i - x_{i-1}) \to 0$, then we say that f is integrable with respect to G on $[a, b]$ in the sense of Riemann–Stieltjes and write the limit l as

$$\int_a^b f(x)\, dG(x).$$

Rather more precisely, the existence of the limit l means that, to any $\epsilon > 0$, there is a $\delta > 0$ such that $|S - l| < \epsilon$ for any finite dissection

$$a = x_0 < x_1 < \ldots < x_r = b$$

of $[a, b]$ provided only that $\max\limits_{1 \leqslant i \leqslant r} (x_i - x_{i-1}) < \delta$.

We shall now see that this Riemann–Stieltjes integral agrees with the Lebesgue–Stieltjes integral in the classical case where f is continuous and G is increasing on $[a, b]$. For the purpose of interpreting the latter integral we extend G so that

$$G(a-0) = G(a) \quad \text{and} \quad G(b+0) = G(b).$$

A. Let f be continuous and G increasing and continuous on the right. Show that the Riemann–Stieltjes integral

$$\int_a^b f(x)\, dG(x) = \int_{[a,b]} f\, dG.$$

B. Show that Ex. A remains true if G is no longer assumed to be continuous on the right.

9.2 Integrals on $\mathbf{R}^k$

The ideas of the last section extend with very little difficulty to $\mathbf{R}^k$. The only trouble arises in finding the k-dimensional analogue of the increasing function $F_0: \mathbf{R} \to \mathbf{R}$. To do this we take our lead from the special case of Lebesgue measure. As a temporary notation denote by I_x the interval of type $(\,,\,]^k$ with one vertex at the origin and the diagonally opposite vertex at the point $x = (x_1, \ldots, x_k)$; we agree that $I_x = \varnothing$ if any $x_i = 0$. Then the Lebesgue measure

$$m(I_x) = |x_1 \ldots x_k|.$$

(In Fig. 6, $x_1 < 0$, $x_2 > 0$.)

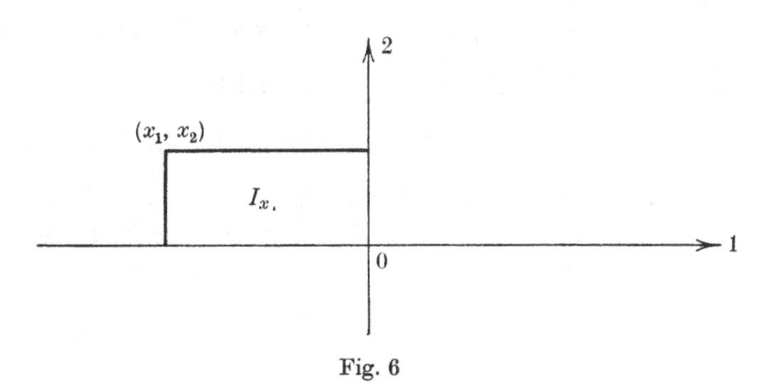

Fig. 6

In particular $$m(I_x) = x_1 \dots x_k,$$
when all the x's are positive. This suggests that we define

$$F_0(x) = x_1 \dots x_k$$

in the case of Lebesgue measure. If the interval I is given by the inequalities

$$a_i < x_i \leqslant b_i \tag{1}$$

for $i = 1, \dots, k$, then

$$m(I) = (b_1 - a_1) \dots (b_k - a_k).$$

(We do not want to use the special property that m is invariant under translations, so we ignore the fact that $m(I) = F_0(b-a)$: but see Ex. 4.) This product may be expanded to show that

$$m(I) = b_1 b_2 \dots b_k - a_1 b_2 \dots b_k + \dots + (-1)^k a_1 a_2 \dots a_k,$$

i.e.

$$m(I) = F_0(b_1, b_2, \dots, b_k) - F_0(a_1, b_2, \dots, b_k) + \dots + (-1)^k F_0(a_1, a_2, \dots, a_k),$$

where there are 2^k terms on the right hand side and the sign in front of any term is $(-1)^n$ if there are n a's present.

With all this in mind we may now return to the Daniell integral $\int$ on $\mathbf{R}^k$, or the corresponding measure μ, and define

$$F_0(x) = \operatorname{sgn}(x_1 \dots x_k) \int \chi_{I_x} = \operatorname{sgn}(x_1 \dots x_k)\, \mu(I_x). \tag{2}$$

Recall that the *sign* function on $\mathbf{R}$ satisfies

$$\begin{aligned} \operatorname{sgn} r &= 1 \quad \text{if} \quad r > 0, \\ &= 0 \quad \text{if} \quad r = 0, \\ &= -1 \quad \text{if} \quad r < 0. \end{aligned}$$

This, of course, gives $F_0(x) = x_1 \dots x_k$ in the Lebesgue case. If I is the interval defined by the inequalities (1), it is not difficult to verify that

$$\mu(I) = F_0(b_1, b_2, \ldots, b_k) - F_0(a_1, b_2, \ldots, b_k) + \ldots + (-1)^k F_0(a_1, a_2, \ldots, a_k), \tag{3}$$

where the sign is $(-1)^n$ if there are n a's present (see Ex. 3 and Fig. 7). We may also verify, just as in § 9.1, that the function F_0 defined by (2)

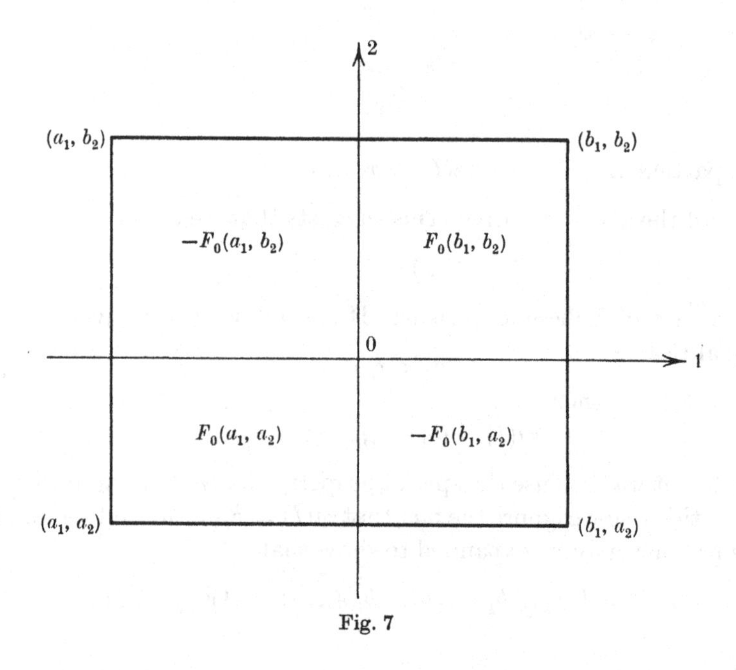

Fig. 7

is continuous on the right as a function of each variable x_i separately; in other words,

$$F_0(x_1, \ldots, x_i + t, \ldots, x_k) \to F_0(x_1, \ldots, x_i, \ldots, x_k) \quad \text{as} \quad t \downarrow 0$$

$(x_1, \ldots, x_k \in \mathbf{R}, i = 1, \ldots, k)$.

Suppose that the function $F: \mathbf{R}^k \to \mathbf{R}$ is continuous on the right in each variable and that

$$\mu_F(I) = F(b_1, b_2, \ldots, b_k) - F(a_1, b_2, \ldots, b_k) + \ldots + (-1)^k F(a_1, a_2, \ldots, a_k) \tag{4}$$

is positive for every interval I (defined by inequalities (1)); then we call F a *Stieltjes function* on $\mathbf{R}^k$. The argument of the last section, based on § 3.2 and § 4.2, may now be extended to prove the k-dimensional analogue of Theorem 9.1.1 (see Ex. 5).

Theorem 1. *The existence of any one of the following implies the existence of the other two satisfying the equations* (5):

(i) *a Daniell integral* $\int$ *for which* L^1 *contains all the step functions on* $\mathbf{R}^k$,

(ii) *a measure* μ *which is finite on the bounded intervals in* $\mathbf{R}^k$,

(iii) *a Stieltjes function* F: $\mathbf{R}^k \to \mathbf{R}$, *where*

$$\int \chi_I = \mu(I) = F(b_1, b_2, \ldots, b_k) - F(a_1, b_2, \ldots, b_k) + \ldots + (-1)^k F(a_1, a_2, \ldots, a_k). \quad (5)$$

In equation (5) the interval I is defined by the inequalities (1) and the sign in front of any of the 2^k terms on the right hand side is $(-1)^n$ if n a's are present.

As before, the integral in (i) is usually written $\int dF$ and called the *Lebesgue–Stieltjes integral* with respect to F, and the measure in (ii) is written μ_F and called the *Lebesgue–Stieltjes measure* with respect to F.

Exercises

1. Let
$$F(x_1, x_2) = 1 \quad \text{for} \quad x_1 \geqslant 0, x_2 \geqslant 0,$$
$$= 0 \quad \text{otherwise.}$$

Show that F is a Stieltjes function on $\mathbf{R}^2$ and find the corresponding measure μ_F.

2. Let
$$F(x_1, x_2) = \max\{0, x_1+x_2+1\} \quad \text{for} \quad x_1+x_2 < 0,$$
$$= 1 \quad \text{for} \quad x_1+x_2 \geqslant 0.$$

Show that F is continuous and increasing in each variable separately, but is not a Stieltjes function on $\mathbf{R}^2$.

3. If μ is a Lebesgue–Stieltjes measure and F_0 is defined by equation (2), verify equation (3). Also verify that F_0 is continuous on the right in each variable x_i separately.

4. If the Lebesgue–Stieltjes measure μ_F is invariant under translations, show that μ_F is a constant multiple of Lebesgue measure. (In the one-dimensional case show that F is continuous, and, if $F(0) = 0$, satisfies $F(a+b) = F(a)+F(b)$ for all a, b in $\mathbf{R}$.)

5. Outline a proof of Theorem 1.

9.3 Topological characterisation

Throughout this chapter the Lebesgue–Stieltjes integrals have been discussed in terms which emphasise their apparent dependence on the axes in $\mathbf{R}^k$: this is particularly true of the rather complicated Stieltjes functions discussed in § 9.2. But now recall from § 4.2 that any function f which is continuous on a bounded closed interval I and vanishes outside I is the limit (everywhere) of an increasing bounded sequence of step functions. (This was proved by dividing I into smaller intervals by successive halving. At the stage where I was expressed as the union of the disjoint intervals $I_1, \ldots, I_r$ the corresponding step function was

$$\phi = c_1\chi_{I_1} + \ldots + c_r\chi_{I_r},$$

where $$c_i = \inf\{f(x): x \in I_i\}$$

for $i = 1, 2, \ldots, r$.) It follows by the Monotone Convergence Theorem that f is integrable with respect to any Lebesgue–Stieltjes integral on $\mathbf{R}^k$.

In the more general setting of topology it is convenient to discuss functions which are continuous on the whole space. If a function $f: \mathbf{R}^k \to \mathbf{R}$ vanishes outside a compact set C we shall say that f *has compact support* (or that f *is a function of compact support*). For the moment regard this as a simple shorthand; the term 'support' will be defined in the next chapter. According to the Heine–Borel Theorem, the compact sets in $\mathbf{R}^k$ are precisely the bounded closed subsets of $\mathbf{R}^k$. Therefore any compact set C may be enclosed in a compact interval I and it follows from the preceding paragraph that a continuous function of compact support is integrable with respect to any Lebesgue–Stieltjes integral on $\mathbf{R}^k$.

We also recall from § 6.3 that any open set A in $\mathbf{R}^k$ is the union of a sequence of disjoint bounded intervals. (This was proved by considering for each $n \geqslant 0$ the collection $\mathscr{S}_n$ of cubes of type $[\,,\,)^k$ of side $1/2^n$ and with vertices at the rational points with denominator 2^n. If S_n is the union of the cubes of $\mathscr{S}_n$ that are contained in A, then A is the union of $S_0, S_1 \setminus S_0, S_2 \setminus S_1, \ldots$.) Thus any open set is measurable with respect to any Lebesgue–Stieltjes measure; moreover, any closed set is measurable and any compact set C has finite measure (because C may be enclosed in a compact interval which obviously has finite measure).

In fact these two topological properties characterise the Lebesgue–Stieltjes integrals and measures on $\mathbf{R}^k$.

Theorem 1. *A Daniell integral $\int$ on $\mathbf{R}^k$ is a Lebesgue–Stieltjes integral if and only if the space L^1 of integrable functions contains all continuous functions of compact support.*

A measure μ on $\mathbf{R}^k$ is a Lebesgue–Stieltjes measure if and only if μ gives finite measure to all compact subsets.

Proof. We have already proved half of this theorem.

Suppose that the space L^1 contains all continuous functions f on $\mathbf{R}^k$ of compact support, and let I be a compact interval. Then we may find a decreasing sequence $\{f_n\}$ of such functions which converges to χ_I. This is obvious enough in $\mathbf{R}^1$ when $I = [a, b]$ as Fig. 8 makes clear.

In $\mathbf{R}^k$ we may use the distance $d(x, y) = |x-y|$ introduced in § 4.1 to construct such a function f_n. Let I_n be the open interval concentric with I whose sides are all $2/n$ longer than the corresponding sides of I (see Fig. 9).

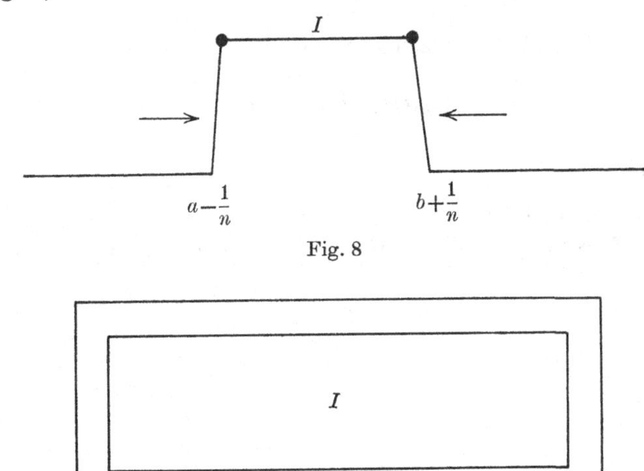

Fig. 8

Fig. 9

Then the function $f_n \colon \mathbf{R}^k \to [0, 1]$ defined by

$$f_n(x) = \frac{d(x, I_n{}^c)}{d(x, I_n{}^c) + d(x, I)}$$

for x in $\mathbf{R}^k$, is easily seen to be continuous and to vanish outside I_n, a fortiori outside $\bar{I}_n$. It is also clear that $f_n(x) = 1$ if and only if $x \in I$, so $f_n \to \chi_I$. In view of this, χ_I is integrable for any compact interval I. Any step function is a finite linear combination of such characteristic functions (cf. Ex. 4.1.6) and so belongs to L^1.

Finally, suppose that any compact subset of $\mathbf{R}^k$ has finite measure with respect to μ. We may express any bounded interval I as the union of an increasing sequence of compact intervals (all contained in $\bar{I}$), and so I has finite measure with respect to μ. This completes the proof.

The next chapter follows up these topological ideas and guarantees an important place for the Lebesgue–Stieltjes integrals and measures in the mainstream of mathematics.

Exercises

1. Let f_n be the function defined in the proof of Theorem 1. Show that f_n is continuous and that $f_n \to \chi_I$. Is it true that $f_n \downarrow \chi_I$? Show that $f_1{}^n \downarrow \chi_I$.

2. Let L be the set of real valued continuous functions on $\mathbf{R}$ of compact support. Show that L is a linear lattice. For what increasing functions F is

$$\int \phi \, dF = \phi(0)$$

for all ϕ in L? For what increasing functions F is

$$\int \phi \, dF = \int x^2 \phi(x) \, dx$$

for all ϕ in L? (Cf. Ex. 9.1.5.)

3. Use the Comparison Theorem 8.3.1 to show that a Lebesgue–Stieltjes integral on $\mathbf{R}^k$ is uniquely determined by its restriction to the continuous functions of compact support.

10

THE RIESZ REPRESENTATION THEOREM

The original Representation Theorem of F. Riesz refers to the linear space $C[a,b]$ of continuous real valued functions on the compact interval $[a,b]$, and states that any positive linear functional P on $C[a,b]$ may be 'represented' by means of a Lebesgue–Stieltjes integral:

$$P(\phi) = \int \phi \, dF$$

for all ϕ in $C[a,b]$. Now it is a remarkable fact that an arbitrary positive linear functional P on $C[a,b]$ is not only 'representable' by means of an integral, it *is* an integral! More precisely, $C[a,b]$ is a linear lattice and the positive linear functional P satisfies the Daniell condition

$$\phi_n \downarrow 0 \Rightarrow P(\phi_n) \to 0,$$

so that *P is an elementary integral* in the sense of Daniell as defined at the beginning of § 8.1. We shall prove this quite simply in a moment. It follows that P may be extended by the standard construction to a Daniell integral $\int$ which obviously must satisfy

$$P(\phi) = \int \phi$$

for all ϕ in $C[a,b]$. (This is just what we mean by 'extending'.) At this stage we already know that P has the character of an integral, but we may refer to Theorem 9.3.1 (or, rather, a simplified version in which everything is restricted to $[a,b]$ in $\mathbf{R}$) to see that the Daniell integral is, in this case, a Lebesgue–Stieltjes integral on $[a,b]$. From the point of view of integration theory this is extremely satisfactory; the original theorem only went this far. But another glance at Theorem 9.3.1 reveals that the measure μ corresponding to the Daniell integral $\int$ gives finite measure to all the open and all the closed subsets of $[a,b]$. Thus μ is a *finite Borel measure* in present day terminology (see § 15.1). The modest beginning with a positive linear functional on $C[a,b]$ has led to an integral and a measure which are custom built to deal with the functions and sets in the topology! This beautiful theorem has been generalised in several significant ways and has had a profound influence on the whole study of integration since its announcement in 1909 [16].

In § 10.1 we establish Riesz's Theorem for positive linear functionals on a locally compact metric space. This is further generalised in § 10.2 to the case of bounded linear functionals by taking differences, and in § 10.3 to the case of complex linear functionals by considering 'real' and 'imaginary' parts.

10.1 Positive linear functionals

As a first step towards a more general statement of Riesz's Theorem let us replace the compact interval $[a, b]$ by an arbitrary *compact* topological space X and let $C(X)$ denote the set of all continuous functions $\phi: X \to \mathbf{R}$. It is immediately verified that $C(X)$ is a linear lattice, i.e. a linear space (over $\mathbf{R}$) closed with respect to absolute values. Thus we may take $C(X)$ as the space L of elementary functions in the Daniell theory. Now suppose that P is an arbitrary positive linear functional on L. We shall prove that P satisfies the Daniell condition

$$\phi_n \downarrow 0 \Rightarrow P(\phi_n) \to 0.$$

In the present situation this follows from a theorem of Dini. It is convenient to introduce a *norm* in $C(X)$, viz.

$$\|\phi\| = \sup_{x \in X} |\phi(x)|.$$

This supremum exists as the continuous function ϕ carries the compact set X onto a compact set in $\mathbf{R}$.

***Theorem 1* (*Dini*).** *Let X be a compact topological space and let $C(X)$ be the set of continuous real valued functions on X. If $\phi_n \in C(X)$ for all n and $\phi_n(x) \downarrow 0$ for all x in X, then $\|\phi_n\| \to 0$.*

(If $\|\phi_n\| \to 0$ we say that $\{\phi_n\}$ converges *uniformly* to 0.)

Proof. For any given $\epsilon > 0$, define

$$S_n^\epsilon = \{x \in X: \phi_n(x) < \epsilon\}.$$

As ϕ_n is continuous, S_n^ϵ is an open subset of X, and as $\phi_n \downarrow 0$ pointwise, $\{S_n^\epsilon\}$ is increasing and $\bigcup S_n^\epsilon = X$. By the compactness of X it follows that there is an integer N for which $S_N^\epsilon = X$. Thus

$$0 \leqslant \phi_n(x) < \epsilon$$

for $n \geqslant N$ and x in X. Hence $\|\phi_n\| \leqslant \epsilon$ for $n \geqslant N$. This establishes Dini's Theorem.

Now return to the positive linear functional P, and assume that $\phi_n \downarrow 0$. Dini's Theorem shows that

$$0 \leqslant \phi_n \leqslant \epsilon 1$$

for $n \geqslant N$, whence
$$0 \leqslant P(\phi_n) \leqslant \epsilon P(1)$$

for $n \geqslant N$, and the Daniell condition for P follows at once. Here, and subsequently, we use 1 as a shorthand for the function χ_X which takes the constant value 1 on X.

Now suppose that (X, d) is a metric space, i.e. there is a distance function, or metric, d on X. The point of this assumption is to ensure a simple proof of the following result.

***Theorem 2* (*Urysohn's Lemma*).** *Let X be a metric space and A, B two disjoint non-empty closed subsets of X. Then there is a continuous function $\psi: X \to [0, 1]$ such that*

$$\psi(x) = 0 \quad \textit{if and only if} \quad x \in A,$$
$$\psi(x) = 1 \quad \textit{if and only if} \quad x \in B.$$

Proof. As A is closed, the distance $d(x, A) = 0$ if and only if $x \in A$. We need only set
$$\psi(x) = \frac{d(x, A)}{d(x, A) + d(x, B)}$$

as in the proof of Theorem 9.3.1.

We are now in a position to state and prove the famous Representation Theorem.

***Theorem 3* (*Riesz*).** *Let X be a compact metric space and $C(X)$ the linear space of continuous real valued functions on X. Any positive linear functional P on $C(X)$ extends to a Daniell integral $\int$ on a linear space L^1 containing $C(X)$. Thus*
$$P(\phi) = \int \phi \tag{1}$$

for all ϕ in $C(X)$. The corresponding measure μ, defined by

$$\mu(A) = \int \chi_A \tag{2}$$

for all χ_A in L^1, gives finite measure to all open and all closed subsets of X.

As $1 \in C(X)$, Stone's Theorem 8.5.1 may be applied to show that

$$\int f = \int f d\mu \tag{3}$$

for all f in L^1 and so equation (1) may be written in the more usual form

$$P(\phi) = \int \phi \, d\mu \tag{4}$$

for all ϕ in $C(X)$. The point of stating the first part of Theorem 3 as we have done, is that the character of P as an integral may be stated (and proved) without reference to the measure μ. This is in the spirit of Riesz's original proof. But, having said this, it is most important to realise the significance of the measure μ and its intimate connection with the Daniell integral expressed by equations (2) and (3).

Note that, as X is a compact metric space, the compact subsets of X are exactly the same as the closed subsets of X (Ex. 1).

Proof. The whole construction of § 8.1 applies and gives a Daniell integral $\int$ extending P to a space L^1 of integrable functions. As $1 \in L$, Stone's Condition is certainly satisfied; thus all sets

$$A_c = \{x \in X: \phi(x) \geqslant c\}$$

($\phi \in C(X)$, $c \in \mathbf{R}$) are measurable. In fact they have finite measure, because X has finite measure $P(1)$.

It is sufficient to prove that any closed subset of X is an A_c for suitable c and ϕ; the result for open subsets follows by taking complements in X. Now we use the metric d in X. Let E be a closed subset of X. The special case $E = X$ is already settled as $1 \in L$, and the case $E = \varnothing$ is trivial. We may therefore assume that E is non-empty and that there is a point p in $X \setminus E$. Take $A = \{p\}$, $B = E$ in Urysohn's Lemma to obtain a continuous function ψ for which $E = \{x \in X: \psi(x) \geqslant 1\}$.

Suppose now that X is *locally compact*, i.e. to each point x of X there is an open set G_x containing x whose closure $\bar{G}_x$ is compact. For example, $\mathbf{R}^k$ is locally compact as we may take G_x to be the open ball, centre x, radius 1. We continue to assume that X is a metric space, though the following lemma does not require this assumption.

Lemma 1. *Let X be a locally compact topological space. Any compact subset K of X may be enclosed in an open subset G whose closure is compact.*

(Cf. the open interval I_n used in the proof of Theorem 9.3.1.)

Proof. To each x in K there is an open set G_x containing x with compact closure $\bar{G}_x$. The open sets G_x form an open covering of K and so there is a finite selection $G_{x_1}, \ldots, G_{x_n}$, say, which cover K. Their union G is open and has compact closure $\bar{G}_{x_1} \cup \ldots \cup \bar{G}_{x_n}$.

If $\phi: X \to \mathbf{R}$ and $S = \{x \in X: \phi(x) \neq 0\}$ then the closure K of S is called the *support* of ϕ.

Suppose that ϕ has support K where K is a compact subset of X. Then ϕ vanishes outside S and a fortiori outside K. (The expression 'ϕ vanishes outside A' really means that $S \subset A$. With this understanding an arbitrary function $\phi: X \to \mathbf{R}$ may be said to 'vanish outside X'; the condition is vacuous.) Conversely, if ϕ vanishes outside some compact set C then $S \subset C$. As C is closed (Ex. 1), the closure K of S is contained in C; now K is a closed subset of the compact set C and as such, K is compact (Ex. 1). Thus ϕ has compact support. This justifies the shorthand used in the previous chapter in which 'function of compact support' was written for 'function which vanishes outside a compact set'.

Proposition 1. *Let X be a locally compact metric space and K a compact subset of X. Then there is a continuous function $\psi: X \to [0, 1]$ of compact support such that $\psi(x) = 1$ if and only if $x \in K$. Moreover,*

$$\psi^n \downarrow \chi_K.$$

(Cf. the function f_n constructed in the proof of Theorem 9.3.1.)

Proof. Let G be the open subset of Lemma 1 and let $A = X \setminus G$, $B = K$. Then certainly A, B are closed and disjoint; suppose for the moment that A, B are both non-empty, then we apply Urysohn's Lemma and deduce the result immediately.

The special case $K = \varnothing$ is trivial for we may put $\psi = 0$ (the zero function) in this case. If $G = X$ then $\bar{G} = X$ is compact; we could have $K = X$, in which case $\psi = 1$, but if $K \neq X$, we find a point p of X not in K and take $A = \{p\}$, $B = K$ in Urysohn's Lemma as in the proof of Theorem 3.

***Theorem 4* (Riesz).** *Let X be a locally compact metric space and $C_c(X)$ the linear space of continuous real valued functions on X of compact support. Any positive linear functional P on $C_c(X)$ extends to a Daniell integral $\int$ on a linear space L^1 containing $C_c(X)$. Thus*

$$P(\phi) = \int \phi \tag{5}$$

for all ϕ in $C_c(X)$. The corresponding measure μ is defined on all open and all closed subsets of X and gives finite measure

$$\mu(K) = \int \chi_K$$

to any compact subset K of X.

Once again we may appeal to Stone's Theorem 8.5.1, because the condition

$$\min\{1, \phi\} \in L \quad \text{for any positive } \phi \text{ in } L$$

is clearly satisfied for $L = C_c(X)$. Thus (5) may be written in the form

$$P(\phi) = \int \phi d\mu \tag{6}$$

for all ϕ in $C_c(X)$.

Proof. First of all, $C_c(X)$ is a linear lattice. In proving this we use the fact that the union of two compact subsets of X is compact; the rest is the same as for $C(X)$. The Daniell condition

$$\phi_n \downarrow 0 \Rightarrow P(\phi_n) \to 0$$

may be verified almost as for Theorem 3. In place of the function 1, which may not lie in $C_c(X)$, we use the function ψ_1 of Proposition 1 which corresponds to the support K_1 of ϕ_1. Let $\epsilon > 0$ be given, and suppose that $\phi_n \downarrow 0$. We may restrict all the ϕ_n's to K_1 and apply Dini's Theorem to see that

$$0 \leqslant \phi_n \leqslant \epsilon\psi_1$$

for $n \geqslant N$. From this it follows that

$$0 \leqslant P(\phi_n) \leqslant \epsilon P(\psi_1)$$

for $n \geqslant N$ and so $P(\phi_n) \to 0$.

The Daniell construction now extends P to an integral $\int$ on a space L^1 of integrable functions. Let K be any compact subset of X and ψ the corresponding function provided by Proposition 1. Then $\psi^n \downarrow \chi_K$ and the Monotone Convergence Theorem shows that $\chi_K \in L^1$, i.e. K has finite measure $\mu(K) = \int \chi_K$.

Let B be any closed subset of X, and ϕ any positive function of $C_c(X)$ with compact support K. Then

$$\min\{\chi_B, \phi\} = \min\{\chi_{B\cap K}, \phi\}.$$

But $B \cap K$ is a closed subset of the compact set K and as such is compact. Thus $\chi_{B\cap K}$ is integrable and so also is

$$\min\{\chi_{B\cap K}, \phi\}.$$

It follows that χ_B is a measurable function, i.e. B is a measurable set. The measurability of open sets follows by taking complements in X.

This fundamental theorem may be extended in two significant ways. The condition that X should be a metric space may be drastically weakened to the condition that X should be a *Hausdorff* topological space, i.e. to any two distinct points x, y of X there exist disjoint open sets U, V such that $x \in U$, $y \in V$. This condition of 'separation' together with the local compactness is enough to prove a variation of Urysohn's Lemma in which one of the sets A, B is also assumed to be compact, and we can only conclude that

$$\psi(x) = 0 \quad \text{if} \quad x \in A,$$
$$\psi(x) = 1 \quad \text{if} \quad x \in B.$$

The Daniell construction carries through and the proofs apply almost verbatim, but the weakening of the conclusion of Urysohn's Lemma means that we cannot apply the $\psi^n \downarrow \chi_K$ argument, and there may be some non-measurable compact (let alone open or closed) subsets of X. We shall say more about this in Chapter 15 when we discuss Baire and Borel measures.

The second significant extension is that the *positive* linear functionals may be replaced by *bounded* linear functionals provided we allow *differences* of two Daniell integrals and *differences* of the corresponding measures. This is worked out in the following section.

Exercises

1. Show that any closed subset of a compact topological space is compact. If X is a Hausdorff topological space, and in particular if X is a metric space, show that a compact subset of X is closed.

2. Let X be a compact topological space. Show that an upper semicontinuous function $f: X \to \mathbf{R}$ is bounded above. Extend Dini's Theorem 1 to upper semicontinuous functions (cf. Lemma 9.1.1).

3. In the notation of Theorem 4 show that the measure of a compact set K is given by

$$\mu(K) = \inf\left\{\int \phi : \phi \in C_c(X),\ \phi \geqslant \chi_K\right\}.$$

4. Let $X = [0,1]$ and let $d(a,b) = 1$ if $a \neq b$, $d(a,b) = 0$ if $a = b$ $(a, b \in X)$. Identify the open, closed and compact subsets of the metric space (X, d) and show that X is locally compact.

Let P be the zero linear functional on $C_c(X)$ and extend P to a Daniell integral on L^1 by the standard construction. Identify the elements of L^1 and describe the corresponding measure μ.

5. Let $X = \mathbf{R}$, $K = [a,b]$ and let F_0 be an increasing function on $\mathbf{R}$. Show that $F_0(b-0) - F_0(a+0)$ is the smallest real number M_K satisfying

$$\left|\int \phi \, dF_0\right| \leqslant M_K \|\phi\|$$

for all ϕ in $C_K(X)$. (See p. 60 for the definition of $C_K(X)$.)

10.2 Bounded linear functionals

Let X be a compact metric space and P a positive linear functional on $C(X)$. Write $\|\phi\| = \sup_{x\in X} |\phi(x)|$ and 1 for the constant function χ_X. As P is positive, the inequality

$$|\phi| \leqslant \|\phi\|\,1 \tag{1}$$

yields $$|P(\phi)| \leqslant P(|\phi|) \leqslant \|\phi\|\,P(1).$$

We shall say that a linear functional F is *bounded* if there is a positive real number M such that

$$|F(\phi)| \leqslant M\|\phi\|$$

for all ϕ. Thus *any positive linear functional on $C(X)$ is bounded.*

Let X be a locally compact metric space and P a positive linear functional on $C_c(X)$. We again write

$$\|\phi\| = \sup_{x\in X} |\phi(x)|.$$

This is legitimate as the continuous function ϕ carries the support of ϕ onto a compact, and therefore bounded, subset of $\mathbf{R}$. It is not true, in general, that $1 \in C_c(X)$, but if K is a compact subset of X we may find an element ψ_K of $C_c(X)$, as in Proposition 10.1.1 which satisfies

$$\chi_K \leqslant \psi_K.$$

(Here, as always, χ_K is the characteristic function of K.) Then, if ϕ is any element of $C_c(X)$ which vanishes outside K, we have, in place of (1), the inequality

$$|\phi| \leqslant \|\phi\|\,\psi_K,$$

which yields $$|P(\phi)| \leqslant P(|\phi|) \leqslant \|\phi\|\,P(\psi_K).$$

Let $C_K(X)$ denote the subspace of $C_c(X)$ consisting of the elements that vanish outside the compact set K. Thus we have shown that *a positive linear functional on $C_c(X)$ is bounded on every $C_K(X)$.*

Both of these results depend on the existence of a norm and of certain special elements such as $1, \psi_K$ in the space in question. As so often happens, it is simpler, and much more powerful, to consider the quite general situation. Let X be an arbitrary set, L a linear lattice of functions $\phi: X \to \mathbf{R}$ and P a positive linear functional on L. Then the inequality

$$|\phi| \leqslant \psi$$

yields $$|P(\phi)| \leqslant P(|\phi|) \leqslant P(\psi).$$

Following Bourbaki [4], Ch. II, we shall say that a linear functional F on L is *relatively bounded* if the set

$$\{F(\phi): |\phi| \leqslant \psi\}$$

is bounded for each positive ψ in L. Thus *any positive linear functional on L is relatively bounded.* It is now a simple exercise to verify that a linear functional F on $C_c(X)$ is relatively bounded if and only if F is bounded on every $C_K(X)$ (Ex. 1).

Lemma 1. *Let F be a relatively bounded linear functional on L and define*

$$T(\psi) = \sup\{F(\phi): |\phi| \leqslant \psi\}$$

for any positive ψ in L. Then

$$T(\psi_1+\psi_2) = T(\psi_1)+T(\psi_2)$$

for all positive ψ_1, ψ_2 in L.

Proof. If $|\phi_1| \leqslant \psi_1$, $|\phi_2| \leqslant \psi_2$ then $|\phi_1+\phi_2| \leqslant \psi_1+\psi_2$ and so

$$F(\phi_1)+F(\phi_2) = F(\phi_1+\phi_2) \leqslant T(\psi_1+\psi_2).$$

Taking suprema for ϕ_1, ϕ_2 independently (as in Ex. 1.2.2) we see that

$$T(\psi_1)+T(\psi_2) \leqslant T(\psi_1+\psi_2).$$

On the other hand, suppose that $|\phi| \leqslant \psi_1+\psi_2$. We may express ϕ as $\phi_1+\phi_2$, where $|\phi_1| \leqslant \psi_1$, $|\phi_2| \leqslant \psi_2$ (for example, by taking $\phi_1 = \text{mid}\{-\psi_1, \phi, \psi_1\}$, $\phi_2 = \phi-\phi_1$). Then

$$F(\phi) = F(\phi_1)+F(\phi_2) \leqslant T(\psi_1)+T(\psi_2)$$

and taking the supremum for ϕ gives

$$T(\psi_1+\psi_2) \leqslant T(\psi_1)+T(\psi_2).$$

Proposition 1. *Let X be an arbitrary set, L a linear lattice of functions $\phi: X \to \mathbf{R}$ and F a relatively bounded linear functional on L. Then there is a positive linear functional T on L satisfying*

$$T(\psi) = \sup\{F(\phi): |\phi| \leqslant \psi\} \tag{2}$$

for all positive ψ in L. Moreover, there are positive linear functionals P, N on L which satisfy

$$F = P-N, \quad T = P+N.$$

Proof. Define $T(\psi)$ for positive ψ by means of equation (2). As F is linear it is clear that $T(\psi) \geqslant 0$ and $T(c\psi) = cT(\psi)$ for all positive c in $\mathbf{R}$ and all positive ψ in L.

If $\psi = \theta_1 - \phi_1 = \theta_2 - \phi_2$, where $\theta_1, \theta_2, \phi_1, \phi_2$ are positive elements of L, then $\theta_1 + \phi_2 = \theta_2 + \phi_1$ and so by Lemma 1,

$$T(\theta_1) + T(\phi_2) = T(\theta_2) + T(\phi_1),$$

i.e.

$$T(\theta_1) - T(\phi_1) = T(\theta_2) - T(\phi_2).$$

Any ψ in L may be expressed as $\psi = \theta - \phi$, where θ, ϕ are positive elements of L (e.g. $\theta = \psi^+$, $\phi = \psi^-$) and we may therefore define T unambiguously on L by

$$T(\psi) = T(\theta) - T(\phi).$$

It is now easy to check, as in the proof of Theorem 8.1.1, that T is a linear functional on L. This completes the first part of the proof.

For the remainder we only need to set

$$P = \tfrac{1}{2}(T + F),$$
$$N = \tfrac{1}{2}(T - F),$$

which are both positive linear functionals on L.

The notation T, P, N is suggested by the Total, Positive and Negative variations of a function of bounded variation (see Exx. 3.5.1–4). If we think of T as the absolute value of F then it is clear that the relatively bounded linear functionals on L form a linear lattice (Ex. 3). With this in mind, the alternative notation

$$|F| = T, \quad F^+ = P, \quad F^- = N$$

is perhaps more appropriate; but we shall often revert to T in preference to $|F|$, because the latter notation is awkward in practice, particularly when $|T(\phi)|$ or $\|T\|$ is in question. The positive and negative parts of F may be expressed directly in terms of F:

$$F^+(\psi) = \sup\{F(\phi) : 0 \leqslant \phi \leqslant \psi\}, \tag{3}$$

$$F^-(\psi) = \sup\{-F(\phi) : 0 \leqslant \phi \leqslant \psi\} \tag{4}$$

for any positive ψ in L. Equation (3) follows from the relation

$$2F^+ = |F| + F,$$

from which we deduce that

$$2F^+(\psi) = \sup_\theta \{F(\theta) + F(\psi) : -\psi \leqslant \theta \leqslant \psi\}$$
$$= \sup_\phi \{F(\phi) : 0 \leqslant \phi \leqslant 2\psi\},$$

by writing $\phi = \theta + \psi$; equation (4) is proved in almost the same way by writing $\phi = \psi - \theta$.

We have stated and proved Proposition 1 in terms of $T = |F|$. It is customary, and possibly a little simpler, to frame the proof in terms of $P = F^+$. This is followed up in Ex. 5. Our reason for emphasising $|F|$ is that this notion carries over to the complex case where the functionals F^+, F^- have no meaning (cf. §16.2).

Up to this point we have only allowed as integrals, either the elementary integrals or the Daniell integrals of §8.1, and these are *positive* linear functionals. If P, Q are Daniell integrals on spaces $L^1(P)$, $L^1(Q)$ then we may define

$$F(f) = P(f) - Q(f)$$

for all functions f in $L^1(P) \cap L^1(Q)$ and refer to $F = P - Q$ as a *signed Daniell integral*. Let μ, ν be the measures corresponding to P, Q and let $\mathcal{M}$, $\mathcal{N}$ be their domains of definition. *Provided μ, ν are finite*, i.e. take only finite values, we may define

$$\lambda(A) = \mu(A) - \nu(A)$$

for all sets A in $\mathcal{M} \cap \mathcal{N}$ and refer to $\lambda = \mu - \nu$ as a *signed measure*. The restriction of this term to differences of finite measures is in line with our caution about the value ∞. It is imperative to avoid the 'difference' $\infty - \infty$; many authors allow *one* of the measures μ, ν to take the value ∞. (In Chapter 16 we shall also refer to a finite signed measure as a *real measure*; this is harmonious with the idea of a *complex measure* which takes its values in C.) The Riesz Representation Theorems are now the obvious extensions of Theorems 10.1.3, 10.1.4.

Theorem 1. *Let X be a compact metric space and $C(X)$ the linear space of continuous real valued functions on X. Any bounded linear functional F on $C(X)$ extends to a signed Daniell integral $\int$ on a linear space L^1 containing $C(X)$. Thus*

$$F(\phi) = \int \phi$$

for all ϕ in $C(X)$. The corresponding signed measure μ, defined by

$$\mu(A) = \int \chi_A$$

for all χ_A in L^1, gives finite measure to all open and all closed subsets of X.

Theorem 2. *Let X be a locally compact metric space and $C_c(X)$ the linear space of continuous real valued functions on X of compact support. If F is a linear functional on $C_c(X)$ which is bounded on every subspace*

$C_K(X)$ (K compact in X), then F extends to a signed Daniell integral $\int$ on a linear space L^1 containing $C_c(X)$. Thus

$$F(\phi) = \int \phi$$

for all ϕ in $C_c(X)$. If A is any open or any closed subset of X then χ_A is measurable (with respect to the positive and negative parts of $\int$) and if A is compact, $\chi_A \in L^1$.

Clearly Theorem 1 is a special case of Theorem 2. We have stated Theorem 1 separately because it is a little shorter and is the one which is most often associated with the name of Riesz.

Proof of Theorem 2. If $|\phi| \leqslant \psi$ and ψ has support K, then $\phi \in C_K(X)$ and

$$|F(\phi)| \leqslant M_K \|\phi\| \leqslant M_K \|\psi\|.$$

Thus F is relatively bounded and Proposition 1 applies. We may use Dini's Theorem as in the proof of Theorem 10.1.4 to check the Daniell condition for F^+, F^-. The Daniell construction now extends F^+, F^- to $L^1(F^+)$, $L^1(F^-)$ and yields the signed Daniell integral $\int = F^+ - F^-$ on $L^1 = L^1(F^+) \cap L^1(F^-)$. Theorem 10.1.4 applies to F^+, F^- separately and shows that χ_A is measurable for all open and all closed subsets of X and χ_A is integrable for compact A. This completes the proof.

An important illustration of Theorem 2 is given by the case where $X = \mathbf{R}$. If $\int$ is the signed Daniell integral of the theorem, then $\int$ is the difference of two Lebesgue–Stieltjes integrals (in view of Theorem 9.3.1). As in § 9.1 we let

$$F_0(x) = \int \chi_{(0,x]} \quad \text{for} \quad x \geqslant 0,$$

$$= -\int \chi_{(x,0]} \quad \text{for} \quad x < 0.$$

This defines a function $F_0\colon \mathbf{R} \to \mathbf{R}$ which is continuous on the right – but, of course, F_0 may not be increasing. Let $P = F^+$, $N = F^-$ be the Daniell integrals given by Theorem 2, let $T = P + N$ and define Stieltjes functions P_0, N_0, T_0 as for F_0: e.g.

$$P_0(x) = P(\chi_{(0,x]}) \quad \text{for} \quad x \geqslant 0,$$

$$= -P(\chi_{(x,0]}) \quad \text{for} \quad x < 0.$$

Then

$$F_0 = P_0 - N_0,$$

$$T_0 = P_0 + N_0.$$

From the first of these equations it follows at once that F_0 has *bounded variation* on any bounded interval (see Ex. 3.5.1 and Ex. 6 below). The signed Daniell integral in this case may be written $\int dF_0$ and called the (*signed*) *Lebesgue–Stieltjes integral* with respect to F_0. (For the practical evaluation of these integrals see Exx. 7, 8.)

Theorem 2 makes no reference to a signed measure μ. We could define measures μ^+, μ^- corresponding to the positive and negative parts of the Daniell integral, but these may both take infinite values and so we cannot define $\mu = \mu^+ - \mu^-$. A natural way to resolve this difficulty is to assume a little more about the linear functional F. Suppose that F is bounded on the whole of $C_c(X)$, rather than merely being bounded on every $C_K(X)$. Can we not then show that the corresponding measures μ^+, μ^- are both finite and hence that the signed measure $\mu = \mu^+ - \mu^-$ exists? Perhaps a little surprisingly, we cannot do this. For example, if $X = [0, 1]$ and d is the discrete metric on X (which sets $d(a, b) = 1$ whenever $a \neq b$, and $d(a, a) = 0$), we may check that all subsets of X are both open and closed, the compact subsets are precisely the finite subsets, and X is locally compact. The zero linear functional P on $C_c(X)$ is obviously bounded, and extends by the Daniell construction to the zero integral on L^1, where L^1 consists of all $f: X \to \mathbf{R}$ which vanish outside a countable subset of X. Now X is not countable; thus χ_X, though measurable, is not integrable, and the corresponding measure $\mu(X) = \infty$. This example shows that we cannot presume that μ is finite although the given functional P is bounded (even zero). Roughly speaking, the trouble in this example is that we cannot get near X by means of a sequence of finite (or even countable) sets which we know have finite measure. To deal with this problem, let us further assume that X is *σ-compact*, i.e. X may be expressed as the union of a sequence $\{K_n\}$ of compact subsets of X. We may replace K_n by the compact set $K_1 \cup K_2 \cup \ldots \cup K_n$, if necessary, and assume that

$$K_n \uparrow X.$$

A familiar example of a σ-compact space is $\mathbf{R}^k$ which is the union of the compact intervals I_n defined by $|x_i| \leqslant n$ for $i = 1, 2, \ldots, k$.

Theorem 3. *Let X be a σ-compact, locally compact metric space and $C_c(X)$ the linear space of continuous real valued functions on X of compact support. Any bounded linear functional F on $C_c(X)$ extends to a signed Daniell integral $\int$ on a linear space L^1 containing $C_c(X)$. Thus*

$$F(\phi) = \int \phi$$

for all ϕ in $C_c(X)$. The corresponding signed measure defined by

$$\mu(A) = \int \chi_A$$

for all χ_A in L^1, gives finite measure to all open and all closed subsets of X.

Proof. There is a positive constant M such that

$$|F(\phi)| \leqslant M \|\phi\|$$

for all ϕ in $C_c(X)$. If $|\phi| \leqslant \psi$, this gives

$$|F(\phi)| \leqslant M \|\psi\|$$

so that
$$T(\psi) \leqslant M \|\psi\| \tag{5}$$

for all positive ψ in $C_c(X)$ (see equation (2) which defines $T = |F|$).

For any compact subset K of X, Proposition 10.1.1 provides a positive element ψ of $C_c(X)$ which satisfies $\|\psi\| \leqslant 1$ and $\chi_K \leqslant \psi$. In view of inequality (5),

$$T(\chi_K) \leqslant M$$

for any compact subset K of X. Now we use the σ-compactness of X. As $K_n \uparrow X$ it follows, by the Monotone Convergence Theorem, that

$$T(\chi_X) \leqslant M,$$

which is equivalent to
$$|\mu| (X) \leqslant M,$$

where
$$|\mu| = \mu^+ + \mu^-.$$

The rest of the theorem is now obvious as μ^+, μ^- and μ are all dominated by $|\mu|$.

Once again we remark that Stone's Theorem 8.5.1 applies in the case of Theorems 1 and 3 and the signed Daniell integral may be written as $\int d\mu$, but Theorem 2 shows that there is some advantage in being able to express these ideas purely in terms of integration.

Let X denote as in Theorem 3 a σ-compact, locally compact metric space and F a bounded linear functional on $C_c(X)$. Recall from § 7.4 that F has the norm

$$\|F\| = \sup_{\|\phi\| \leqslant 1} F(\phi)$$

which in the present setting is the same as

$$\|F\| = \sup_{|\phi| \leqslant 1} F(\phi)$$

(because of the particular norm $\|\phi\|$ in $C_c(X)$). A comparison with equation (2) which defines $T = |F|$ shows that

$$\|F\| = |F| (1)$$

or, in other words, $$\|F\| = |\mu|\,(X).$$

If we define a norm $$\|\mu\| = |\mu|\,(X)$$

on the linear space of finite signed measures on X, then

$$\|F\| = \|\mu\|. \tag{6}$$

Theorem 3 provides a one–one linear mapping of the bounded linear functionals on $C_c(X)$, which form the *dual space* of $C_c(X)$, onto certain signed measures on X, the *signed Borel measures*, and equation (6) shows that this mapping is in fact an *isometry*.

Exercises

1. Let X be a locally compact metric space. Show that a linear functional F on $C_c(X)$ is relatively bounded if and only if F is bounded on every $C_K(X)$ (K compact).

2. Let L be a linear lattice of functions $\phi: X \to \mathbf{R}$ and F a linear functional on L. Show that F is relatively bounded if and only if $F = G-H$ where G, H are positive linear functionals on L. In this case show that the positive linear functionals P, N of Proposition 1 are 'best possible' in the sense that $P \leqslant G$ and $N \leqslant H$ (i.e. $G-P$ and $H-N$ are positive linear functionals).

3. In the notation of Proposition 1 show that the relatively bounded linear functionals on L form a linear lattice in which $|F| = T$ is defined by equation (2). Let

$$A = \max\{F, G\} = \tfrac{1}{2}(F+G)+\tfrac{1}{2}|F-G|,$$

$$B = \min\{F, G\} = \tfrac{1}{2}(F+G)-\tfrac{1}{2}|F-G|.$$

Show that A is the smallest linear functional on L that is greater than or equal to F and G. Prove a similar result for B. (As in Ex. 2, $A \geqslant F$ means that $A-F$ is a positive linear functional.)

4. In the notation of Ex. 3 show that, for any positive ψ in L,

$$A(\psi) = \sup\{F(\theta)+G(\phi)\text{: positive } \theta, \phi \in L,\ \theta+\phi = \psi\},$$

and find a similar expression for $B(\psi)$.

The functionals F, G are said to be *mutually singular* if

$$\min\{|F|, |G|\} = 0.$$

Show that the linear functionals P, N of Proposition 1 are mutually singular. For any $\epsilon > 0$ and any positive ψ in L, show that there exist positive θ, ϕ in L such that $\theta+\phi = \psi$ and $P(\theta) < \epsilon$, $N(\phi) < \epsilon$.

5. Let X be an arbitrary set, L a linear lattice of functions $\phi: X \to \mathbf{R}$ and F a relatively bounded linear functional on L. For any positive ψ in L define

$$P(\psi) = \sup\{F(\phi)\text{: } 0 \leqslant \phi \leqslant \psi\}.$$

Without reference to Proposition 1, show that

$$P(\psi_1+\psi_2) = P(\psi_1)+P(\psi_2)$$

for any positive ψ_1, ψ_2 in L, and that P extends to a positive linear functional P on L.

Let $N = P-F$ and $T = 2P-F$. Show that N and T are positive linear functionals on L and that

$$N(\psi) = \sup\{-F(\phi): 0 \leqslant \phi \leqslant \psi\},$$
$$T(\psi) = \sup\{F(\phi): |\phi| \leqslant \psi\}$$

for any positive ψ in L.

6. Let L' denote the linear lattice of step functions on $\mathbf{R}$ whose intervals are of type (,]. Let F be a relatively bounded linear functional on L' and T, P, N the corresponding positive linear functionals on L' given by Proposition 1. Let the function $F_0: \mathbf{R} \to \mathbf{R}$ be defined by

$$\begin{aligned} F_0(x) &= F(\chi_{(0,x]}) \quad \text{for} \quad x \geqslant 0, \\ &= -F(\chi_{(x,0]}) \quad \text{for} \quad x < 0, \end{aligned}$$

with similar definitions for T_0, P_0, N_0. Show that

$$F_0 = P_0 - N_0, \quad T_0 = P_0 + N_0,$$

and that the total variation of F_0 on $[a,b]$ is $T_0(b)-T_0(a)$ for any $a < b$ in $\mathbf{R}$. What are the positive and negative variations of F_0 on $[a,b]$? (See Exx. 3.5.1–3.)

7. Extend Proposition 9.1.1 to the case where $G: \mathbf{R} \to \mathbf{R}$ has a continuous derivative G', and hence evaluate

$$\int_{-1}^{2} x^3 \, dx^2.$$

8. Extend Ex. 9.1.11 to the case where $g \in L^1(I)$ for every bounded interval I on $\mathbf{R}$.

The Riemann–Stieltjes Integral

Recall the definition of the Riemann–Stieltjes integral given before Exx. 9.1 A, B.

A. Let f, G be real valued functions on the compact interval $[a,b]$ and suppose that f is continuous and G has bounded variation on $[a,b]$. Show that the Riemann–Stieltjes integral

$$\int_a^b f(x)\,dG(x) = \int_{[a,b]} f\,dG.$$

(In evaluating the Lebesgue–Stieltjes integral on the right hand side we repeat the convention that $G(a-0) = G(a)$ and $G(b+0) = G(b)$.)

10.3 Complex linear functionals

This is an appropriate point at which to mention linear functionals which take complex values. Denote by $\mathbf{C}$ the field of complex numbers $z = x+iy$ where x, y are real numbers and $i^2 = -1$. The reader is assumed to be familiar with these numbers and their construction from the reals. As in the previous section X is a locally compact metric space. We shall distinguish between the two spaces $C_c(X, \mathbf{R})$, $C_c(X, \mathbf{C})$: the former consists of continuous real valued functions of compact support on X, and is a linear space over $\mathbf{R}$; the latter consists of continuous complex valued functions of compact support on X and is a *linear space over* $\mathbf{C}$. In the present setting this last statement means that if ζ_1, $\zeta_2 \in C_c(X, \mathbf{C})$ and $z_1, z_2 \in \mathbf{C}$ then the function $z_1\zeta_1 + z_2\zeta_2$ (defined by $(z_1\zeta_1+z_2\zeta_2)(x) = z_1\zeta_1(x)+z_2\zeta_2(x)$ for all x in X) also belongs to $C_c(X, \mathbf{C})$.

In the interests of simplicity, we shall also denote these two spaces by $L_\mathbf{R}$, $L_\mathbf{C}$, respectively. (In one of our most important examples, X will be a curve C and the space $L_\mathbf{C}$ will then be $C_c(C, \mathbf{C})$!) The spaces $L_\mathbf{R}$, $L_\mathbf{C}$ are related in a particularly simple way. If $\xi, \eta \in L_\mathbf{R}$ then we may define $\zeta = \xi + i\eta$ by the obvious rule

$$\zeta(x) = \xi(x) + i\eta(x)$$

for all x in X, and verify immediately that $\zeta \in L_\mathbf{C}$. Conversely, if ζ in $L_\mathbf{C}$ is given, we may define real valued functions ξ, η by the same equation and check that ξ, $\eta \in L_\mathbf{R}$.

If F, G are two linear functionals on $L_\mathbf{R}$ then there is a function $H = F+iG$ defined on $L_\mathbf{C}$ by

$$(F+iG)(\xi+i\eta) = F(\xi) - G(\eta) + i(F(\eta)+G(\xi)) \tag{1}$$

for all ξ, η in $L_\mathbf{R}$. It is clear from this definition that H is linear over $\mathbf{R}$, i.e.

$$H(a_1\zeta_1 + a_2\zeta_2) = a_1H(\zeta_1) + a_2H(\zeta_2)$$

for any ζ_1, ζ_2 in $L_\mathbf{C}$ and a_1, a_2 in $\mathbf{R}$. But we also check at once that

$$H(i\zeta) = iH(\zeta)$$

and so H is linear over $\mathbf{C}$, i.e. H is a linear functional on $L_\mathbf{C}$. On the other hand, if we are given a linear functional H on $L_\mathbf{C}$, we may restrict H to $L_\mathbf{R}$ and define real valued functions F, G by

$$H(\xi) = F(\xi) + iG(\xi)$$

($\xi \in L_\mathbf{R}$). The linearity of H over $\mathbf{C}$ now ensures that $H = F+iG$, as defined by (1). Thus the linear functionals on $L_\mathbf{R}$, $L_\mathbf{C}$ are also related in a simple way.

If F, G are bounded linear functionals on $L_{\mathbf{R}}$ then we may find $M \geqslant 0$ such that

$$|F(\xi)| \leqslant M\,\|\xi\|,$$
$$|G(\xi)| \leqslant M\,\|\xi\|$$

for all ξ in $L_{\mathbf{R}}$. But this implies by (1) that

$$|H(\zeta)| \leqslant 4M\,\|\zeta\|$$

for all ζ in $L_{\mathbf{C}}$, i.e. H is a bounded linear functional on $L_{\mathbf{C}}$. Here we have written

$$\|\zeta\| = \sup_{x \in X} |\zeta(x)|$$

for the *norm* of ζ in $L_{\mathbf{C}}$. Recall that $|x+iy| = (x^2+y^2)^{\frac{1}{2}}$; if we compose the functions $\zeta\colon X \to \mathbf{C}$, $|\ |\colon \mathbf{C} \to \mathbf{R}$ and note that ζ has compact support, it follows, just as in the real case, that this supremum exists (i.e. is finite, in our terminology).

Conversely, if H is a bounded linear functional on $L_{\mathbf{C}}$ and

$$|H(\zeta)| \leqslant M\,\|\zeta\|$$

for all ζ in $L_{\mathbf{C}}$, we may restrict ζ to $L_{\mathbf{R}}$ and deduce that

$$|F(\xi)| \leqslant M\,\|\xi\|,$$
$$|G(\xi)| \leqslant M\,\|\xi\|.$$

This shows that *H is a bounded linear functional on $L_{\mathbf{C}}$ if and only if F, G are bounded linear functionals on $L_{\mathbf{R}}$.*

The extension of Theorem 10.2.3 to the complex case now has an air of inevitability.

Theorem 1. *Let X be a σ-compact, locally compact metric space and $C_c(X, \mathbf{C})$ the linear space of continuous complex valued functions on X of compact support. Any bounded linear functional H on $C_c(X, \mathbf{C})$ extends to a complex Daniell integral $\int$ on a complex linear space $L_{\mathbf{C}}^1$. Thus*

$$H(\zeta) = \int \zeta$$

for all ζ in $C_c(X, \mathbf{C})$. The corresponding complex measure defined by

$$\lambda(A) = \int \chi_A$$

for all χ_A in $L_{\mathbf{C}}^1$, gives finite measure to all open and all closed subsets of X.

The terms *complex Daniell integral* and *complex measure* still have to be defined; this will be done in the following proof, but should now be quite obvious in view of the above discussion.

Proof. Let $H = F+iG$ as described above. Then F, G are bounded linear functionals on $C_c(X, \mathbf{R})$ which therefore extend by Theorem 10.2.3 to finite signed Daniell integrals F_1, G_1, say, on real linear spaces $L^1(F_1)$, $L^1(G_1)$. Let $L_{\mathbf{R}}^1$ denote the intersection of these spaces and $L_{\mathbf{C}}^1$ the 'complexification' of $L_{\mathbf{R}}^1$ consisting of all $f+ig$ with f, g in $L_{\mathbf{R}}^1$ (aping the above construction of $L_{\mathbf{C}}$ from $L_{\mathbf{R}}$). Clearly $L_{\mathbf{R}}^1$ contains $L_{\mathbf{R}} = C_c(X, \mathbf{R})$ and so $L_{\mathbf{C}}^1$ contains $L_{\mathbf{C}} = C_c(X, \mathbf{C})$. The linear functional $H_1 = F_1+iG_1$ on $L_{\mathbf{C}}^1$ is defined by an equation analogous to (1). We naturally call H_1 a *complex Daniell integral* and write $H_1 = \int$ for consistency in stating our theorem. The finite signed measures μ, ν corresponding to F_1, G_1 are defined by

$$\mu(A) = F_1(\chi_A),$$
$$\nu(A) = G_1(\chi_A)$$

for all χ_A in $L_{\mathbf{R}}^1$ (or equivalently, for all χ_A in $L_{\mathbf{C}}^1$, as χ_A is real valued).

The *complex measure* $\lambda = \mu+i\nu$ is now defined by

$$\lambda(A) = \mu(A)+i\nu(A) = \int \chi_A$$

for all A satisfying $\chi_A \in L_{\mathbf{R}}^1$, and certainly therefore for all open and all closed subsets A of X.

The extension of Theorem 10.2.3 to the complex case and the discussion of the last few paragraphs are fairly obvious; they say, more or less, that the functions ζ and the linear functionals H may be described in terms of their 'real and imaginary parts'. But there is one other property that is not quite so obvious. Suppose that the set of complex numbers

$$\{H(\zeta): |\zeta| \leqslant \psi\}$$

is bounded for every positive ψ in $L_{\mathbf{R}}$; we shall say in this case that H is *relatively bounded*. It is easy to verify that H is relatively bounded if and only if F, G are relatively bounded (Ex. 1). The real valued functionals F, G are dominated by the positive functionals $|F|$, $|G|$, and so H is dominated by the positive functional $2(|F|+|G|)$, i.e.

$$|H(\zeta)| \leqslant 2(|F|+|G|)(|\zeta|)$$

for all ζ in $L_{\mathbf{C}}$. This follows at once from (1). It is a little more surprising that there is a *smallest* positive linear functional T dominating H and defined by

$$T(\psi) = \sup\{|H(\zeta)|: |\zeta| \leqslant \psi\}$$

for any positive ψ in $L_{\mathbf{R}}$. This is almost the same as the definition in Proposition 10.2.1 but we have inserted an absolute value on the right hand side. The complex analogue of Proposition 10.2.1 is true, at least for the special linear space $L_{\mathbf{C}} = C_c(X, \mathbf{C})$.

Proposition 1. *Let H be a relatively bounded linear functional on $L_{\mathbf{C}}$. Then there is a positive linear functional T on $L_{\mathbf{R}}$ satisfying*

$$T(\psi) = \sup\{|H(\zeta)| : |\zeta| \leqslant \psi\}$$

for all positive ψ in $L_{\mathbf{R}}$.

As before we shall write $|H| = T$.

Proof. For any ζ in $L_{\mathbf{C}}$, there is a complex number c of unit modulus such that

$$cH(\zeta) = |H(\zeta)|,$$

whence

$$H(c\zeta) = cH(\zeta) = |H(\zeta)|.$$

Suppose that ψ_1, ψ_2 are positive functions in $L_{\mathbf{R}}$. If $|\zeta_1| \leqslant \psi_1, |\zeta_2| \leqslant \psi_2$, then $|\zeta_1 + \zeta_2| \leqslant |\zeta_1| + |\zeta_2| \leqslant \psi_1 + \psi_2$. By the above observation, we may assume, without loss of generality, that $H(\zeta_1)$ and $H(\zeta_2)$ are positive. Thus

$$|H(\zeta_1)| + |H(\zeta_2)| = |H(\zeta_1 + \zeta_2)| \leqslant T(\psi_1 + \psi_2).$$

Taking suprema for ζ_1, ζ_2 independently, gives

$$T(\psi_1) + T(\psi_2) \leqslant T(\psi_1 + \psi_2).$$

To prove the opposite inequality we use the fact that any ζ in $L_{\mathbf{C}}$ satisfying $|\zeta| \leqslant \psi_1 + \psi_2$ may be expressed as $\zeta = \zeta_1 + \zeta_2$, where $\zeta_1, \zeta_2 \in L_{\mathbf{C}}$ and $|\zeta_1| \leqslant \psi_1$, $|\zeta_2| \leqslant \psi_2$. Let us assume this for the moment. It follows that

$$|H(\zeta)| \leqslant |H(\zeta_1)| + |H(\zeta_2)| \leqslant T(\psi_1) + T(\psi_2).$$

Taking the supremum for $|\zeta| \leqslant \psi_1 + \psi_2$ gives

$$T(\psi_1 + \psi_2) \leqslant T(\psi_1) + T(\psi_2).$$

This establishes the additive property

$$T(\psi_1 + \psi_2) = T(\psi_1) + T(\psi_2)$$

for positive ψ_1, ψ_2. Now extend T, exactly as in the proof of Proposition 10.2.1, to a positive linear functional on the space $L_{\mathbf{R}}$.

It only remains to prove the crucial property that *any ζ in $L_{\mathbf{C}}$ which is dominated by $\psi_1 + \psi_2$ is the sum of two elements ζ_1, ζ_2 in $L_{\mathbf{C}}$ which are dominated by ψ_1, ψ_2, respectively.*

In this special case where $L_{\mathbf{C}} = C_c(X, \mathbf{C})$, a standard method is to set

$$\zeta_1 = \frac{\zeta\psi_1}{\psi_1 + \psi_2}, \quad \zeta_2 = \frac{\zeta\psi_2}{\psi_1 + \psi_2}. \tag{2}$$

(As usual these quotients are defined to be zero where the denominators vanish.) The conditions $\zeta = \zeta_1 + \zeta_2$, $|\zeta_1| \leqslant \psi_1$, $|\zeta_2| \leqslant \psi_2$ are obviously satisfied and the verification of the continuity of ζ_1, ζ_2, even at points x where $\psi_1(x) + \psi_2(x) = 0$, is a simple exercise (Ex. 2).

This last part of the proof is dependent on special properties of the space $C_c(X, \mathbf{C})$ and is quite different from the lattice theoretic proof given for the real case in Lemma 10.2.1. There we had $|\phi| \leqslant \psi_1 + \psi_2$ and defined

$$\phi_1 = \operatorname{mid}\{-\psi_1, \phi, \psi_1\}, \quad \phi_2 = \phi - \phi_1.$$

Now the linear space $L_{\mathbf{C}}$ is closed with respect to absolute values, i.e.

$$\zeta \in L_{\mathbf{C}} \Rightarrow |\zeta| \in L_{\mathbf{C}}. \tag{3}$$

The expressions $\frac{1}{2}(|\zeta| + \zeta)$, $\frac{1}{2}(|\zeta| - \zeta)$ are no longer helpful in the way they were for real valued functions, but we can define a simple analogue

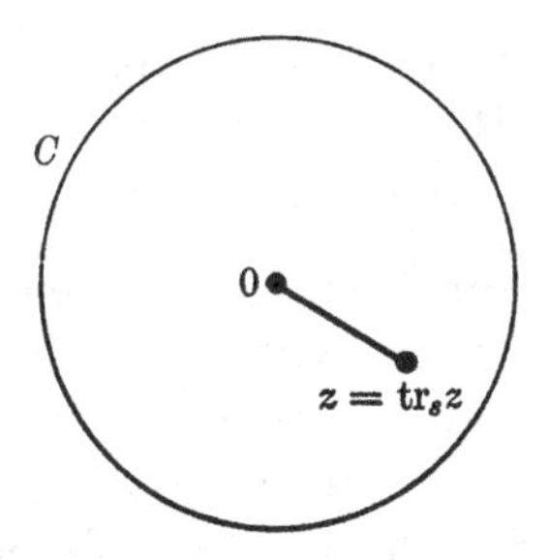

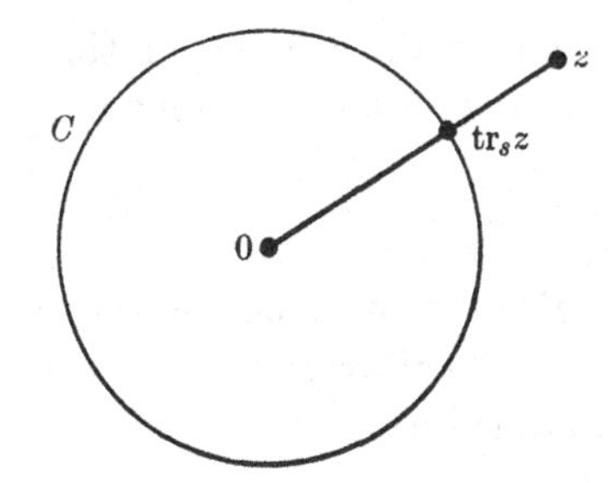

Fig. 10

for complex valued functions of the idea of truncation. Let the complex number z be written in the familiar polar form

$$z = r(\cos\theta + i\sin\theta)$$

and define

$$\begin{aligned} \operatorname{sgn} z &= \cos\theta + i\sin\theta \quad \text{if} \quad r > 0 \\ &= 0 \qquad\qquad\qquad \text{if} \quad r = 0. \end{aligned}$$

In other words,

$$\operatorname{sgn} z = z/|z|,$$

with the usual understanding that the right hand side is 0 if $|z| = 0$. In any case

$$z = |z| \operatorname{sgn} z$$

holds for all z in $\mathbf{C}$, though the function sgn is not continuous at the origin. If s is any positive real number we may 'truncate z radially by means of s': let

$$\operatorname{tr}_s z = \min\{|z|, s\} \operatorname{sgn} z.$$

As illustrated in Fig. 10, let C denote the circle, centre 0, radius s; if z lies inside (or on) C, then $\operatorname{tr}_s z = z$; but if z lies outside C, then $\operatorname{tr}_s z$ is the intersection of C and the radial segment joining 0 to z. It is now easily verified (Ex. 3) that $L_{\mathbf{C}}$ has the truncation property:

$$\zeta \in L_{\mathbf{C}} \Rightarrow \mathrm{tr}_{\psi}\zeta \in L_{\mathbf{C}} \tag{4}$$

for any positive function ψ in $L_{\mathbf{R}}$. Here, of course, $\mathrm{tr}_{\psi}\zeta$ is shorthand for the function on X whose value at x is $\mathrm{tr}_{\psi(x)}\zeta(x)$.

With this in mind, we return to the situation where $|\zeta| \leqslant \psi_1 + \psi_2$ and define

$$\zeta_1 = \mathrm{tr}_{\psi_1}\zeta, \quad \zeta_2 = \zeta - \zeta_1.$$

Clearly, $\zeta_1, \zeta_2 \in L_{\mathbf{C}}$ and $\zeta = \zeta_1 + \zeta_2$. If $|\zeta(x)| \leqslant \psi_1(x)$ then $\mathrm{tr}_{\psi_1(x)}\zeta(x) = \zeta(x)$ and $\zeta_2(x) = 0$; if $|\zeta(x)| > \psi_1(x)$ then

$$|\zeta_1(x)| = \psi_1(x) \quad \text{and} \quad |\zeta_2(x)| = |\zeta(x)| - \psi_1(x) \leqslant \psi_2(x).$$

In either case $|\zeta_1| \leqslant \psi_1$, $|\zeta_2| \leqslant \psi_2$. This may be compared with the previous method (2) in which '0ζ is divided in the ratio $\psi_1 : \psi_2$'.

As a nice illustration of the ideas of this section, suppose that $C: [a, b] \to \mathbf{C}$ is a continuous function; in other words

$$C(t) = A(t) + iB(t) \quad \text{for} \quad a \leqslant t \leqslant b, \tag{5}$$

where A, B are continuous real valued functions on the compact interval $[a, b]$. We shall refer to C as a *curve* (or *path*) in the complex plane $\mathbf{C}$. Recall from the Appendix to Volume 1, Theorem 8 that the continuous image of a compact set is compact: thus the set of points $C(t)$ $(t \in [a, b])$ is a compact subset X of $\mathbf{C}$. Strictly speaking, the curve C is a function whose values are points in $\mathbf{C}$ though we shall often think of C rather vaguely as the set of points in the complex plane traced out in a certain way which is specified by the given 'parametrisation' (5). To avoid anomalies which occur when the points are traced out more than once, it is sometimes important to restrict attention to arcs rather than curves: an *arc* (in the complex plane) is a homeomorphism of a compact interval $[a, b]$ onto a subset of the complex plane (Figs. 11, 12).

Let C be a curve in $\mathbf{C}$ and suppose further that the functions A, B of (5) have bounded variation on $[a, b]$. This allows us to define

$$\int_C f(z)\,dz = \int_a^b f(C(t))\,dC(t),$$

where the right hand side is a complex Lebesgue–Stieltjes integral which is evaluated, just as for (1), by multiplying out the real and imaginary parts. The linear functional H defined by

$$H(f) = \int_C f(z)\,dz,$$

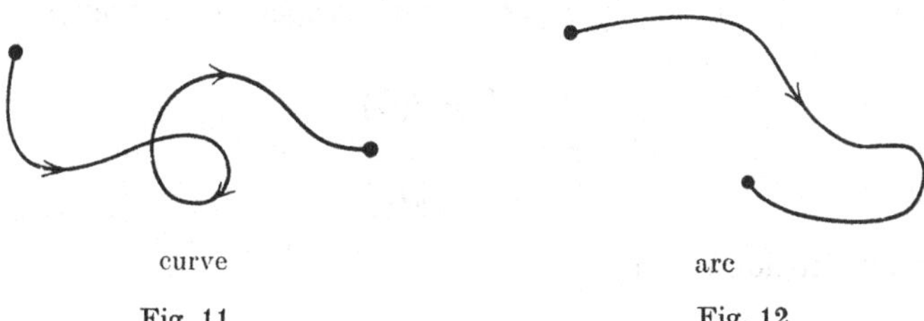

Fig. 11 Fig. 12

which associates with any continuous function f on X its 'complex integral along C', is, in fact, bounded. Our assumption that A, B should be of bounded variation is necessary and sufficient to ensure that C is *rectifiable*, i.e. has finite length $l(C)$ given by the supremum of the lengths of inscribed polygons (Ex. 5); and it is quite a simple exercise to show that $\|H\| \leqslant l(C)$ (Ex. 6). This gives the familiar inequality

$$\left|\int_C f(z)\,dz\right| \leqslant l(C)\,\|f\|,$$

where

$$\|f\| = \sup_{z\in X} |f(z)|$$

is the 'maximum modulus' of f on X. For suitably perverse curves C, we may have $\|H\| < l(C)$. For example, as t increases from a to b the

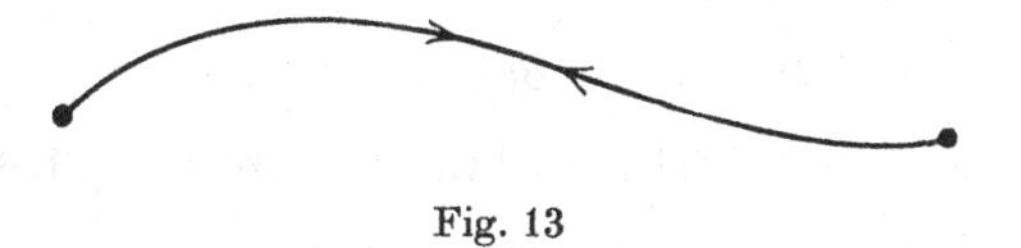

Fig. 13

point $C(t)$ might trace out the set X twice – once there and back, so to speak (Fig. 13) – and

$$\int_C f(z)\,dz$$

would then be zero for all continuous f on X (Ex. 4).

From now on let us assume that C is an arc. The reader is invited to verify in this case that $\|H\| = l(C)$ (Ex. 7). In books on complex functions, $|H|\,(f)$ is often written

$$\int_C f(z)\,|dz|$$

or

$$\int_C f\,ds$$

and called the integral of f along C with respect to *arc length*. As we should expect,

$$\int_C 1\,ds = l(C)$$

because

$$|H|\,(1) = \|H\|,$$

from the definition of $|H| = T$.

Exercises

1. Let F, G be linear functionals on $L_{\mathbf{R}}$ and define $H = F + iG$ as in (1). Show that H is relatively bounded if and only if F, G are relatively bounded.

2. Verify that the functions ζ_1, ζ_2 defined in (2) are elements of $C_c(X, \mathbf{C})$.

3. Verify that $L_{\mathbf{C}} = C_c(X, \mathbf{C})$ has the truncation property (4).

4. Let

$$C(t) = e^{it} \quad \text{for} \quad 0 \leqslant t \leqslant 2\pi,$$
$$= e^{-it} \quad \text{for} \quad 2\pi \leqslant t \leqslant 4\pi.$$

Describe what happens to the point $C(t)$ as t increases from 0 to 4π. Show that

$$\int_C f(z)\,dz = 0$$

for any function f which is continuous on the unit circle $|z| = 1$.

5. Let C be the curve in the z-plane given by

$$C(t) = A(t) + iB(t) \quad (a \leqslant t \leqslant b),$$

where A, B are continuous real valued functions on $[a, b]$. Let

$$a = t_0 < t_1 < \ldots < t_r = b.$$

The length of the inscribed polygon with vertices $C(t_0), C(t_1), \ldots, C(t_r)$ is

$$\sum_{i=1}^{r} |C(t_i) - C(t_{i-1})|.$$

Show that (as the polygon varies) this length is bounded above if and only if A and B have bounded variation on $[a, b]$. (In this case the curve C is said to be *rectifiable* and the supremum of the lengths of the inscribed polygons is the *length* $l(C)$ of C.)

6. Show that the length $l(C)$ of the rectifiable curve C of Ex. 5 is greater than or equal to the norm $\|H\|$, where

$$H(f) = \int_C f(z)\,dz.$$

7. In Ex. 6 let C be an arc and show that $\|H\| = l(C)$.

8. Let $C: [a,b] \to \mathbf{C}$ be a curve. Suppose that G is a strictly increasing continuous mapping of $[c,d]$ onto $[a,b]$ and define the curve C_1 by the equations

$$C_1(s) = A(G(s)) + iB(G(s)) = A_1(s) + iB_1(s)$$

for $c \leqslant s \leqslant d$. Show that the functions A_1, B_1 have bounded variation if and only if A, B have bounded variation, and that

$$\int_C f(z)\,dz = \int_{C_1} f(z_1)\,dz_1.$$

[In other words, from the point of view of integration, the curves C, C_1 may be regarded as equivalent.]

11

GENERAL MEASURES

In Volume 1 we constructed Lebesgue measure on $\mathbf{R}^k$ from the primitive idea of the measure of an interval. This was done by establishing an elementary integral on the step functions, extending this elementary integral by means of the Daniell construction to the Lebesgue integral on $\mathbf{R}^k$ and finally deducing Lebesgue measure from the Lebesgue integral. In this chapter we follow the same pattern, starting from a primitive measure μ on a collection $\mathscr{C}$ of sets which is very nearly arbitrary except that we want any simple function

$$\phi = a_1\chi_{A_1} + \ldots + a_m\chi_{A_m},$$

with $a_1, \ldots, a_m$ in $\mathbf{R}$ and $A_1, \ldots, A_m$ in $\mathscr{C}$, to be expressible in terms of *disjoint* sets in $\mathscr{C}$ (as in Propositions 3.1.1 and 4.1.1). In this way we extend the primitive measure μ to a 'complete' measure which is analogous to Lebesgue measure and has all the properties of the measure described in Theorem 8.4.2.

11.1 Rings of sets

According to Theorem 8.4.2 the collection $\mathscr{R}$ of 'measurable sets' satisfies the following conditions

$$S, T \in \mathscr{R} \Rightarrow S \cup T, S \cap T, S \setminus T, S \triangle T \in \mathscr{R}.$$

In other words $\mathscr{R}$ is closed with respect to the formation of the union, intersection, difference and symmetric difference of any two subsets. These four conditions are not independent. For example, if $\mathscr{R}$ is closed with respect to unions and differences (of pairs of subsets), then the other two conditions follow from the general relations

$$A \triangle B = (A \setminus B) \cup (B \setminus A),$$
$$A \cap B = A \setminus (A \setminus B).$$

Again, if $\mathscr{R}$ is closed with respect to symmetric differences and intersections (of pairs of subsets), the relations

$$A \cup B = (A \triangle B) \triangle (A \cap B),$$
$$A \setminus B = A \triangle (A \cap B)$$

show that $\mathscr{R}$ satisfies all four conditions. On the other hand, if $\mathscr{R}$ is closed with respect to unions and intersections (of pairs of subsets), the other two conditions do not follow: we need only consider two distinct non-empty sets A, B where $A \subset B$ and let $\mathscr{R} = \{A, B\}$.

Now let X be an arbitrary set; a collection $\mathscr{R}$ of subsets of X will be called a *ring* if $\mathscr{R}$ is non-empty and is closed with respect to unions and differences (of pairs of subsets). As $\mathscr{R}$ is non-empty, there is at least one set A in $\mathscr{R}$ and so $A \setminus A = \varnothing$ belongs to $\mathscr{R}$. In view of this, we might as well have assumed that $\varnothing \in \mathscr{R}$ in our definition. If the 'universal' set X itself belongs to the ring $\mathscr{R}$ then we shall say that $\mathscr{R}$ is an *algebra*. Note that this term is ambiguous unless there is a clear understanding about the universal set X. For example, $\{\varnothing\}$ is a perfectly good algebra provided $X = \varnothing$, but it is only a ring if X is any larger! There are good historic reasons for calling $\mathscr{R}$ a ring, or Boolean ring. A short discussion is given in Exx. A–H; for a detailed treatment the reader is referred to the delightful little book by Halmos on Boolean Algebras [10].

A plentiful supply of rings and algebras is guaranteed by the following existence theorem.

Theorem 1. *Let $\mathscr{C}$ be an arbitrary collection of subsets of X. Then there is a unique smallest ring of subsets of X containing $\mathscr{C}$ and a unique smallest algebra of subsets of X containing $\mathscr{C}$.*

Proof. Let $\mathscr{F}$ be the family of all rings (of subsets of X) containing $\mathscr{C}$ and define

$$\mathscr{R}_0 = \bigcap (\mathscr{R} : \mathscr{R} \in \mathscr{F}),$$

i.e. $\mathscr{R}_0$ is the intersection of all rings $\mathscr{R}$ containing $\mathscr{C}$. This family $\mathscr{F}$ is not empty as it contains the ring $\mathscr{P}(X)$ of all subsets of X. We verify at once that $\mathscr{R}_0$ is itself a ring: for if $A, B \in \mathscr{R}_0$, then $A, B \in \mathscr{R}$ for every $\mathscr{R}$ in $\mathscr{F}$, whence $A \cup B, A \setminus B \in \mathscr{R}$ for every $\mathscr{R}$ in $\mathscr{F}$, i.e. $A \cup B, A \setminus B \in \mathscr{R}_0$. It is also obvious that $\mathscr{R}_0$ contains $\mathscr{C}$. If $\mathscr{R}$ is any ring containing $\mathscr{C}$ then the above construction shows that $\mathscr{R}_0 \subset \mathscr{R}$; and this property determines $\mathscr{R}_0$ uniquely. The same argument shows that there is a unique smallest algebra $\mathscr{A}_0$ containing $\mathscr{C}$.

We shall also refer to $\mathscr{R}_0$ as *the ring generated by* $\mathscr{C}$ and $\mathscr{A}_0$ as *the algebra generated by* $\mathscr{C}$. In the case of $\mathscr{A}_0$ the universal set X must be given beforehand or clearly understood from the context.

The proof of Theorem 1 gives essentially no clue to the structure of the ring $\mathscr{R}_0$. In practice we are often interested in a *finitely generated subring* of $\mathscr{R}_0$, i.e. the ring generated by a finite collection $\{A_1, \ldots, A_m\}$ of sets in $\mathscr{C}$, and here we can be quite specific. As a convenient notation we write $A^1 = A$, $A^c = X \setminus A$ and use ϵ for either 1 or c.

Lemma 1. *If $A_1, \ldots, A_m$ are subsets of X, then A_i is the union of the disjoint sets*

$$A_1^{\epsilon_1} \cap \ldots \cap A_m^{\epsilon_m},$$

with $\epsilon_i = 1$ (Fig. 14).

Proof. Select $x \in A_i$ and let

$$\begin{aligned} \epsilon_j &= 1 \quad \text{if} \quad x \in A_j, \\ &= c \quad \text{if} \quad x \notin A_j. \end{aligned}$$

Then $\epsilon_i = 1$ and $x \in A_1^{\epsilon_1} \cap \ldots \cap A_m^{\epsilon_m}$. The rest is obvious.

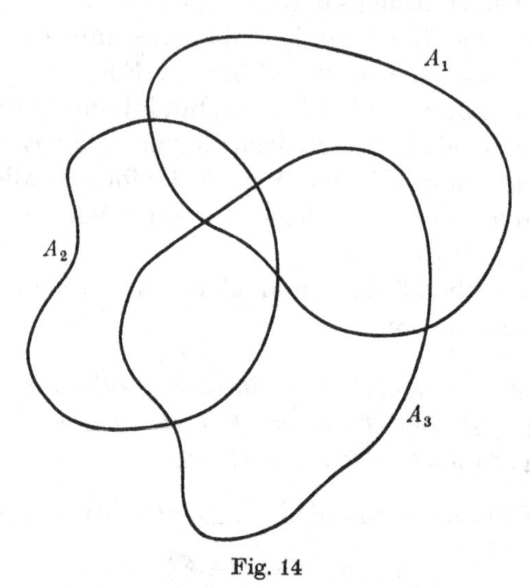

Fig. 14

Lemma 2. *If $A_1, \ldots, A_m$ belong to a ring $\mathscr{R}$ then the $2^m - 1$ disjoint sets*

$$A_1^{\epsilon_1} \cap \ldots \cap A_m^{\epsilon_m}$$

with not all $\epsilon_i = c$, also belong to $\mathscr{R}$.

Proof. The result is obvious for $m = 1$. For $m = 2$ it follows from the definition of $\mathscr{R}$ if we observe that $A \cap B^c$ is just another way of writing $A \setminus B$. The same observation for $m \geqslant 3$ gives an immediate proof by induction.

To avoid possible misunderstanding, note that the $2^m - 1$ sets of Lemma 2 need not be distinct; as an extreme case, they could all be empty!

Lemma 3. *Let $\mathscr{D}$ be a collection of disjoint subsets of X. Then the finite unions*

$$D_1 \cup \ldots \cup D_s,$$

with $D_1, \ldots, D_s$ in $\mathscr{D}$, form a ring.

It is an important convention that we include among these finite unions the *empty union* $\varnothing$, corresponding to $s = 0$.

Proof. This follows immediately from the definition of a ring.

Proposition 1. *The ring generated by* $\{A_1, \ldots, A_m\}$ *consists of unions of the* $2^m - 1$ *disjoint sets*
$$A_1^{\epsilon_1} \cap \ldots \cap A_m^{\epsilon_m},$$
with not all $\epsilon_i = c$. *The algebra generated by* $\{A_1, \ldots, A_m\}$ *consists of unions of the* 2^m *disjoint sets*
$$A_1^{\epsilon_1} \cap \ldots \cap A_m^{\epsilon_m}.$$

Proof. This follows at once from Lemmas 2 and 3.

Exercises

1. Find the ring generated by $\mathscr{C} = \{[1,2), [1,3)\}$.

2. Find the ring generated by $\mathscr{C} = \{[1,3), [2,4)\}$.

3. Let $\mathscr{C}$ be a non-empty collection of subsets of X. Show that $\mathscr{C}$ is an algebra if and only if
$$A, B \in \mathscr{C} \Rightarrow A \cup B, A^c \in \mathscr{C}$$
(i.e. $\mathscr{C}$ is closed with respect to unions and complements).

4. Let $\mathscr{R}$ be a ring of subsets of X. Show that the algebra generated by $\mathscr{R}$ consists of all subsets A of X for which either A or A^c belongs to $\mathscr{R}$.

5. According to Theorem 8.4.2, the measurable sets with respect to a Daniell integral form a ring. Show that the sets of finite measure form a ring. When are these rings also algebras?

6. Let $f: X \to Y$. If $\mathscr{R}$ is a ring of subsets of Y show that
$$f^{-1}(\mathscr{R}) = \{f^{-1}(A): A \in \mathscr{R}\}$$
is a ring of subsets of X. If $\mathscr{R}$ is the ring generated by $\mathscr{C}$ show that $f^{-1}(\mathscr{R})$ is the ring generated by $f^{-1}(\mathscr{C})$.

7. Let $f: X \to Y$. If $\mathscr{R}$ is a ring of subsets of X, can we deduce that
$$f(\mathscr{R}) = \{f(A): A \in \mathscr{R}\}$$
is a ring of subsets of Y?

Boolean Rings and Algebras

A *ring* is a non-empty set R together with two operations which associate with any elements a, b of R a sum $a+b$ and a product ab, both in R, which satisfy the following conditions for all a, b, c in R:

R1. $a+b = b+a$,
R2. $(a+b)+c = a+(b+c)$,

R3. there is an element 0 in R such that $a+0 = a$,
R4. there is an element $-a$ in R such that $a+(-a) = 0$,
R5. $(ab)c = a(bc)$,
R6. $a(b+c) = ab+ac$,
R7. $(a+b)c = ac+bc$.

It is usual to write a^2 for the product aa. A *Boolean ring* is a ring R in which the equation

$$a^2 = a$$

is satisfied for all a in R.

A. Let R be a Boolean ring. By expanding $(a+b)^2$ show that $ab+ba = 0$ for all a, b in R and deduce that $a+a = 0$ and $ab = ba$ for all a, b in R. (In other words R is a commutative ring of characteristic 2.)

B. Let $\mathbf{F}_2$ be the field consisting of two elements 0, 1 in which addition and multiplication are as for the integers 0, 1, except that $1+1 = 0$ (Ex. 1.3.1). We may regard the characteristic function χ_A of a set A as taking its values in $\mathbf{F}_2$: with this understanding show that

$$\chi_A+\chi_B = \chi_{A \triangle B}, \quad \chi_A\chi_B = \chi_{A \cap B}.$$

C. Let X be a non-empty set and denote by $\mathscr{P}(X)$ the collection of all subsets of X and by $P(X)$ the collection of all functions $f: X \to \mathbf{F}_2$. Show that $P(X)$ is a Boolean ring with respect to (pointwise) addition and multiplication of functions. Use the mapping

$$A \to \chi_A$$

to show that $\mathscr{P}(X)$ is a Boolean ring with respect to the operations $\triangle$, $\cap$ (in place of addition and multiplication) and that $P(X)$, $\mathscr{P}(X)$ are isomorphic.

[The direct verification of the ring axioms for $\mathscr{P}(X)$ is tedious: e.g. the associativity of $\triangle$.]

D. In a Boolean ring R we may introduce the analogue of inclusion (cf. Ex. C) by writing

$$a \leqslant b \quad \text{if and only if} \quad ab = a.$$

Verify that (i) $a \leqslant a$ (ii) $a \leqslant b, b \leqslant a \Rightarrow a = b$ (iii) $a \leqslant b, b \leqslant c \Rightarrow a \leqslant c$. (In other words $\leqslant$ is a partial ordering on R.)

E. In the terminology of Ex. D, $0 \leqslant a$ for all a in R. Thus 0 is the analogue of the empty set $\varnothing$. An element 1 of a Boolean ring R is a *unit* if $1a = a$ for all a in R. This means that $a \leqslant 1$ for all a in R. Thus 1 is the analogue of the universal set.

[A ring of subsets of X as described in the text is a subring of $\mathscr{P}(X)$ and and algebra of subsets of X with universal set X is a subring of $\mathscr{P}(X)$ with unit X. Stone has proved the following significant theorem: *Any Boolean ring is isomorphic to a ring of subsets of some set* X.]

F. In a Boolean ring R with 1 we may introduce the analogues of intersection and union (cf. Ex. C) by writing

$$a \wedge b = ab, \quad a \vee b = a+b+ab.$$

Show that

L1. $a \vee b = b \vee a, \quad a \wedge b = b \wedge a;$

L2. $(a \vee b) \vee c = a \vee (b \vee c), \quad (a \wedge b) \wedge c = a \wedge (b \wedge c);$

L3. $a \vee a = a, \quad a \wedge a = a;$

L4. $(a \vee b) \wedge c = (a \wedge c) \vee (b \wedge c), \quad (a \wedge b) \vee c = (a \vee c) \wedge (b \vee c);$

L5. $a \vee 0 = a, \quad a \wedge 1 = a,$

$a \vee 1 = 1, \quad a \wedge 0 = 0;$

L6. to any element a in R there is an element a' in R such that

$$a \vee a' = 1, \quad a \wedge a' = 0.$$

Show that the element a' (the *complement* of a) is uniquely determined by a and that

$$(a')' = a, \quad (a \vee b)' = a' \wedge b', \quad (a \wedge b)' = a' \vee b'.$$

[This is one classical formulation of the axioms for a Boolean algebra which brings out the duality between the lattice operations $\vee$, $\wedge$. In brief these axioms state that R is a *distributive lattice* (L1–L4) *with* 0 *and* 1 (L5) which is *complemented* (L6).]

G. Let $(R, \vee, \wedge)$ be a distributive complemented lattice with 0 and 1 as in Ex. F. Define

$$a+b = (a \wedge b') \vee (a' \wedge b), \quad ab = a \wedge b$$

for all a, b in R. Show that R is a Boolean ring with unit 1 with respect to these operations of addition and multiplication.

H. Combine Exx. F, G.

11.2 Simple functions

Let $\mathscr{C}$ be a non-empty, but otherwise arbitrary, collection of subsets of X and denote by $L(\mathscr{C})$ the linear space of *simple functions* (strictly, *$\mathscr{C}$-simple functions*)

$$\phi = a_1\chi_{A_1} + \dots + a_m\chi_{A_m},$$

where $a_1, \dots, a_m \in \mathbf{R}$ and $A_1, \dots, A_m \in \mathscr{C}$. These are analogous to the step functions of §3.1, §4.1 and the simple functions of §6.2. The case where $\mathscr{C}$ is a ring is of particular importance, and we consider this first.

Proposition 1. *Let $\mathscr{R}$ be a ring of subsets of X. Then any element ϕ of $L(\mathscr{R})$ has an expression in terms of disjoint sets in $\mathscr{R}$. In fact, if $\phi \neq 0$, then ϕ has unique expression as*

$$\phi = c_1\chi_{S_1} + \dots + c_r\chi_{S_r},$$

where $c_1 < c_2 < \dots < c_r$ are the distinct non-zero values assumed by ϕ and $S_1, \dots, S_r$ are disjoint elements of $\mathscr{R}$.

Proof. Let $\phi = a_1\chi_{A_1} + \ldots + a_m\chi_{A_m}$ and denote by $B_1, \ldots, B_n$ ($n = 2^m - 1$) the disjoint sets described in Lemma 11.1.2. Each χ_{A_i} is the sum of certain χ_{B_j} (Lemma 11.1.1) and so by the obvious gathering of terms,

$$\phi = b_1\chi_{B_1} + \ldots + b_n\chi_{B_n}.$$

More precisely,

$$b_j = \sum_{A_i \supset B_j} a_i. \tag{1}$$

Now if $c_1 < \ldots < c_r$ are the non-zero values assumed by ϕ, another gathering of terms gives

$$\phi = c_1\chi_{S_1} + \ldots + c_r\chi_{S_r},$$

where

$$S_k = \bigcup_{b_j = c_k} B_j.$$

This final expression for ϕ is obviously unique. (It is interesting to compare it with the one given in §6.2 after Proposition 1, where we allowed zero as one of the c's; that was fine when the sets involved – the Lebesgue measurable subsets of $\mathbf{R}^k$ – formed an algebra, but in the present context it is simpler to ignore the complement of $S_1 \cup \ldots \cup S_r$ which may not lie in $\mathscr{R}$.)

As an immediate consequence of Proposition 1 we have:

Corollary. *$L(\mathscr{R})$ is a linear lattice satisfying the stronger form of Stone's Condition*, viz.

$$\min\{1, \phi\} \in L(\mathscr{R}) \quad \textit{for any positive } \phi \textit{ in } L(\mathscr{R}).$$

Proof. If $\phi = b_1\chi_{B_1} + \ldots + b_n\chi_{B_n}$, as in the above proof, then

$$|\phi| = |b_1|\,\chi_{B_1} + \ldots + |b_n|\,\chi_{B_n},$$

because $B_1, \ldots, B_n$ are disjoint. Moreover,

$$\min\{1, \phi\} = d_1\chi_{B_1} + \ldots + d_n\chi_{B_n},$$

where $d_j = \min\{1, b_j\}$ for each j.

We draw attention to the fact that the proof of this corollary only depends on the first part of Proposition 1, viz., that any $\mathscr{R}$-simple function may be expressed in terms of disjoint sets in $\mathscr{R}$. This 'disjointness condition' was used several times in the construction of the Lebesgue integral for step functions: the key results were Propositions 3.1.1 and 4.1.1, which state that if $\mathscr{C}$ is the collection of all bounded intervals in $\mathbf{R}$, $\mathbf{R}^k$, respectively, then any $\mathscr{C}$-simple function may be expressed in terms of disjoint sets in $\mathscr{C}$.

We shall now explore this condition more generally. Let $\mathscr{C}$ be a non-empty collection of subsets of X and, as a temporary notation, let $\overline{\mathscr{C}}$ denote the collection of sets

$$E_1 \cup \ldots \cup E_s,$$

where $E_1, \ldots, E_s$ are disjoint elements of $\mathscr{C}$; as for Lemma 11.1.3, we agree to include in $\overline{\mathscr{C}}$ the empty union $\varnothing$, corresponding to $s = 0$.

Proposition 2. *The following three disjointness conditions are equivalent.*

(i) *Any element ϕ of $L(\mathscr{C})$ may be expressed as*

$$\phi = e_1\chi_{E_1} + \ldots + e_s\chi_{E_s},$$

where $E_1, \ldots, E_s$ are disjoint elements of $\mathscr{C}$.

(ii) *The finite disjoint unions of elements of $\mathscr{C}$ form a ring, i.e. $\overline{\mathscr{C}}$ is a ring.*

(iii) *If A, B are elements of $\mathscr{C}$ then $A \cap B$, $A \setminus B$ are finite disjoint unions of elements of $\mathscr{C}$, i.e.*

$$A, B \in \mathscr{C} \Rightarrow A \cap B, A \setminus B \in \overline{\mathscr{C}}.$$

Proof. (i) $\Rightarrow$ (ii) As for the corollary to Proposition 1, $L(\mathscr{C})$ is a linear lattice. Let $\mathscr{R} = \{A\colon \chi_A \in L(\mathscr{C})\}$. If $\chi_A, \chi_B \in L(\mathscr{C})$ then

$$\chi_{A \cup B} = \max\{\chi_A, \chi_B\} \quad \text{and} \quad \chi_{A \setminus B} = (\chi_A - \chi_B)^+$$

both belong to $L(\mathscr{C})$. Thus $\mathscr{R}$ is a ring. But $\chi_A \in L(\mathscr{C})$ if and only if $A \in \overline{\mathscr{C}}$. In other words $\mathscr{R} = \overline{\mathscr{C}}$.

(ii) $\Rightarrow$ (iii) $\overline{\mathscr{C}}$ is a ring containing $\mathscr{C}$.

(iii) $\Rightarrow$ (i) Suppose that $\phi = a_1\chi_{A_1} + \ldots + a_m\chi_{A_m}$ with $A_1, \ldots, A_m$ in $\mathscr{C}$. As in the proof of Proposition 1,

$$\phi = b_1\chi_{B_1} + \ldots + b_n\chi_{B_n}, \tag{2}$$

where $B_1, \ldots, B_n$ are the disjoint sets $A_1^{\epsilon_1} \cap \ldots \cap A_m^{\epsilon_m}$ with not all $\epsilon_i = c$. It will be enough to prove that $B_1, \ldots, B_n \in \overline{\mathscr{C}}$ for (2) will expand to give an expression for ϕ in terms of disjoint sets from $\mathscr{C}$.

If $m = 1$ then obviously $A_1 \in \overline{\mathscr{C}}$. If $m = 2$ the result is given (recall that $A \setminus B = A \cap B^c$). Suppose that $m \geqslant 2$ and

$$A_1^{\epsilon_1} \cap \ldots \cap A_m^{\epsilon_m} = E_1 \cup \ldots \cup E_s,$$

where $E_1, \ldots, E_s$ are disjoint elements of $\mathscr{C}$. Then

$$(E_1 \cup \ldots \cup E_s) \cap A^\epsilon = (E_1 \cap A^\epsilon) \cup \ldots \cup (E_s \cap A^\epsilon),$$

where each $E_i \cap A^\epsilon$ is in $\overline{\mathscr{C}}$ and $E_i \cap A^\epsilon$, $E_j \cap A^\epsilon$ for $i \neq j$ are disjoint as they lie in E_i, E_j, respectively. A simple induction now completes the proof.

If $\mathscr{C}$ satisfies any one of the equivalent conditions of Proposition 2, we shall say that $\mathscr{C}$ *satisfies the Disjointness Condition*. For example, if $\mathscr{C}$ consists of the bounded intervals in $\mathbf{R}^k$ it is easy to verify condition (iii); Proposition 2 therefore gives an independent proof of Proposition 4.1.1 and the fact that the elementary figures, defined at the end of § 4.1, form a ring. In this particular case, $\mathscr{C}$ satisfies the stronger condition:

$$A, B \in \mathscr{C} \Rightarrow A \cap B \in \mathscr{C} \quad \text{and} \quad A \setminus B \in \overline{\mathscr{C}}.$$

In the literature such a collection is usually known as a *semi-ring*†. In $\mathbf{R}^k$ we may consider the *simplexes* (of dimension not greater than k). These do not form a semi-ring though they do satisfy the Disjointness Condition; their finite disjoint unions, which we might call *polyhedral sets*, then form a ring. See Exx. 5, 6 for some details.

If $\mathscr{C}$ satisfies the Disjointness Condition, and in particular if $\mathscr{C}$ is a semi-ring, then the proof of Proposition 1, Corollary shows that $L(\mathscr{C})$ is a linear lattice. In Ex. 2 we give necessary and sufficient conditions for $L(\mathscr{C})$ to be a linear lattice: one of these shows that if $L(\mathscr{C})$ is a linear lattice then there is a ring $\mathscr{R}$ containing $\mathscr{C}$ such that $L(\mathscr{C}) = L(\mathscr{R})$ and so $L(\mathscr{C})$ automatically satisfies the stronger Stone condition by Proposition 1, Corollary. The simple case $\mathscr{C} = \{[1,3),[2,4)\}$ shows that $L(\mathscr{C})$ need not be a linear lattice, and the case $\mathscr{C} = \{[1,3),[2,4),[2,3)\}$, where $\mathscr{C}$ is closed with respect to intersections, shows that $L(\mathscr{C})$ may be a linear lattice without satisfying condition (i) of Proposition 2. In the next two sections where we construct an elementary integral on $L(\mathscr{C})$ starting with a measure μ on $\mathscr{C}$, we shall see that it is not enough to assume that $L(\mathscr{C})$ is a linear lattice (satisfying the Stone condition), and it is in this context that the Disjointness Condition comes into its own.

Exercises

1. If $\mathscr{C} = \{[1,3),[2,4)\}$ show that $L(\mathscr{C})$ is not a linear lattice.

2. Prove the ***Proposition:*** *For any collection $\mathscr{C}$ of subsets of X the following conditions are equivalent.*

(i) *$L(\mathscr{C})$ is a linear lattice,*
(ii) *$\{A: \chi_A \in L(\mathscr{C})\}$ is a ring,*
(iii) *$L(\mathscr{C})$ is closed with respect to multiplication,*
(iv) *$A, B \in \mathscr{C} \Rightarrow \chi_{A \cap B} \in L(\mathscr{C})$.*

3. If $L(\mathscr{C})$ is a linear lattice show that $\{A: \chi_A \in L(\mathscr{C})\}$ is the ring generated by $\mathscr{C}$.

† It is customary to insist that a semi-ring includes the empty set.

4. A sufficient condition for $L(\mathscr{C})$ to be a linear lattice is that $\mathscr{C}$ should be closed with respect to intersections, i.e.

$$A, B \in \mathscr{C} \Rightarrow A \cap B \in \mathscr{C}.$$

A sufficient condition for $L(\mathscr{C})$ to be a linear lattice is that $\mathscr{C} \cup \{\varnothing\}$ should be closed with respect to symmetric differences; equivalently,

$$A, B \in \mathscr{C}, A \neq B \Rightarrow A \triangle B \in \mathscr{C}.$$

Is either of these conditions necessary?

[Let $a_0, a_1, \ldots, a_r$ be elements of $\mathbf{R}^k$ for which $a_1 - a_0, \ldots, a_r - a_0$ are linearly independent. The set of all points

$$x = \lambda_0 a_0 + \lambda_1 a_1 + \ldots + \lambda_r a_r,$$

with $\lambda_0, \lambda_1, \ldots, \lambda_r > 0$ and $\lambda_0 + \lambda_1 + \ldots + \lambda_r = 1$, is the *simplex* (of dimension r) with vertices $a_0, a_1, \ldots, a_r$. For example, in $\mathbf{R}^2$ the simplexes are the open triangles, open line segments and single points.

In $\mathbf{R}^2$ an *open half-plane* is defined by a strict linear inequality. An *open convex polygon* is the bounded (non-empty) intersection of finitely many open half-planes. You may assume from elementary combinatorial topology the fact that an open convex polygon may be *triangulated*, i.e. expressed as the union of a finite set of disjoint simplexes in $\mathbf{R}^2$.]

5. Show that the simplexes in $\mathbf{R}^2$ satisfy the Disjointness Condition.

6. Extend Ex. 5 to $\mathbf{R}^k$ ($k \geqslant 3$).

11.3 Additive set functions

By a *set function* we simply mean a function whose domain of definition is a collection of sets. Let $\mathscr{C}$ be a non-empty collection of subsets of X and $\int$ a linear functional on $L(\mathscr{C})$. We may define a set function $\mu: \mathscr{C} \to \mathbf{R}$ by means of the equation

$$\mu(A) = \int \chi_A \tag{1}$$

for all A in $\mathscr{C}$. From the linearity of $\int$ we deduce at once that μ is (*finitely*) *additive*, i.e.

$$\mu(A_1 \cup \ldots \cup A_m) = \mu(A_1) + \ldots + \mu(A_m) \tag{2}$$

for any disjoint sets $A_1, \ldots, A_m$ in $\mathscr{C}$, *provided their union is also in* $\mathscr{C}$. The main purpose of this section is to solve the converse problem: given $\mathscr{C}$ and an additive function $\mu: \mathscr{C} \to \mathbf{R}$; to find a linear functional $\int$ on $L(\mathscr{C})$ satisfying (1).

First of all, we consider the case where $\mathscr{C}$ is a ring $\mathscr{R}$. The awkward proviso in the definition of additivity may be dropped in this case because $\mathscr{R}$ is closed with respect to finite unions; moreover, equation (2) may be replaced by the simpler equation

$$\mu(A_1 \cup A_2) = \mu(A_1) + \mu(A_2)$$

for any disjoint sets A_1, A_2 in $\mathscr{R}$.

Theorem 1. *Let $\mu: \mathscr{R} \to \mathbf{R}$ be an additive function. Then there exists a unique linear functional $\int$ on $L(\mathscr{R})$ satisfying*

$$\int \chi_A = \mu(A)$$

for all A in $\mathscr{R}$.

If it is desired to stress the fact that this integral depends on μ we may write $\int \phi \, d\mu$ in place of $\int \phi$ and call it 'the integral of ϕ with respect to μ'.

Proof. If $\phi = a_1\chi_{A_1} + \ldots + a_m\chi_{A_m}$, where $a_1, \ldots, a_m \in \mathbf{R}$, $A_1, \ldots, A_m \in \mathscr{R}$, then we define

$$\int \phi = a_1\mu(A_1) + \ldots + a_m\mu(A_m).$$

As usual there is the question of consistency. If we refine the given expression for ϕ as in the proof of Proposition 11.2.1 to obtain

$$\phi = b_1\chi_{B_1} + \ldots + b_n\chi_{B_n},$$

and use the additivity of μ each time we gather terms, we find that

$$a_1\mu(A_1) + \ldots + a_m\mu(A_m) = b_1\mu(B_1) + \ldots + b_n\mu(B_n).$$

More formally: $$A_i = \bigcup_{B_j \subset A_i} B_j,$$

and so by additivity, $$\mu(A_i) = \sum_{B_j \subset A_i} \mu(B_j).$$

Hence

$$\begin{aligned} \sum_i a_i\mu(A_i) &= \sum_i \sum_{B_j \subset A_i} a_i\mu(B_j) \\ &= \sum_j \sum_{A_i \supset B_j} a_i\mu(B_j) \\ &= \sum_j b_j\mu(B_j) \end{aligned}$$

by equation (1) of § 11.2.

This argument shows that any refinement in terms of disjoint sets will give the same value for $\int \phi$. But now if any two expressions are given for ϕ we may apply Lemma 11.1.2 to the sets involved in both expressions to obtain a common refinement in terms of disjoint sets and hence deduce the consistency. This is just the familiar argument of § 3.1 and § 4.1.

Now that we are satisfied that the definition of $\int\phi$ is unequivocal, it is quite obvious that $\int$ is a linear functional on the linear space $L(\mathscr{R})$. (If we had chosen to define $\int\phi$ in terms of the unique expression for ϕ given by Proposition 11.2.1, then the proof of the linearity of $\int$ would have presented essentially the same kind of difficulty that we have just faced.)

Theorem 1 is sufficient for almost all that follows, but we must give some idea of what can happen in a more general setting. We start with a non-empty collection $\mathscr{C}$ and an additive function $\mu: \mathscr{C} \to \mathbf{R}$. If $\mathscr{C}$ is sufficiently sparse the condition of additivity may say nothing at all. For example, if $\mathscr{C} = \{[1,3), [2,4)\}$, *any* function $\mu: \mathscr{C} \to \mathbf{R}$ is additive! In this case it is a trivial matter to construct an integral satisfying (1).

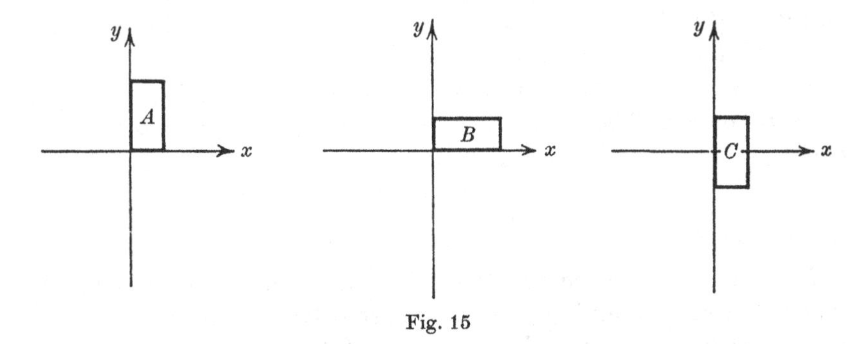

Fig. 15

But consider the more interesting case of three distinct sets A, B, C for which $A \cap B = A \cap C = B \cap C \neq \varnothing$ and let

$$\mathscr{C} = \{A, B, C, A \triangle B, A \triangle C\}.$$

To be quite specific, we could take A, B, C as the subsets of $\mathbf{R}^2$ defined respectively by the inequalities

$$\begin{aligned} 0 \leqslant x \leqslant 1, &\qquad 0 \leqslant y \leqslant 2; \\ 0 \leqslant x \leqslant 2, &\qquad 0 \leqslant y \leqslant 1; \\ 0 \leqslant x \leqslant 1, &\qquad -1 \leqslant y \leqslant 1; \end{aligned}$$

(see Fig. 15).

It is easy to check that the five sets constituting $\mathscr{C}$ are distinct and that *any* function $\mu: \mathscr{C} \to \mathbf{R}$ is additive. If we give $\mu(A)$, $\mu(B)$, $\mu(C)$ equal values and $\mu(A \triangle B)$, $\mu(A \triangle C)$ unequal values, then it is impossible to define a linear functional $\int$ on $L(\mathscr{C})$ satisfying (1) because the relations

$$\chi_{A \cap B} = \tfrac{1}{2}(\chi_A + \chi_B - \chi_{A \triangle B}),$$

$$\chi_{A \cap C} = \tfrac{1}{2}(\chi_A + \chi_C - \chi_{A \triangle C}),$$

would require different values for

$$\int \chi_{A \cap B}, \quad \int \chi_{A \cap C},$$

despite the fact that $A \cap B = A \cap C$. Incidentally, it is easy to verify that $L(\mathscr{C})$ is a linear lattice in this case (see Ex. 2) and so we require some more stringent condition on $\mathscr{C}$ than this. In the light of our previous experience in constructing the Lebesgue integral on the linear space of step functions and the discussion at the end of the last section, it is natural to assume that $\mathscr{C}$ satisfies the Disjointness Condition.

***Theorem* 2.** *Let $\mathscr{C}$ be a non-empty collection of subsets of X and suppose that $\mathscr{C}$ satisfies the Disjointness Condition: if $A, B \in \mathscr{C}$ then $A \cap B$, $A \setminus B$ are finite disjoint unions of sets in $\mathscr{C}$.*

Then any additive function $\mu: \mathscr{C} \to \mathbf{R}$ extends uniquely to the ring $\mathscr{R}$ generated by $\mathscr{C}$ and there is a unique functional $\int$ on $L(\mathscr{R}) = L(\mathscr{C})$ such that

$$\int \chi_A = \mu(A)$$

for all A in $\mathscr{C}$.

Proof. By Proposition 11.2.2 (ii), $\mathscr{R}$ consists of the finite disjoint unions of sets in $\mathscr{C}$ and $L(\mathscr{R}) = L(\mathscr{C})$. Let A be an element of $\mathscr{R}$ and suppose that A is expressed as the union of disjoint sets in two ways:

$$A = A_1 \cup \ldots \cup A_m = B_1 \cup \ldots \cup B_n.$$

Each $A_i \cap B_j$ is in turn expressed as the union of disjoint sets from $\mathscr{C}$: denote by m_{ij} the sum of their measures. (At this stage $\mu(A_i \cap B_j)$ is not defined unless $A_i \cap B_j$ belongs to $\mathscr{C}$!) For each i, A_i is the union of the disjoint sets $A_i \cap B_j$ as j varies, and so by the additivity of μ on $\mathscr{C}$, it follows that

$$\mu(A_i) = \sum_j m_{ij}.$$

In exactly the same way $\mu(B_j) = \sum_i m_{ij}$

so that $$\sum_i \mu(A_i) = \sum_{i,j} m_{ij} = \sum_j \mu(B_j).$$

We may therefore define $\mu(A)$ *unambiguously* as

$$\mu(A_1) + \ldots + \mu(A_m) = \mu(B_1) + \ldots + \mu(B_n).$$

The uniqueness of this extension of μ to $\mathscr{R}$ is obvious from the additivity, and the rest of the proof follows from Theorem 1.

Exercises

1. Let $\mu: \mathscr{C} \to \mathbf{R}$ be additive. If $\varnothing \in \mathscr{C}$ show that $\mu(\varnothing) = 0$. If μ is positive and $\mu(A \triangle B) = 0$, show that $\mu(A) = \mu(B)$.

2. Let A, B, C be the sets described in Fig. 15 and let

$$\mathscr{C} = \{A, B, C, A \triangle B, A \triangle C\}.$$

Show that $L(\mathscr{C})$ is a linear lattice.

Show that any function $\mu: \mathscr{C} \to \mathbf{R}$ is additive. If $\mu(A) = 3$, $\mu(B) = 3$, $\mu(C) = 3$, $\mu(A \triangle B) = 2$, $\mu(A \triangle C) = 1$, show that μ cannot be extended to an additive function on a ring $\mathscr{R}$ containing $\mathscr{C}$.

3. Let $\mathscr{C} = \{\varnothing, [0,1), [0,\frac{1}{2}), [0,\frac{1}{4}), [0,\frac{3}{4}), [\frac{1}{4},\frac{3}{4})\}$ and define

$$\mu(\varnothing) = 0, \mu[0,1) = 4, \mu[0,\tfrac{1}{2}) = 2, \mu[0,\tfrac{1}{4}) = 2, \mu[0,\tfrac{3}{4}) = 4, \mu[\tfrac{1}{4},\tfrac{3}{4}) = 2.$$

Show that μ is additive on $\mathscr{C}$. Can μ be extended to an additive set function on the ring generated by $\mathscr{C}$?

4. Let $\mu: \mathscr{R} \to \mathbf{R}$ be an additive set function on the ring $\mathscr{R}$. If

$$A_1, A_2, A_3, \ldots \in \mathscr{R}$$

show that

$$\mu(A_1 \cup A_2) = \mu(A_1) + \mu(A_2) - \mu(A_1 \cap A_2),$$

$$\mu(A_1 \cup A_2 \cup A_3) = \mu(A_1) + \mu(A_2) + \mu(A_3) - \mu(A_1 \cap A_2) - \mu(A_1 \cap A_3) - \mu(A_2 \cap A_3) + \mu(A_1 \cap A_2 \cap A_3).$$

What is the formula for n sets?

11.4 Measures

At this point we make the decision to concentrate on set functions whose values are positive real numbers or ∞. It is convenient to denote by $[0,\infty]$ the 'closed interval' $[0,\infty) \cup \{\infty\}$, although this is admittedly a relaxation of our previous strict attitude to the point ∞ (cf. § 1.1 and § 6.2). The discussion of real and complex valued set functions is postponed until Chapter 16, though this material would fit quite logically at the end of § 11.3.

Let $\mathscr{C}$ be a non-empty collection of sets in X. A function $\mu: \mathscr{C} \to [0,\infty]$ is (*finitely*) *additive* if

$$\mu(A_1 \cup \ldots \cup A_m) = \mu(A_1) + \ldots + \mu(A_m) \tag{1}$$

for any disjoint sets $A_1, \ldots, A_m$ in $\mathscr{C}$, provided their union is also in $\mathscr{C}$. Equation (1) may be compared with equation (2) at the beginning of the previous section, but with the obvious interpretation for infinite values, viz. the left hand side is infinite if and only if at least one term on the right hand side is infinite.

A *countably additive* (or *σ-additive*) function $\mu: \mathscr{C} \to [0, \infty]$ is an additive function which satisfies

$$\mu(\bigcup A_n) = \Sigma\mu(A_n) \tag{2}$$

for any sequence $\{A_n\}$ of disjoint sets A_n in $\mathscr{C}$, *provided their union is also in* $\mathscr{C}$. If any $\mu(A_n) = \infty$, or if $\Sigma\mu(A_n)$ is a divergent series of positive real numbers, we write $\Sigma\mu(A_n) = \infty$: thus (2) has an obvious interpretation for infinite values.

A σ-additive function $\mu: \mathscr{C} \to [0, \infty]$ which takes at least one finite value will be called a *measure* on $\mathscr{C}$. If the empty set $\varnothing$ belongs to $\mathscr{C}$, it follows at once that $\mu(\varnothing) = 0$ (cf. Ex. 1).

The natural domain of definition for a measure μ is a *σ-ring*, viz. a ring $\mathscr{R}$ which is closed with respect to countable unions, or a *σ-algebra*, viz. an algebra $\mathscr{A}$ which is closed with respect to countable unions; for in these cases the embarrassing proviso at the end of the definition of σ-additivity is quite unnecessary. The primitive measure of intervals was recognised as additive many centuries before it was realised that it is also σ-additive. Jordan content (§ 13.5) is a more sophisticated example of a measure whose σ-additivity is in practice reduced to additivity because the ring of contented sets is not closed with respect to countable unions.

The following existence theorem is analogous to Theorem 11.1.1.

Theorem 1. *Let* $\mathscr{C}$ *be an arbitrary collection of subsets of* X. *Then there is a unique smallest σ-ring* $\mathscr{S}$ *of subsets of* X *containing* $\mathscr{C}$, *and a unique smallest σ-algebra* $\mathscr{A}$ *of subsets of* X *containing* $\mathscr{C}$.

Proof. As for Theorem 11.1.1, let $\mathscr{S}$ be the intersection of all σ-rings of subsets of X containing $\mathscr{C}$, and let $\mathscr{A}$ be the intersection of all σ-algebras of subsets of X containing $\mathscr{C}$.

We refer to $\mathscr{S}$ as the *σ-ring generated by* $\mathscr{C}$ and $\mathscr{A}$ as the *σ-algebra generated by* $\mathscr{C}$.

The main purpose of the present section is to extend a measure from a restricted domain (such as the bounded intervals or the elementary figures in $\mathbf{R}^k$) to a σ-algebra (such as the Lebesgue measurable sets in $\mathbf{R}^k$). As a first step in the extension programme we have the following quite elementary result.

Proposition 1. *Let* $\mathscr{C}$ *be a non-empty collection of subsets of* X *satisfying the Disjointness Condition* (*of Theorem* 11.3.2) *and let* $\mathscr{R}$ *be the ring* *generated by $\mathscr{C}$. Then any measure on $\mathscr{C}$ may be extended uniquely to a measure on $\mathscr{R}$.*

Proof. The first part of the proof of Theorem 11.3.2 applies verbatim, even allowing for possible infinite values, and so μ extends uniquely to an *additive* function μ on $\mathscr{R}$ and $\mu(\varnothing) = 0$. It remains to prove that μ is σ-additive on $\mathscr{R}$.

Let $\{A_n\}$ be a sequence of disjoint elements of $\mathscr{R}$ whose union A is also in $\mathscr{R}$. We shall prove that

$$\mu(A) = \sum_{n=1}^{\infty} \mu(A_n).$$

Now
$$A = E_1 \cup \ldots \cup E_s,$$

where $E_1, \ldots, E_s$ are disjoint elements of $\mathscr{C}$, and

$$\mu(A) = \mu(E_1) + \ldots + \mu(E_s)$$

by the definition of μ on $\mathscr{R}$. The proof therefore reduces to the case where $A \in \mathscr{C}$. Each A_n is in $\mathscr{R}$ and so is a union

$$A_{n1} \cup \ldots \cup A_{ns_n}$$

of disjoint elements of $\mathscr{C}$, where

$$\mu(A_n) = \mu(A_{n1}) + \ldots + \mu(A_{ns_n}).$$

Now A is the union of the disjoint sets A_{ij} which may be written as an array

$$\begin{array}{llll} A_{11}, & A_{12}, & \ldots, & A_{1s_1}, \\ A_{21}, & A_{22}, & \ldots, & A_{2s_2}, \\ \ldots & & & \end{array}$$

in which the rows are finite. If we enumerate this array by going along the rows, then the countable additivity of μ on $\mathscr{C}$ gives

$$\mu(A) = \Sigma \mu(A_{ij}).$$

An elementary proposition on series with positive terms, suitably interpreted for infinite values (Ex. 2), allows us to gather the terms in each row and deduce that

$$\mu(A) = \sum_{n=1}^{\infty} \mu(A_n)$$

as required.

Suppose now that μ is a *finite* measure on a ring $\mathscr{R}$, i.e. $\mu(A)$ is finite for all A in $\mathscr{R}$. Then § 11.3 suggests that we define a linear functional $\int$ on the linear lattice $L(\mathscr{R})$ of simple functions in such a way that

$$\int \chi_A = \mu(A)$$

for all A in $\mathscr{R}$. As μ is positive it follows that $\int$ is a positive linear functional, i.e.

$$\phi \geqslant 0 \Rightarrow \int \phi \geqslant 0.$$

For if

$$\phi = c_1 \chi_{S_1} + \ldots + c_r \chi_{S_r},$$

with $S_1, \ldots, S_r$ disjoint, as in Proposition 11.2.1, then $c_1, \ldots, c_r \geqslant 0$ and

$$\int \phi = c_1 \mu(S_1) + \ldots + c_r \mu(S_r) \geqslant 0.$$

The discussion in § 11.3 only used the additivity of μ. How can we interpret the σ-additivity of μ in terms of $\int$? The following result gives the gratifying answer that the σ-additivity of μ is equivalent to the Daniell condition for $\int$, and this of course now means that $\int$ *is an elementary integral on* $L(\mathscr{R})$.

Proposition 2. *Let* $\mu: \mathscr{R} \to [0, \infty)$ *be an additive function. (Note that the values of* μ *are finite and positive.) Then the following four conditions are equivalent.*

(i) μ *is* σ*-additive.*

(ii) $S_n \uparrow S \in \mathscr{R} \Rightarrow \mu(S_n) \to \mu(S)$.

(iii) $S_n \downarrow \varnothing \Rightarrow \mu(S_n) \to 0$.

(iv) $\phi_n \downarrow 0 \Rightarrow \int \phi_n \to 0$.

$$(S_n \in \mathscr{R}, \quad \phi_n \in L(\mathscr{R}).)$$

In (ii) and (iii) we have introduced a fairly obvious use of arrows. They are interpreted as follows.

(ii) *If* $\{S_n\}$ *is an increasing sequence of sets of* $\mathscr{R}$ *whose union* S *also belongs to* $\mathscr{R}$*, then the sequence* $\{\mu(S_n)\}$ *converges to* $\mu(S)$.

(iii) *If* $\{S_n\}$ *is a decreasing sequence of sets of* $\mathscr{R}$ *whose intersection is empty, then the sequence* $\{\mu(S_n)\}$ *converges to zero.*

Proof. In proving the equivalence of (i) and (ii) we shall use a notation which brings out the analogy with sequences and series (cf. § 1.1). If $\{A_n\}$ is a sequence of disjoint sets in $\mathscr{R}$ then we may define an increasing sequence $\{S_n\}$ of 'partial sums':

$$S_n = A_1 \cup \ldots \cup A_n.$$

On the other hand, if $\{S_n\}$ is an increasing sequence of sets in $\mathscr{R}$ then we may define

$$A_1 = S_1, \quad A_n = S_n \setminus S_{n-1} \quad (n \geqslant 2)$$

so that the A's are disjoint and

$$S_n = A_1 \cup \ldots \cup A_n$$

for all n. It is now clear that

$$S_n \uparrow S = \bigcup A_n$$

in either case. By the additivity of μ

$$\mu(S_n) = \mu(A_1) + \ldots + \mu(A_n)$$

so that $\quad \mu(S_n) \to \mu(S) \quad$ if and only if $\quad \Sigma\mu(A_n) = \mu(S)$.

This shows that (i) and (ii) are equivalent.

Suppose that (ii) holds and that $S_n \downarrow \varnothing$. Let $T_n = S_1 \setminus S_n$. Then $T_n \uparrow S_1$ and so $\mu(T_n) \to \mu(S_1)$. As μ only assumes finite values we deduce from the additivity that $\mu(T_n) = \mu(S_1) - \mu(S_n)$ and so $\mu(S_n) \to 0$. Conversely, if (iii) holds and $S_n \uparrow S \in \mathscr{R}$, let $T_n = S \setminus S_n$. Then $T_n \downarrow \varnothing$ and $\mu(T_n) \to 0$. But $\mu(T_n) = \mu(S) - \mu(S_n)$ (again using the finiteness of μ) and so $\mu(S_n) \to \mu(S)$.

It is clear that (iii) is a particular case of the Daniell condition (iv) with $\phi_n = \chi_{S_n}$. Conversely, suppose that (iii) holds and that $\{\phi_n\}$ is a decreasing sequence of positive simple functions which converges (everywhere) to 0 (cf. the proof of Lemma 3.2.1). Let $A_1 = \{x: \phi_1(x) > 0\}$ so that $A_1 \in \mathscr{R}$ and $\mu(A_1)$ is a (finite) real number. Also let

$$M_1 = \max_{x \in X} \phi_1(x)$$

(well-defined as ϕ_1 takes only finitely many values). For any $\epsilon > 0$, define

$$S_n^{\epsilon} = \{x: \phi_n(x) \geqslant \epsilon\}.$$

As $0 \leqslant \phi_n \leqslant \phi_1$ it follows that

$$\phi_n \leqslant \epsilon\chi_{A_1} + M_1\chi_{S_n^{\epsilon}}$$

and

$$\int \phi_n \leqslant \epsilon\mu(A_1) + M_1\mu(S_n^{\epsilon}).$$

But $\phi_n(x) \downarrow 0$ for all x, which implies that

$$S_n^{\epsilon} \downarrow \varnothing$$

and so $\mu(S_n^{\epsilon}) \to 0$ by condition (iii). We may therefore find a positive integer N such that $\mu(S_N^{\epsilon}) < \epsilon$ and hence

$$0 \leqslant \int \phi_n \leqslant \epsilon(\mu(A_1) + M_1)$$

for $n \geqslant N$. This shows finally that $\int \phi_n \to 0$ and completes the proof.

If μ takes the value ∞ it is no longer true, in general, that (ii) and (iii) are equivalent. For example, if $S_n = [n, \infty)$ and μ is Lebesgue measure on $\mathbf{R}$, then $S_n \downarrow \varnothing$ although $\mu(S_n) = \infty$ for all n. The positiveness of μ is nowhere used in proving the equivalence of (i), (ii) and (iii) but the final part of the proof is vitally dependent on this assumption (Ex. 13).

Let us sum up what has been achieved so far in this chapter. Let $\mathscr{R}$ be a ring of subsets of X and μ a *finite* measure on $\mathscr{R}$. The simple functions

$$\phi = a_1\chi_{A_1} + \ldots + a_m\chi_{A_m},$$

where $a_1, \ldots, a_m \in \mathbf{R}$, $A_1, \ldots, A_m \in \mathscr{R}$, form a linear lattice L which satisfies the strong Stone condition

$$\min\{1, \phi\} \in L \quad \text{for any positive } \phi \text{ in } L$$

(Proposition 11.2.1, Corollary). There is a unique linear functional $\int$ on L defined by

$$\int \phi = a_1\mu(A_1) + \ldots + a_m\mu(A_m)$$

(Theorem 11.3.1). This linear functional is an elementary integral on L (Proposition 2). In view of all this, we may apply the standard Daniell construction of § 8.1 to obtain a Daniell integral and a corresponding collection $\mathscr{M}$ of measurable functions. According to Theorem 8.4.2 (i), (iii) $\mathscr{M}$ is a σ-ring; but Stone's condition implies that X is a measurable set and so $\mathscr{M}$ is a σ-algebra. Let $\bar{\mu}$: $\mathscr{M} \to [0, \infty]$ be defined by

$$\bar{\mu}(S) = \int \chi_S \quad \text{if} \quad \chi_S \text{ is integrable,}$$
$$= \infty \qquad \text{otherwise.}$$

Then $\bar{\mu}$ is a measure on the σ-algebra $\mathscr{M}$ (Theorem 8.4.2 (iii)) and clearly $\bar{\mu}$ extends the given finite measure μ on $\mathscr{R}$. We have used a different symbol for the extended measure as it is important in the present context to distinguish the extended measure from the given one.

Before we state the fundamental extension theorem for measures we must mention one quite obvious property of null sets as defined in the Daniell construction, viz. that *any subset of a null set is null.* According to Theorem 8.4.2 (i), the set A is null if and only if $A \in \mathscr{M}$ and $\bar{\mu}(A) = 0$. In other words $\bar{\mu}$ has the following property of *completeness:* if $A \in \mathscr{M}$ and $\bar{\mu}(A) = 0$, then any subset B of A also belongs to $\mathscr{M}$ (and hence $\bar{\mu}(B) = 0$). Incomplete measures certainly exist: for example, if $\mathscr{R}$ is the ring of elementary figures in $\mathbf{R}^2$ (the smallest ring containing the bounded intervals) then $\mathscr{R}$ contains the interval I defined by $0 \leqslant x \leqslant 1, y = 0$ which has (plane) measure zero, but there are plenty of subsets of I that are not elementary figures. Thus the restriction of Lebesgue measure to $\mathscr{R}$ is not a complete measure. The very important Borel measures to be studied in Chapter 15 give further examples of incomplete measures.

Theorem 2 (Extension Theorem). *Any measure μ on a ring $\mathscr{R}$ extends to a complete measure $\bar{\mu}$ on a σ-algebra $\mathscr{M}$ containing $\mathscr{R}$.*

Proof. When μ is finite this has all been stated in the last two paragraphs. In general if $A, C \in \mathscr{R}$, $C \subset A$ and $\mu(A)$ is finite, then the equation

$$\mu(A) = \mu(C) + \mu(A \setminus C)$$

shows that $\mu(C)$, $\mu(A \setminus C)$ are finite and

$$\mu(C) \leqslant \mu(A).$$

Now if $A, B \in \mathscr{R}$ and $\mu(A)$, $\mu(B)$ are finite, then $\mu(A \setminus B)$ is finite and

$$\mu(A \cup B) = \mu(A) + \mu(B) - \mu(A \cap B)$$

is finite. From this it follows that the sets A in $\mathscr{R}$ for which $\mu(A)$ is finite form a subring $\mathscr{R}_0$ of $\mathscr{R}$. Let μ_0 be the restriction of μ to $\mathscr{R}_0$. There is a complete measure $\bar{\mu}$ extending μ_0 to a σ-algebra $\mathscr{M}$. All that remains to prove is that $\bar{\mu}(E) = \infty$ for any E in $\mathscr{R}$ for which $\mu(E) = \infty$. (In other words $\bar{\mu}$ does not embarrass us by giving a finite measure, or no measure at all, to one of the sets which we discarded to obtain $\mathscr{R}_0$. As an extreme case we could have $\mu(E) = \infty$ for all non-empty E in $\mathscr{R}$ so that $\mathscr{R}_0 = \{\varnothing\}$ and we discard all but the empty set! The situation here is saved by what we might call a predisposition of the Daniell construction to give as large a measure as possible. This vague statement will be made precise in Proposition 13.4.1: the final part of our present proof is only a special case.)

Suppose that $E \in \mathscr{R}$. For any positive ϕ in $L(\mathscr{R}_0)$,

$$S = \{x \in X \colon \phi(x) > 0\}$$

belongs to $\mathscr{R}_0$, and so

$$\min\{\chi_E, \phi\} = \min\{\chi_{E \cap S}, \phi\}$$

belongs to $L(\mathscr{R}_0)$. Hence χ_E is certainly measurable, i.e. $E \in \mathscr{M}$. Suppose further that $\chi_E \in L^1$, i.e. $\bar{\mu}(E)$ is finite. By Lemma 8.2.1 there exist positive functions g, h in L^{inc} such that $\chi_E = g - h$. Thus

$$g \geqslant \chi_E.$$

By the definition of L^{inc} there is an increasing sequence $\{\phi_n\}$ of positive functions in $L(\mathscr{R}_0)$ such that $\phi_n \to g$ outside a null set A. By the definition of null sets, there is an increasing sequence $\{\psi_n\}$ of positive functions in $L(\mathscr{R}_0)$ which diverges at each point of A, and so the increasing sequence $\{\theta_n\}$ defined by

$$\theta_n = \phi_n + \psi_n$$

satisfies

$$\lim \theta_n \geqslant \chi_E. \tag{3}$$

(Stretch the notation temporarily to allow ∞ as a limit, or truncate θ_n by 1 if you prefer.) The sets

$$S_n = \{x \in X : \theta_n(x) > 0\}$$

belong to $\mathscr{R}_0$ and by (3) we have

$$\bigcup S_n \supset E.$$

Thus $$S_n \cap E \uparrow E.$$

Finally, μ and $\bar{\mu}$ agree on the sets $S_n \cap E$ which belong to $\mathscr{R}_0$ and so

$$\mu(S_n \cap E) \to \mu(E) = \bar{\mu}(E)$$

by the σ-additivity of both μ and $\bar{\mu}$. Thus $\mu(E)$ is finite whenever $\bar{\mu}(E)$ is finite, and this completes the proof.

Exercises

1. Let μ be a measure on $\mathscr{C}$. If $\varnothing \in \mathscr{C}$ show that $\mu(\varnothing) = 0$. Can we deduce this from Ex. 11.3.1?

2. Let Σa_n be a series of positive terms and let Σb_n be a series obtained by inserting parentheses in Σa_n (e.g.

$$a_1 + (a_2 + a_3) + (a_4 + a_5 + a_6) + (a_7 + a_8 + a_9 + a_{10}) + \ldots).$$

Show that Σa_n, Σb_n either converge to the same sum or both diverge. (This question relates to the proof of Proposition 1.)

3. Show that the collection of all finite subsets of an infinite set X is a ring, but not a σ-ring.

4. Show that the collection of all countable subsets of an uncountable set X is a σ-ring but not a σ-algebra.

5. Show that a ring which is closed under countable disjoint unions is a σ-ring, but that a ring which is closed under countable intersections is not necessarily a σ-ring.

6. Let $\mathscr{R}$ be a ring of subsets of X and denote by $\mathscr{A}$ the collection of all subsets B of X such that $A \cap B \in \mathscr{R}$ for all A in $\mathscr{R}$. Show that (i) $\mathscr{A}$ is an algebra containing $\mathscr{R}$, (ii) if $\mathscr{R}$ is an algebra then $\mathscr{A} = \mathscr{R}$, (iii) if $\mathscr{R}$ is a σ-ring, then $\mathscr{A}$ is a σ-algebra.

7. Let $\mathscr{C}$ be a non-empty collection of subsets of X and $\mathscr{R}, \mathscr{S}$ respectively the ring, σ-ring generated by $\mathscr{C}$. Show that every set in $\mathscr{R}$ is a subset of a finite union of sets in $\mathscr{C}$, and every set in $\mathscr{S}$ is a subset of a countable union of sets in $\mathscr{C}$.

8. Find two semi-rings whose intersection is not a semi-ring.

9. Let X be the set of rational numbers in $[0,1]$ and let $\mathscr{C}$ be the set of 'intervals' of the form $\{x \in X: a < x \leqslant b\}$ where $0 \leqslant a \leqslant b \leqslant 1$ and $a, b \in X$. Show that $\mathscr{C}$ is a semi-ring and every set in $\mathscr{C}$ is either empty or infinite. Show that the σ-ring generated by $\mathscr{C}$ contains all subsets of X.

10. Let $X = (0,1]$ and let $\mathscr{C}$ consist of the half-open intervals $(a,b]$ where $0 \leqslant a < b \leqslant 1$. Let $\mu(a,b] = b-a$ if $a \neq 0$ and $\mu(0,b] = \infty$. Show that μ is additive but not σ-additive.

11. Let X be an infinite set and $\mathscr{P}$ the collection of all subsets of X. Let $\{x_n\}$ be a sequence of distinct points of X and $\{p_n\}$ a sequence of positive real numbers. For any set A in $\mathscr{P}$ let

$$\mu(A) = \sum_{x_n \in A} p_n,$$

where the sum extends over all integers n for which $x_n \in A$ (and may of course be infinite). Show that μ is a measure on $\mathscr{P}$.

12. Let $\int$ be a Daniell integral on the space L^1 with σ-algebra $\mathscr{M}$ of measurable sets and let h be a positive function in L^1. Show that

$$\nu(A) = \int_A h \quad (A \in \mathscr{M})$$

defines a measure ν on $\mathscr{M}$.

13. Let $\{A_n\}$ be a strictly decreasing sequence of subsets of X whose intersection is not empty. Find the ring $\mathscr{R}$ generated by $\{A_1, A_2, \ldots\}$ and show that there is an additive function $\mu: \mathscr{R} \to \mathbf{R}$ satisfying $\mu(A_n) = n$ for all n.

Let $\phi_n = \dfrac{1}{n} \chi_{A_n}$. What is the relevance of the sequence $\{\phi_n\}$ to Proposition 2?

14. Let $\{A_n\}$ be the sequence of Ex. 13 and take $X = A_1$. Find the σ-ring and the σ-algebra generated by $\{A_1, A_2, \ldots\}$.

15. Let X be an uncountable set and let $\mathscr{C}$ consist of the 'singleton' sets $\{x\}$ where $x \in X$. Show that $\mathscr{C}$ satisfies the disjointness condition of Theorem 11.3.2 and find the ring $\mathscr{R}$ generated by $\mathscr{C}$. Show that any set function $\mu: \mathscr{C} \to [0, \infty]$ is σ-additive and may be extended to a measure on $\mathscr{R}$.

If $\mu(A) = c$ (possibly 0 or ∞) for all A in $\mathscr{C}$, what is the corresponding measure on $\mathscr{R}$?

16. Let $\mathscr{C}$, $\mathscr{R}$ be as in Ex. 15. Find the σ-ring and σ-algebra generated by $\mathscr{C}$. If $\mu(A) = c \in [0, \infty]$ for all A in $\mathscr{C}$, what is the measure obtained by extending first to $\mathscr{R}$ and then using Theorem 2?

17. Let $\mathscr{C}$ be as in Ex. 15. What is the algebra $\mathscr{A}$ generated by $\mathscr{C}$? If μ is identically zero on $\mathscr{C}$ show how to extend μ to a measure on $\mathscr{A}$ by taking (i) $\mu(X) = 0$ (ii) $\mu(X) = 1$ (iii) $\mu(X) = \infty$.

In each case find the measure $\bar{\mu}$ given by Theorem 2 with $\mathscr{R} = \mathscr{A}$.

18. Let μ be a measure on a collection $\mathscr{C}$ of subsets of X. A set E in $\mathscr{C}$ is an *atom* of μ if $\mu(E) \neq 0$, and for any subset A of E $(A \in \mathscr{C})$, either $\mu(A) = 0$ or $\mu(A) = \mu(E)$.

Find the atoms, if any, of the measures described in Exx. 8.4.1–4 and 11.4.11. What are the atoms of the Lebesgue–Stieltjes measure in Theorem 9.1.1?

19. *Theorem.* *Let μ be a finite measure on a σ-algebra $\mathscr{A}$ of subsets of X and suppose that μ has no atoms. Then μ takes every value between* 0 *and* $\mu(X)$.

For the definition of atom see Ex. 18. Prove this theorem using the following steps.

(i) Given E in $\mathscr{A}$ with $\mu(E) > 0$ and any $\epsilon > 0$; there is a set B in $\mathscr{A}$ satisfying $B \subset E$ and $0 < \mu(B) < \epsilon$.

(ii) Let $0 < t < \mu(X)$. According to (i) there are sets A in $\mathscr{A}$ such that $0 < \mu(A) < t$. Let

$$s_1 = \sup\{\mu(A) \colon A \in \mathscr{A}, \mu(A) \leqslant t\}$$

(where obviously $0 < s_1 \leqslant t$). Then there exists a set A_1 in $\mathscr{A}$ such that

$$s_1/2 < \mu(A_1) \leqslant s_1.$$

Let

$$s_2 = \sup\{\mu(A) \colon A \in \mathscr{A}, A \supset A_1, \mu(A) \leqslant t\}.$$

Then there exists a set A_2 in $\mathscr{A}$ such that $A_2 \supset A_1$ and

$$s_2 - s_1/2^2 < \mu(A_2) \leqslant s_2.$$

Continue this construction inductively to find a decreasing sequence $\{s_n\}$ of real numbers and an increasing sequence $\{A_n\}$ of sets in $\mathscr{A}$ such that $A = \bigcup A_n \in \mathscr{A}$ and $s_n \downarrow \mu(A)$.

(iii) Deduce from (i) that $\mu(A) = t$.

(iv) *Post Script*: in the light of (iii) what can you now say about the sequence $\{s_n\}$?

[This theorem has an interesting history including contributions from Sierpiński, Fréchet and Hahn: see [9], pp. 51–3. The construction given in (ii) is due to D. Newton.]

12

THE CLASSICAL APPROACH

Following in the steps of Lebesgue, most exponents of integration theory in the present century have begun by constructing a measure μ, have proceeded from there to define measurable functions and then, finally, have introduced the idea of integration with respect to the measure μ. We refer to this as the classical approach. In this chapter we shall step aside briefly to compare the classical approach with what we have done so far; very little subsequently depends on this discussion, but it may be valuable in linking up with other treatments of the subject.

The extension of a measure μ from a ring $\mathscr{R}$ to a σ-algebra $\mathscr{M}$ was achieved in the last chapter by means of the Daniell construction. Exactly the same extension may be achieved by the purely measure theoretic construction of Carathéodory: the details of this are fairly complicated and are postponed until § 13.5. For the present we assume the existence of a measure μ on a σ-algebra $\mathscr{A}$ and explore the consequent notions of measurability with respect to $\mathscr{A}$, integrability with respect to μ and measurability with respect to μ.

12.1 $\mathscr{A}$-measurable functions

Let us begin tentatively by supposing that μ is a measure on a collection $\mathscr{C}$ of subsets of X. In the light of Propositions 6.2.1 and 8.4.2 we shall say that $f: X \to \mathbf{R}$ is *measurable* with respect to $\mathscr{C}$ if

$$A_c = \{x \in X: f(x) \geqslant c\}$$

belongs to $\mathscr{C}$ for all c in $\mathbf{R}$. In particular, if $f = \chi_A$, then

$$\begin{aligned} A_c &= X \quad \text{for} \quad c \leqslant 0, \\ &= A \quad \text{for} \quad 0 < c \leqslant 1, \\ &= \varnothing \quad \text{for} \quad 1 < c. \end{aligned}$$

We certainly hope that all characteristic functions χ_A with A in $\mathscr{C}$ will be measurable and so it is reasonable to assume that $\varnothing$ and X both belong to $\mathscr{C}$. With this assumption, χ_A is measurable if and only if $A \in \mathscr{C}$. We also hope that the measurable functions will form a linear space. As

$$A \cup B = \{x \in X: \chi_A(x) + \chi_B(x) \geqslant 1\},$$

$$A \setminus B = \{x \in X: \chi_A(x) - \chi_B(x) \geqslant 1\},$$

this leads to the assumption that $\mathscr{C}$ is an *algebra* of subsets of X. (In fact this is still not enough to ensure that the measurable functions form a linear space: see Ex. 1.) One other sensible requirement is that the class of measurable functions should be closed with respect to pointwise convergence of sequences. In particular, if $\{A_n\}$ is an increasing sequence of sets in $\mathscr{C}$ whose union is A, then

$$\chi_{A_n} \to \chi_A$$

would imply that A belongs to $\mathscr{C}$. This finally drives us to the assumption that $\mathscr{C}$ is a *σ-algebra* of subsets of X.

It should perhaps be noted at this point that Halmos [11], and several authors following him, define measurable functions with respect to a *σ-ring* $\mathscr{S}$. If $\mathscr{S}$ is not a σ-algebra, i.e. if $X \notin \mathscr{S}$, our definition would imply that the constant functions are not measurable. It is bad enough to exclude the constant function 1 from the class of measurable functions, but it is intolerable to exclude the zero function! Halmos deals with this elegantly by the simple expedient of ignoring the points x for which $f(x) = 0$: more precisely, he defines

$$N(f) = \{x \in X : f(x) \neq 0\}$$

and says that f is measurable if and only if

$$N(f) \cap A_c \in \mathscr{S}$$

for all c in $\mathbf{R}$. This ensures that the zero function is measurable and that χ_A is measurable if and only if $A \in \mathscr{S}$, but it leaves the awkward situation where the constant functions $c1 (c \neq 0)$ are not measurable. It is more natural for us to assume that $X \in \mathscr{S}$, i.e. that $\mathscr{S}$ is a σ-algebra, as this is the exact counterpart of Stone's Axiom in the present setting.

So much by way of introduction. Let us now suppose that $\mathscr{A}$ is a σ-algebra of subsets of X and say that $f: X \to \mathbf{R}$ is *measurable with respect to $\mathscr{A}$*, or *$\mathscr{A}$-measurable*, if

$$A_c = \{x \in X : f(x) \geqslant c\}$$

belongs to $\mathscr{A}$ for all c in $\mathbf{R}$. Note that this definition depends on the σ-algebra $\mathscr{A}$, but that there is no mention of a measure μ.

If

$$B_c = \{x \in X : f(x) > c\}$$

then

$$B_c = \bigcup_{n=1}^{\infty} A_{c+1/n}$$

and

$$A_c = \bigcap_{n=1}^{\infty} B_{c-1/n}.$$

It follows that f is measurable with respect to $\mathscr{A}$ if and only if $B_c \in \mathscr{A}$ for all c in $\mathbf{R}$.

The following basic result should be compared with Theorems 6.1.1 and 8.4.1.

Theorem 1. *Let $\mathscr{A}$ be a σ-algebra of subsets of X. The class of $\mathscr{A}$-measurable functions is a linear lattice which is closed with respect to pointwise convergence of sequences. Explicitly:*

(i) *if f, g are $\mathscr{A}$-measurable, then so are $|f|$, f^+, f^-, $af+bg$ $(a, b \in \mathbf{R})$, $\max\{f, g\}$ and $\min\{f, g\}$;*

(ii) *if $f_n \to f$ and f_n is $\mathscr{A}$-measurable for $n = 1, 2, \ldots$ then f is $\mathscr{A}$-measurable.*

It should be noted that there is no 'almost everywhere' in (ii); this is hardly surprising as there is no mention of a measure in the theorem!

Proof. It is convenient to begin by establishing part (ii). (For a direct proof of part (i) see Ex. 5.) If

$$l_n = \inf\{f_n, f_{n+1}, \ldots\}$$

then
$$l_n(x) \geqslant c \Leftrightarrow f_m(x) \geqslant c \quad \text{for} \quad m \geqslant n.$$

Thus
$$\{x: l_n(x) \geqslant c\} = \bigcap_{m \geqslant n} \{x: f_m(x) \geqslant c\}$$

belongs to $\mathscr{A}$, as f_m is measurable for $m \geqslant n$. Hence l_n is measurable for $n \geqslant 1$. But $\{l_n\}$ is an increasing sequence which converges to

$$\liminf f_n = f$$

(by Proposition 5.2.1). In other words

$$f = \sup\{l_1, l_2, \ldots\}.$$

Arguing as above,

$$f(x) > c \Leftrightarrow l_n(x) > c \quad \text{for some} \quad n \geqslant 1,$$

i.e.
$$\{x: f(x) > c\} = \bigcup_{n \geqslant 1} \{x: l_n(x) > c\}$$

and so f is measurable.

Before we conclude the proof of Theorem 1 let us note the following results which are the exact counterparts of Propositions 8.4.3 and 8.4.4; their proofs are virtually the same as before. Now that we have part (ii) of Theorem 1, the only point of difference is that we must verify that the process of truncating a measurable function above and below by constant functions yields a measurable function, and this follows immediately using the sets A_c and B_c.

Proposition 1. *A measurable function $f: X \to \mathbf{R}$ may be expressed as the limit, everywhere, of a sequence of simple functions; if f is also positive, then f may be expressed as the limit, everywhere, of an increasing sequence of positive simple functions.*

In this proposition a simple function ϕ is, of course, an element of $L(\mathscr{A})$, i.e.

$$\phi = a_1\chi_{A_1} + \ldots + a_m\chi_{A_m},$$

where $a_1, \ldots, a_m \in \mathbf{R}, A_1, \ldots, A_m \in \mathscr{A}$. Strictly, we should say that f is $\mathscr{A}$-measurable and ϕ is $\mathscr{A}$-simple.

Proposition 2. *If $h: \mathbf{R}^2 \to \mathbf{R}$ is continuous and if $f, g: X \to \mathbf{R}$ are measurable, then the composite function $F: X \to \mathbf{R}$ defined by the equation*

$$F(x) = h(f(x), g(x))$$

for x in X, is again measurable.

To deduce part (i) of Theorem 1 we need only define h by

$$h(x, x_2) = ax_1 + bx_2$$

or

$$h(x_1, x_2) = |x_1|$$

for x_1, x_2 in $\mathbf{R}$. This shows that $af + bg$ and $|f|$ are measurable, and the others follow by the standard linear relations as for Theorem 8.4.1 (ii).

As a consequence of Proposition 2 we have the following fact which combines with Theorem 1 to show that the $\mathscr{A}$-measurable functions form a *linear algebra*.

Corollary *If $f, g: X \to \mathbf{R}$ are measurable, then their product fg is also measurable.*

Proof. Let

$$h(x_1, x_2) = x_1 x_2$$

for x_1, x_2 in $\mathbf{R}$.

Exercises

1. Let $X = \{1, 2, \ldots\}$ and let $\mathscr{A}$ be the algebra consisting of all finite subsets of X and their complements in X. Let

$$f(n) = 0 \quad \text{if} \quad n \text{ is odd,}$$

$$= \frac{1}{n} \quad \text{if} \quad n \text{ is even.}$$

If measurability with respect to $\mathscr{A}$ is defined as in the first paragraph of this section, show that f is measurable but that $-f$ is not measurable.

2. Let $\mathscr{A}$ be a finite algebra of subsets of X. Show that $\mathscr{A}$ is a σ-algebra and that $f: X \to \mathbf{R}$ is $\mathscr{A}$-measurable if and only if f is $\mathscr{A}$-simple (i.e. belongs to $L(\mathscr{A})$).

3. Show that f is $\mathscr{A}$-measurable if and only if $\{x \in X: f(x) < c\}$ belongs to $\mathscr{A}$ for all c in $\mathbf{R}$. Similarly for $\{x \in X: f(x) \leqslant c\}$.

4. Write out proofs of Propositions 1 and 2 (cf. Propositions 8.4.3, 8.4.4).

5. Let $\mathscr{A}$ be a σ-algebra of subsets of X and suppose that f, g are $\mathscr{A}$-measurable functions on X and $k \in \mathbf{R}$. Show that kf and $|f|$ are $\mathscr{A}$-measurable.

For any x in X, show that $f(x)+g(x) > c$ if and only if there exists a rational number r such that $f(x) > r$ and $g(x) > c-r$. Deduce that $f+g$ is measurable. Combine these results to prove Theorem 1 (i).

6. If f is $\mathscr{A}$-measurable show that f^2 is measurable. Use the identity $4fg = \{(f+g)^2-(f-g)^2\}$ and Ex. 5 to deduce Proposition 2, Corollary.

12.2 Integration with respect to μ

In §12.1 we studied properties of a σ-algebra $\mathscr{A}$: now consider a measure μ on the σ-algebra $\mathscr{A}$. Following the pattern of § 11.2 and § 11.3 we may express a simple function ϕ as

$$\phi = c_1\chi_{S_1} + \ldots + c_r\chi_{S_r},$$

where $c_1, \ldots, c_r$ are distinct non-zero elements of $\mathbf{R}$ and $S_1, \ldots, S_r$ are disjoint elements of $\mathscr{A}$, and define

$$\int \phi \, d\mu = c_1\mu(S_1) + \ldots + c_r\mu(S_r),$$

provided the measures $\mu(S_1), \ldots, \mu(S_r)$ are all finite. Suppose that f is a positive $\mathscr{A}$-measurable function and that $\{\phi_n\}$ is an increasing sequence of $\mathscr{A}$-simple functions which converges everywhere to f. (The existence of such sequences is guaranteed by Proposition 12.1.1.) If the integrals

$$\int \phi_n \, d\mu$$

exist and are bounded, we say that the positive function f is *integrable* with respect to μ and define

$$\int f \, d\mu = \lim \int \phi_n \, d\mu.$$

By now it should be a conditioned reflex to check the consistency of this definition. One method, that seems a little like cheating in the present context, is to use the Daniell integral constructed in §11.4 which extends the integral with respect to μ on $L(\mathscr{A})$ and satisfies the Monotone Convergence Theorem: this is just the argument used in proving Stone's Theorem 8.5.1. Without specific reference to the Daniell construction we may extract the following two lemmas.

Lemma 1. *If $\{\phi_n\}$ is a decreasing sequence of integrable simple functions which converges to zero, then*

$$\int \phi_n \, d\mu \to 0.$$

This, of course, is the Daniell condition! The proof is given in Proposition 11.4.2 (taking for $\mathscr{R}$ the subring of $\mathscr{A}$ on which μ is finite).

Lemma 2. *Let $\{\phi_n\}$, $\{\psi_n\}$ be increasing sequences of simple functions whose integrals with respect to μ are bounded and let $\{\phi_n\}$, $\{\psi_n\}$ converge to f, g, respectively. If $f \geqslant g$, then*

$$\lim \int \phi_n \, d\mu \geqslant \lim \int \psi_n \, d\mu.$$

This is merely a repeat of Lemma 8.1.2 where the words 'almost everywhere' are omitted and 'elementary functions' are replaced by 'simple functions'. Lemma 2, with $f = g$, is just what is needed to check the consistency.

It now follows at once that

$$\int (f+g) \, d\mu = \int f \, d\mu + \int g \, d\mu$$

and

$$\int cf \, d\mu = c \int f \, d\mu$$

for all positive integrable functions f, g and all positive real numbers c. By a process we have now used several times (e.g. in the proof of Theorem 8.1.1), $\int d\mu$ extends to the set $L^1(\mu)$ consisting of all $f = g - h$, where g, h are positive integrable functions, and

$$\int f \, d\mu = \int g \, d\mu - \int h \, d\mu.$$

Moreover we have

Theorem 1. *Let μ be a measure on the σ-algebra $\mathscr{A}$. Then $L^1(\mu)$ is a linear lattice on which $\int d\mu$ is a positive linear functional.*

It is noteworthy that the above discussion makes no assumption about the completeness of μ. For an incomplete measure μ we must be more careful about any statement involving *null sets*, i.e. sets A in $\mathscr{A}$ for which $\mu(A) = 0$. As before we say that $\mathfrak{P}(x)$ holds for *almost all x*

(in X) if there is a null set A such that $\mathfrak{P}(x)$ holds for all x in $X \setminus A$. This does not necessarily mean that the set S for which $\mathfrak{P}(x)$ fails to hold is a null set, but rather that S is contained in a null set A. It is possible to have $f(x) = g(x)$ for almost all x, where f is measurable but g is not. For example, if S is a non-measurable subset of a null set A, then $\chi_A = \chi_S$ almost everywhere, χ_A is measurable, but χ_S is not measurable. In the next chapter we shall give an elementary construction which shows how to extend μ to a complete measure; nevertheless, it will emerge, particularly in the context of Borel measures on $\mathbf{R}^k$, that there is much to be gained by studying measures which may be incomplete.

Even without the assumption of completeness we may prove the most fundamental theorem for $L^1(\mu)$.

Theorem 2 (The Monotone Convergence Theorem). *Let $\{f_n\}$ be a monotone sequence of functions in $L^1(\mu)$ whose integrals are bounded. Then $\{f_n\}$ converges almost everywhere to a function f, where f lies in $L^1(\mu)$, and*

$$\int f d\mu = \lim \int f_n d\mu.$$

Proof. Without loss of generality we may assume that $\{f_n\}$ is an *increasing* sequence of *positive* functions and that

$$\int f_n d\mu \leqslant K,$$

where $K > 0$.

(i) (Cf. the proof of Theorem 3.2.1.) For any $\epsilon > 0$ let

$$S_n{}^\epsilon = \left\{x: f_n(x) \geqslant \frac{K}{\epsilon}\right\}.$$

Then $S_n{}^\epsilon$ is measurable, and in fact

$$\chi_{S_n{}^\epsilon} \leqslant \frac{\epsilon}{K} f_n$$

implies that

$$\mu(S_n{}^\epsilon) \leqslant \epsilon$$

for all n. The increasing sequence $\{S_n{}^\epsilon\}$ converges to S^ϵ, say, where

$$\mu(S^\epsilon) \leqslant \epsilon$$

by the σ-additivity of μ. The decreasing sequence $\{S^{1/n}\}$ converges to S, say, where

$$\mu(S) \leqslant 1/n$$

for all n, and so $\mu(S) = 0$. This null set S is exactly the set of points x

in X for which $f_n(x) \to \infty$: we check that $f_n(x) \to \infty$ if and only if $x \in S^{1/m}$ for all $m \geqslant 1$. Now define

$$f(x) = \lim f_n(x) \quad \text{if} \quad x \notin S,$$
$$= 0 \qquad\qquad \text{if} \quad x \in S.$$

(ii) We are now able to use the easier part of the original proof of the Monotone Convergence Theorem. Multiplying all f_n by χ_S, if necessary, we may as well assume that $f_n \to f$ everywhere. For each integer $m \geqslant 1$ let $\{\phi_{mn}\}$ be an increasing sequence of simple functions which converges (everywhere) to f_m and define

$$\phi_n = \max \phi_{ij}$$

for $1 \leqslant i, j \leqslant n$. Then $\{\phi_n\}$ is an increasing sequence of simple functions which satisfies

$$\phi_{mn} \leqslant \phi_n \leqslant f_n$$

for $m \leqslant n$. Keep m fixed and let n tend to infinity: then $\{\phi_n\}$ converges to an integrable function g, where

$$f_m \leqslant g \leqslant f$$

for all m. Now let m tend to infinity and deduce that $g = f$. This shows that f is integrable and

$$\int \phi_n d\mu \to \int f d\mu.$$

But

$$\int \phi_n d\mu \leqslant \int f_n d\mu \leqslant \int f d\mu$$

and the squeezing argument gives

$$\int f_n d\mu \to \int f d\mu.$$

Once again we remark that many authors allow ∞ and $-\infty$ as values of their functions and most of them would therefore define

$$f(x) = \lim f_n(x)$$

for all x, in the above theorem. In any case, f takes finite values outside a null set. We return to this question in Chapter 17 where we define the $\mathscr{L}^p$ spaces.

The Monotone Convergence Theorem leads as before to Lebesgue's main theorem.

Theorem 3 (The Dominated Convergence Theorem). *Let* $\{f_n\}$ *be a sequence of functions in* $L^1(\mu)$ *which converges (everywhere) to a func-*

tion f and which is dominated by the positive function g in $L^1(\mu)$:

$$|f_n| \leqslant g$$

for all n. Then $f \in L^1(\mu)$ and

$$\int f_n \, d\mu \to \int f \, d\mu.$$

We can easily extend this result to the case where $f_n \to f$ almost everywhere, but we must then assume that f is measurable. A similar remark applies to Fatou's Lemma: the details are left to the reader (Exx. 6, 7).

Exercises

1. Let μ be a measure on the σ-algebra $\mathscr{A}$ and let

$$\phi = a_1 \chi_{A_1} + \ldots + a_m \chi_{A_m},$$

where $a_1, \ldots, a_m$ are non-zero real numbers and $A_1, \ldots, A_m$ are disjoint elements of $\mathscr{A}$. Show that ϕ is integrable with respect to μ if and only if $\mu(A_1), \ldots, \mu(A_m)$ are all finite.

2. If $f\colon X \to \mathbf{R}$ is integrable with respect to μ show that

$$\mu\{x \in X : |f| \geqslant c\}$$

is finite for any $c > 0$.

3. Let μ_1, μ_2 be two measures on the σ-algebra $\mathscr{A}$ and let $\mu = \mu_1 + \mu_2$. If f is integrable with respect to μ_1 and μ_2 show that f is integrable with respect to μ and

$$\int f \, d\mu = \int f \, d\mu_1 + \int f \, d\mu_2.$$

4. Let μ be a finite measure on the σ-algebra $\mathscr{A}$. If f is measurable with respect to μ and

$$p \leqslant f(x) \leqslant q$$

for all x in X, show that f is integrable with respect to μ and

$$p\mu(X) \leqslant \int f \, d\mu \leqslant q\mu(X).$$

5. Write out a proof of Theorem 1 on the lines of Theorem 8.1.1.

6. Write out a proof of the Dominated Convergence Theorem as stated in Theorem 3. Extend to the case where $\{f_n\}$ converges almost everywhere to a measurable function f.

7. State and prove Fatou's Lemma for $L^1(\mu)$ with an extension, as in Ex. 6, to the case where $\{f_n\}$ converges almost everywhere to a measurable function f.

12.3 μ-measurable functions

Let μ be a measure on the σ-algebra $\mathscr{A}$ of subsets of X. There are now two possible notions of measurable function. According to the definition in § 12.1, the function $f: X \to \mathbf{R}$ is measurable with respect to $\mathscr{A}$ if

$$\{x \in X : f(x) \geqslant c\}$$

belongs to $\mathscr{A}$ for all c in $\mathbf{R}$. In § 12.2 we constructed a space $L^1(\mu)$ of integrable functions with respect to μ and we may therefore follow the lead of Stone and say that $f: X \to \mathbf{R}$ is *measurable with respect to* μ, or μ*-measurable*, if

$$\text{mid}\{-g, f, g\}$$

belongs to $L^1(\mu)$ for all positive g in $L^1(\mu)$. It is easy to prove the analogue of Theorem 8.4.1 for $L^1(\mu)$: the statement is true verbatim, with the one proviso that we must omit the words 'almost everywhere' if the measure μ is incomplete. Thus the space $M_{\mathscr{A}}$ of $\mathscr{A}$-measurable functions and the space M_μ of μ-measurable functions have many properties in common, but we can hardly expect them to be the same for *all* measures μ defined on $\mathscr{A}$. In fact the former space is always contained in the latter:

Proposition 1. *An $\mathscr{A}$-measurable function is μ-measurable for all μ defined on $\mathscr{A}$.*

As a simple step on the way to Proposition 1 we have another useful result.

Proposition 2. *If f is $\mathscr{A}$-measurable, $g \in L^1(\mu)$ and $|f| \leqslant g$, then $f \in L^1(\mu)$.*

Proof. By considering f^+ and f^-, if necessary, we may assume that f is positive. There is an increasing sequence $\{\phi_n\}$ of simple functions converging to $f \leqslant g$ and so the integrals

$$\int \phi_n \, d\mu$$

exist and are bounded above by $\int g \, d\mu$. Hence $f \in L^1(\mu)$.

Proof of Proposition 1. Let f be $\mathscr{A}$-measurable and let g be a positive integrable function. By the definition of $L^1(\mu)$, g is certainly $\mathscr{A}$-measurable, so the truncated function $\text{mid}\{-g, f, g\}$ is $\mathscr{A}$-measurable (by Theorem 12.1.1) and dominated by g; hence is integrable by Proposition 2.

In the light of Proposition 8.4.2 and Stone's Theorem 8.5.1, these two notions of measurability coincide for the measures $\mu: \mathscr{M} \to [0, \infty]$ derived from the Daniell construction. We close this chapter by establishing a necessary and sufficient condition for $M_{\mathscr{A}}$ and M_{μ} to coincide for a measure μ on a σ-algebra $\mathscr{A}$. Following Royden [18] we shall say that a subset E of X is *locally measurable* with respect to μ if $E \cap A \in \mathscr{A}$ for each A in $\mathscr{A}$ with $\mu(A)$ finite; and the measure μ is *saturated* if every locally measurable set is measurable, i.e. is in $\mathscr{A}$. Almost our first remarks after the definition of Lebesgue measure in § 6.2 are to the effect that Lebesgue measure is saturated. The *σ-finite* measures to be introduced in the following chapter are immediately seen to be saturated. One may verify directly, or appeal to Proposition 3 below, to see that all measures derived from Daniell integrals are saturated. In fact most of the measures that arise in practice are saturated (but see Exx. 3, 4).

Proposition 3. *Let μ be a measure on the σ-algebra $\mathscr{A}$. Then $M_{\mathscr{A}} = M_{\mu}$ if and only if μ is saturated.*

Proof. (i) Assume that every μ-measurable function is $\mathscr{A}$-measurable. Let E be a subset of X such that $E \cap A \in \mathscr{A}$ for every A in $\mathscr{A}$ with $\mu(A)$ finite; we want to prove the $E \in \mathscr{A}$. The condition on E means that

$$\min \{\chi_E, \chi_A\}$$

belongs to $L^1(\mu)$ for all χ_A in $L^1(\mu)$. In the definition of μ-measurability we may replace g by simple functions ϕ, or even by functions $K\chi_A$ with χ_A in $L^1(\mu)$; this reduction is the same as in § 8.3 (or even more explicitly in § 6.1). Thus χ_E is μ-measurable and therefore $\mathscr{A}$-measurable by our assumption. This shows that $E \in \mathscr{A}$ and proves that μ is saturated.

(ii) By the analogue of Theorem 8.4.1 for $L^1(\mu)$, the class M_{μ} of all μ-measurable functions is a linear lattice which is closed with respect to pointwise limits of sequences, and Proposition 1 shows that M_{μ} contains the constant functions. Let f be any function in M_{μ}. As in Proposition 8.4.2 the functions f_n defined by

$$f_n(x) = n(\min \{f(x), c\} - \min \{f(x), c - 1/n\})$$

are in M_{μ}, and

$$f_n \to \chi_{A_c},$$

where

$$A_c = \{x \in X: f(x) \geqslant c\}.$$

Thus χ_{A_c} belongs to M_{μ}, and so

$$\min \{\chi_{A_c}, \chi_A\}$$

belongs to $L^1(\mu)$ for all χ_A in $L^1(\mu)$. In other words $A_c \cap A \in \mathscr{A}$ for all A in $\mathscr{A}$ with $\mu(A)$ finite. Finally, if μ is saturated, this shows that $A_c \in \mathscr{A}$ (for all c in $\mathbf{R}$), i.e. f is measurable with respect to $\mathscr{A}$.

Exercises

1. Let μ be a measure on the σ-algebra $\mathscr{A}$ of subsets of X. We say that μ is *σ-finite* if X is the union of a sequence $\{X_n\}$, where $X_n \in \mathscr{A}$ and $\mu(X_n)$ is finite for all n. Show that such a σ-finite measure μ is saturated.

2. Let μ be a measure on the σ-algebra $\mathscr{A}$. Show that the sets which are locally measurable with respect to μ form a σ-algebra $\mathscr{B}$ containing $\mathscr{A}$. Define

$$\begin{aligned} \bar{\mu}(E) = \mu(E) &\quad \text{if} \quad E \in \mathscr{A} \\ = \infty &\quad \text{if} \quad E \in \mathscr{B} \setminus \mathscr{A}. \end{aligned}$$

Show that $\bar{\mu}$ is a saturated measure on $\mathscr{B}$.

3. Let $\mathscr{A}$ be a σ-algebra of subsets of X and let μ be the 'infinite' measure defined by

$$\begin{aligned} \mu(A) = \infty &\quad \text{if} \quad A \neq \varnothing, \\ = 0 &\quad \text{if} \quad A = \varnothing. \end{aligned}$$

Under what circumstances is μ saturated?

4. Let $X = [0, 1]$ and let $\mathscr{A}$ be the σ-algebra which consists of the countable subsets of X and their complements in X. For A in $\mathscr{A}$ let

$$\begin{aligned} \mu(A) = 0 &\quad \text{if} \quad A \text{ is countable}, \\ = \infty &\quad \text{if} \quad A^c \text{ is countable}. \end{aligned}$$

Show that μ is not saturated.

What happens when μ is extended by the Daniell construction as in Theorem 11.4.2?

13

UNIQUENESS AND APPROXIMATION THEOREMS

The first major problem of measure theory is to extend a measure μ from a ring $\mathscr{R}$ to a σ-ring or σ-algebra containing $\mathscr{R}$. One such extension was achieved in § 11.4 by means of the Daniell construction, but it is natural to ask whether or not any other method of extension would lead to the same measure. It is comforting to know that the most frequently used general method of Carathéodory *does* in fact give the same measure as the Daniell construction. This is proved in Theorem 13.5.1 and constitutes a kind of uniqueness theorem in the sense that each measure μ on a ring $\mathscr{R}$ extends to a complete measure $\bar{\mu}$ on a σ-algebra $\mathscr{M}$ and this measure $\bar{\mu}$ is determined by one of two equivalent fixed procedures associated with the names of Carathéodory and Daniell. But the real uniqueness question remains: given a ring $\mathscr{R}$ and a σ-algebra A containing $\mathscr{R}$, can we be sure that a measure μ on $\mathscr{R}$ extends to one and only one measure λ on $\mathscr{A}$? In general the answer is in the negative: for example, let $\mathscr{R}$ consist of the finite subsets of $[0, 1]$ on the real line and let $\mu(A) = 0$ for all A in $\mathscr{R}$. The smallest σ-algebra $\mathscr{A}$ containing $\mathscr{R}$ consists of the countable subsets of $[0, 1]$ and their complements in $[0, 1]$. We can obviously define λ by setting $\lambda(A) = 0$ for all A in $\mathscr{A}$. But we may also set $\nu(A) = 0$ for all countable A in $\mathscr{A}$ and $\nu(A) = 1$ for their complements. Thus the extension to $\mathscr{A}$ is not unique in this case. It is also interesting to note that the complete measure obtained by the Daniell construction is quite different from both λ and ν as it gives the value ∞ to all uncountable subsets of $[0, 1]$.

In § 13.2 we establish the uniqueness of the extension λ in the case where the universal set X is *σ-finite* with respect to μ, i.e. X is the union of a sequence of sets in $\mathscr{R}$ whose measures are all finite. The first proof of Theorem 13.2.1 using the monotone collections of § 13.1 has the Halmos hallmark of elegance [11]. In § 13.4 the approximation theorems of § 13.3 yield stronger uniqueness theorems by relating all extensions to the measure $\bar{\mu}$: briefly it turns out that $\bar{\mu}$ has a predisposition to give the measure ∞, and that any other measure μ' extending μ agrees with $\bar{\mu}$ whenever $\bar{\mu}$ is finite (and both μ', $\bar{\mu}$ are defined). In view of this it is perhaps not surprising that $\bar{\mu}$ coincides with the extension obtained by Carathéodory using the idea of *outer measure* (§ 13.5).

13.1 Monotone collections of sets

Let $\mathscr{C}$ be an arbitrary collection of subsets of X. As we saw in Theorem 11.4.1, there is a smallest σ-ring $\mathscr{S}$ containing $\mathscr{C}$, viz. the intersection of all σ-rings containing $\mathscr{C}$. We refer to $\mathscr{S}$ as the *σ-ring generated by* $\mathscr{C}$. In a similar way, the intersection of all σ-algebras (of subsets of X) containing $\mathscr{C}$ provides the smallest σ-algebra $\mathscr{A}$ containing $\mathscr{C}$ and we call $\mathscr{A}$ the *σ-algebra generated by* $\mathscr{C}$. These descriptions of $\mathscr{S}$ and $\mathscr{A}$ are admirably brief, but they give very little idea of the structure of $\mathscr{S}$ and $\mathscr{A}$ in terms of $\mathscr{C}$ (cf. the concluding paragraphs of § 11.1). In general this is very complicated, but if $\mathscr{C}$ is a ring there is a simple result (Theorem 1) which relates the σ-ring $\mathscr{S}$ to the idea of monotone convergence. As one might expect, this has many applications in the realm of measure and integration (cf. Berberian [3] who stresses its importance).

A collection $\mathscr{M}$ of subsets of X is said to be *monotone* if $\mathscr{M}$ contains the union of every increasing sequence of sets in $\mathscr{M}$ and the intersection of every decreasing sequence of sets in $\mathscr{M}$. In other words, if $A_n \in \mathscr{M}$ for all n and $A_n \uparrow A$, then $A \in \mathscr{M}$; and similarly with the arrow reversed. (We allow $\mathscr{M}$ to be empty.)

A monotone collection need not be a ring (Ex 1). But we do have the following simple proposition.

Proposition 1. *A ring $\mathscr{R}$ is a σ-ring if and only if $\mathscr{R}$ is monotone.*

Proof. Let $A_n \in \mathscr{R}$ for $n = 1, 2, \ldots$.

(i) Suppose that $\mathscr{R}$ is a σ-ring. If $A_n \uparrow A$ then $A = \bigcup A_n \in \mathscr{R}$. If $A_n \downarrow A$ then $A_1 \setminus A = \bigcup (A_1 \setminus A_n) \in \mathscr{R}$ and so $A \in \mathscr{R}$.

(ii) Suppose that $\mathscr{R}$ is monotone. Let $B_n = A_1 \cup \ldots \cup A_n$ for $n \geqslant 1$. Then $B_n \in \mathscr{R}$ (as $\mathscr{R}$ is a ring) and $B_n \uparrow \bigcup A_n \in \mathscr{R}$, therefore $\mathscr{R}$ is a σ-ring.

Let $\mathscr{C}$ be an arbitrary collection of subsets of X. Once again, as in Theorem 11.4.1, the intersection $\mathscr{M}$ of all monotone collections containing $\mathscr{C}$ is the smallest monotone collection containing $\mathscr{C}$ and is called the *monotone collection generated by* $\mathscr{C}$. Let $\mathscr{S}$ be the σ-ring generated by $\mathscr{C}$; as $\mathscr{S}$ is monotone, $\mathscr{M}$ is always contained in $\mathscr{S}$, but $\mathscr{M}$ may be much smaller. For example, if $\mathscr{C}$ consists of the single points $\{x\}$ in $[0, 1]$ then $\mathscr{M} = \mathscr{C}$, whereas $\mathscr{S}$ consists of the countable subsets of $[0, 1]$. But if $\mathscr{C}$ happens to be a ring, we shall see that $\mathscr{M}$ and $\mathscr{S}$ coincide.

Theorem 1. *Let $\mathscr{R}$ be a ring and $\mathscr{M}$, $\mathscr{S}$ respectively the monotone collection and σ-ring generated by $\mathscr{R}$. Then $\mathscr{M} = \mathscr{S}$.*

Proof. We have already seen that $\mathscr{M} \subset \mathscr{S}$. All we need prove is that $\mathscr{M}$ is a ring, for it will then follow that $\mathscr{M}$ is a σ-ring containing $\mathscr{R}$ and hence containing $\mathscr{S}$. With this in mind let A, B be any two subsets of X and write

$$A \sim B$$

if and only if $A \setminus B$, $B \setminus A$ and $A \cup B$ all belong to $\mathscr{M}$. This relation is clearly symmetric, i.e.

$$A \sim B \Leftrightarrow B \sim A.$$

We begin by noting that $A \sim B$ for any A, B in $\mathscr{R}$, simply because $\mathscr{R}$ is a ring contained in $\mathscr{M}$. Let $\tilde{B}$ denote the collection of all subsets A of X which satisfy $A \sim B$.

Suppose that $B \in \mathscr{R}$; then $\tilde{B}$ certainly contains $\mathscr{R}$. Moreover, $\tilde{B}$ is monotone: for if $\{A_n\}$ is an increasing sequence of sets in $\tilde{B}$ and $A_n \uparrow A$, then

$$(A_n \setminus B) \uparrow (A \setminus B),$$
$$(B \setminus A_n) \downarrow (B \setminus A),$$
$$(A_n \cup B) \uparrow (A \cup B),$$

where the limits $A \setminus B$, $B \setminus A$, $A \cup B$ all belong to $\mathscr{M}$, and so $A \sim B$; similarly for a decreasing sequence $\{A_n\}$ with all arrows reversed. It follows that $\tilde{B}$ contains $\mathscr{M}$, thus

$A \sim B$ *for any* A *in* $\mathscr{M}$ *and any* B *in* $\mathscr{R}$.

Suppose now that $A \in \mathscr{M}$ and consider $\tilde{A}$. The above argument shows that $\tilde{A}$ is monotone and contains $\mathscr{R}$, and hence that $\tilde{A}$ contains $\mathscr{M}$. Thus

$A \sim B$ *for any* A, B *in* $\mathscr{M}$,

which is just another way of saying that $\mathscr{M}$ is a ring!

In practice it is often convenient to use a slightly weaker form of Theorem 1:

Corollary. *Let $\mathscr{R}$ be a ring and $\mathscr{S}$ the σ-ring generated by $\mathscr{R}$. Then any monotone collection containing $\mathscr{R}$ must also contain $\mathscr{S}$.*

Exercises

1. Let X be a non-empty set and let $\mathscr{M}$ consist of the singleton subsets $\{x\}$ where $x \in X$. Show that $\mathscr{M}$ is monotone.

2. For each of the following collections $\mathscr{C}$, describe the σ-ring, σ-algebra and monotone collection generated by $\mathscr{C}$.

(i) X is non-empty, P is a one–one mapping of X onto X, and $\mathscr{C}$ consists of the subsets A of X for which $P(A) = A$.

(ii) $X = \mathbf{R}^3$ and $\mathscr{C}$ consists of all subsets A of X such that

$$(x, y, z) \in A \Rightarrow (x, y, z') \in A$$

for all z' in $\mathbf{R}$.

(iii) $X = \mathbf{R}^2$ and $\mathscr{C}$ consists of all subsets of a countable union of horizontal lines.

13.2 Uniqueness of measures

The fundamental uniqueness theorem that we prove in this section depends on reducing the discussion to the case where all measures are finite. Let μ be a measure on the ring $\mathscr{R}$ and E an element of $\mathscr{R}$ with $\mu(E)$ finite. There are two simple minded ways of producing a finite measure in this situation. The more natural one is to restrict μ to the ring $\mathscr{R} \cap E$ (i.e. the set of all $A \cap E$ with A in $\mathscr{R}$). Equally simple, but not quite so obvious, is to define a measure μ_E *on the given ring* $\mathscr{R}$ by the rule

$$\mu_E(A) = \mu(A \cap E)$$

for all A in $\mathscr{R}$. We refer to μ_E as the *contraction* of μ by E. (The reader is encouraged to check that $\mathscr{R} \cap E$ is indeed a ring and μ_E is indeed a measure: Exx. 1, 2.)

A subset A of X is *σ-finite with respect to* μ if A is the union of a sequence $\{A_n\}$ of sets A_n in $\mathscr{R}$ for each of which $\mu(A_n)$ is finite. By replacing A_n by $A_1 \cup \ldots \cup A_n$, if necessary, we may assume that $\{A_n\}$ is increasing. Alternatively, from an increasing sequence $\{A_n\}$ we may produce a sequence $\{B_n\}$ of disjoint sets in $\mathscr{R}$, where

$$B_1 = A_1, \quad B_n = A_n \setminus A_{n-1} \quad (n \geqslant 2)$$

and clearly $A = \bigcup B_n$. The measure μ is *σ-finite* if every A in $\mathscr{R}$ is σ-finite with respect to μ. In practice it often happens that the 'universal set' X is σ-finite with respect to μ, i.e. there is a sequence $\{X_n\}$ of sets in $\mathscr{R}$ with $X = \bigcup X_n$ and $\mu(X_n)$ finite. In this case μ is σ-finite, because any set A in $\mathscr{R}$ is the union of the sets $A \cap X_n$. A typical example of a σ-finite measure is Lebesgue measure on $\mathbf{R}^k$, for $\mathbf{R}^k$ is the union of the bounded intervals I_n defined by the inequalities $|x_i| \leqslant n$ for $i = 1, \ldots, k$.

Theorem 1. *Let $\mathscr{R}$ be a ring of subsets of X and $\mathscr{A}$ the σ-algebra generated by $\mathscr{R}$. If μ is a measure on $\mathscr{R}$ for which X is σ-finite, then there is one and only one measure λ extending μ to $\mathscr{A}$.*

It is not sufficient in Theorem 1 to assume that μ is σ-finite: with that weaker assumption, the uniqueness of extension only goes as far as the σ-ring generated by $\mathscr{R}$. See Ex. 5.

Proof. The extension procedure of § 11.4 yields at least one such measure λ, viz. the restriction of $\bar{\mu}$ to $\mathscr{A}$. Let ν be another measure on $\mathscr{A}$ which agrees with λ on $\mathscr{R}$.

(i) Suppose first of all that λ and ν are finite. Let $\mathscr{M}$ consist of all sets A in $\mathscr{A}$ for which $\lambda(A) = \nu(A)$. Clearly $\mathscr{M}$ contains the ring $\mathscr{R}$. By the σ-additivity of λ, ν it follows (almost exactly as in the proof of Proposition 11.4.2) that $\mathscr{M}$ is monotone. As X is σ-finite, the σ-ring generated by $\mathscr{R}$ coincides with the σ-algebra $\mathscr{A}$. Combining these facts by means of Theorem 13.1.1, Corollary we see that $\mathscr{M} = \mathscr{A}$ and so $\lambda = \nu$.

(ii) Let $X_n \uparrow X$ where $X_n \in \mathscr{R}$ and $\mu(X_n)$ is finite. We may apply part (i) to the finite measures λ_{X_n}, ν_{X_n} (which are defined on the σ-algebra $\mathscr{A}$) and deduce that

$$\lambda(A \cap X_n) = \nu(A \cap X_n)$$

for all A in $\mathscr{A}$ and all $n \geqslant 1$. But

$$A \cap X_n \uparrow A$$

(as $n \to \infty$) and the σ-additivity of λ, ν gives

$$\lambda(A) = \nu(A)$$

for all A in $\mathscr{A}$.

So much for the extension from $\mathscr{R}$ to $\mathscr{A}$. Suppose now that λ is a measure on the σ-algebra $\mathscr{A}$. The extension of λ to a complete measure may be accomplished in a comparatively simple way; it requires neither integration theory nor any of the sophisticated techniques of § 11.4. In fact, all we need do is to consider the sets E which may be 'squeezed' between sets A, B of $\mathscr{A}$ where $A \subset B$ and $\lambda(B \setminus A) = 0$.

Proposition 1. *Let λ be a measure on the σ-algebra $\mathscr{A}$ and let*

$$\tilde{\mathscr{A}} = \{E: A \subset E \subset B, \quad \textit{where } A, B \in \mathscr{A} \textit{ and } \lambda(B \setminus A) = 0\}.$$

Then $\tilde{\mathscr{A}}$ is a σ-algebra and the equations

$$\lambda(A) = \tilde{\lambda}(E) = \lambda(B)$$

define a complete measure $\tilde{\lambda}$ on $\tilde{\mathscr{A}}$ extending λ.

Proof. It is clear that $\tilde{\mathscr{A}}$ contains $\mathscr{A}$. If E, A, B are as above, then $B^c \subset E^c \subset A^c$, where $\lambda(A^c \setminus B^c) = \lambda(B \setminus A) = 0$; thus $\tilde{\mathscr{A}}$ is closed with respect to complements. In view of the inclusion

$$(\bigcup B_n) \setminus (\bigcup A_n) \subset \bigcup (B_n \setminus A_n)$$

and the σ-additivity of λ, $\tilde{\mathscr{A}}$ is closed with respect to countable unions. Thus $\tilde{\mathscr{A}}$ is a σ-algebra.

To check the consistency of the definition of $\tilde{\lambda}$, suppose that

$$A \subset E \subset B, \quad C \subset E \subset D,$$

where $A, B, C, D \in \mathscr{A}$ and

$$\lambda(B \setminus A) = \lambda(D \setminus C) = 0.$$

It is clear that $\lambda(A) = \lambda(B)$, $\lambda(C) = \lambda(D)$. Moreover,

$$A \setminus C \subset D \setminus C,$$

so that $$\lambda(A \setminus C) = 0$$

and $$\lambda(A) = \lambda(A \cap C);$$

also $$C \setminus A \subset B \setminus A$$

implies that $$\lambda(C \setminus A) = 0$$

and $$\lambda(C) = \lambda(A \cap C).$$

Combining these gives $$\lambda(A) = \lambda(C)$$

and establishes the consistency. The σ-additivity of $\tilde{\lambda}$ follows from the σ-additivity of λ. Finally, if $\tilde{\lambda}(E) = 0$, any subset F of E is squeezed between $\varnothing$ and the null set B which shows that F is in $\tilde{\mathscr{A}}$ and $\tilde{\lambda}(F) = 0$. In other words $\tilde{\lambda}$ is complete.

If λ' is a complete measure which extends λ to an algebra $\mathscr{A}'$, then $\mathscr{A}'$ must contain $\tilde{\mathscr{A}}$ and λ' must coincide with $\tilde{\lambda}$ on $\tilde{\mathscr{A}}$ (Ex. 7). In this sense $\tilde{\lambda}$ is the unique 'most economical' complete measure extending λ. We refer to $\tilde{\lambda}$ as the *completion* of λ. Two other equivalent descriptions of $\tilde{\mathscr{A}}$ are often given, viz.

$$\tilde{\mathscr{A}} = \{A \cup B \colon A \in \mathscr{A}, B \subset C, C \in \mathscr{A}, \lambda(C) = 0\}$$

and $$\tilde{\mathscr{A}} = \{A \triangle B \colon A \in \mathscr{A}, B \subset C, C \in \mathscr{A}, \lambda(C) = 0\}.$$

See Ex. 8.

Exercises

1. Let E be an element of the ring $\mathscr{R}$. Show that

$$\mathscr{R} \cap E = \{A \cap E \colon A \in \mathscr{R}\}$$

is a ring.

2. Let μ be a measure on the ring $\mathscr{R}$. For any E in $\mathscr{R}$ show that the contraction μ_E is a measure on $\mathscr{R}$.

3. Let $\mathscr{R}$ be the ring generated by the non-empty collection $\mathscr{C}$. Let μ be a measure on $\mathscr{R}$ and suppose that the restriction of μ to $\mathscr{C}$ is finite. Show that μ is finite.

4. Let μ be a measure on a σ-ring $\mathscr{S}$. Show that the sets of finite measure form a ring and the sets of σ-finite measure form a σ-ring. Assume further that μ is σ-finite: show that the sets of finite measure form a σ-ring if and only if μ is finite. Is this last statement true if μ is not σ-finite?

5. Prove the following variation of Theorem 1. *Let $\mathscr{R}$ be a ring of subsets of X and $\mathscr{S}$ the σ-ring generated by $\mathscr{R}$. If μ is a σ-finite measure on $\mathscr{R}$, then there is one and only one measure λ extending μ to $\mathscr{S}$.* [Hint: use Ex. 11.4.7.]

Give an example to show that we cannot replace $\mathscr{S}$ in this theorem by the σ-algebra $\mathscr{A}$ generated by $\mathscr{R}$.

6. Let $\mathscr{S}$ be the σ-ring generated by a ring $\mathscr{R}$. Let μ_1, μ_2 be measures on $\mathscr{S}$ such that (i) $\mu_1(A) \leqslant \mu_2(A)$ for all A in $\mathscr{R}$, (ii) the restrictions of μ_1, μ_2 to $\mathscr{R}$ are σ-finite. Show that $\mu_1(A) \leqslant \mu_2(A)$ for all A in $\mathscr{S}$.

7. Let λ be a measure on the σ-algebra $\mathscr{A}$ and let λ' be a complete measure which extends λ to an algebra $\mathscr{A}'$. Show that $\mathscr{A}' \supset \tilde{\mathscr{A}}$ and $\lambda'(E) = \tilde{\lambda}(E)$ for all E in $\tilde{\mathscr{A}}$. (For the notation see Proposition 1.)

8. Verify that the σ-algebra $\tilde{\mathscr{A}}$ of Proposition 1 may also be described as in the last paragraph of this section.

13.3 Approximation theorems

Let $\int$ be the Daniell integral obtained by extending an elementary integral on a linear lattice L and $\bar{\mu}$ the corresponding measure. As usual we assume the Stone condition (1) of p. 22. Recall that the set A is *null* if there is an increasing sequence $\{\psi_n\}$ of positive functions in L such that $\{\psi_n(x)\}$ diverges for all x in A and

$$\int \psi_n \leqslant K$$

for some $K > 0$. We shall re-examine this definition in the light of Theorem 3.2.1 (which inspired the definition in the first place: see p. 31 of Volume 1).

For any $\epsilon > 0$ let

$$S_n^{\epsilon} = \{x \in X : \psi_n(x) \geqslant K/\epsilon\}.$$

Then S_n^{ϵ} is measurable and the inequality

$$\chi_{S_n^{\epsilon}} \leqslant \frac{\epsilon}{K} \psi_n$$

implies that

$$\bar{\mu}(S_n^{\epsilon}) \leqslant \epsilon$$

for all $n \geqslant 1$. Let S^{ϵ} be the union of the increasing sequence $\{S_n^{\epsilon}\}$. Then S^{ϵ} is also measurable and

$$\bar{\mu}(S^{\epsilon}) \leqslant \epsilon.$$

If $x \in A$ then $\psi_n(x) \to \infty$, whence $x \in S_N^\epsilon$ for a sufficiently large N (depending on x and ϵ) and so $x \in S^\epsilon$. In other words

$$A \subset S^\epsilon.$$

If ϕ is an elementary function and $c \in \mathbf{R}$, let us agree to call

$$A_c = \{x \in X : \phi(x) \geqslant c\}$$

an *elementary set*.

In line with the first definition of null set given in Chapter 2, we now see that the null set A may be covered by the increasing sequence $\{S_n^\epsilon\}$ of elementary sets whose measures are bounded above by ϵ. This idea is capable of a remarkably fruitful generalisation.

Theorem 1 (The Approximation Theorem). *Let L be a linear lattice of real valued functions on X and $\bar{\mu}$ the measure obtained from the Daniell integral extending an elementary integral on L. If $\bar{\mu}(E)$ is finite, then to any $\epsilon > 0$, there is an increasing sequence $\{S_n^\epsilon\}$ of elementary sets whose union S^ϵ contains E and satisfies*

$$\bar{\mu}(E) \leqslant \bar{\mu}(S^\epsilon) \leqslant \bar{\mu}(E) + \epsilon. \tag{1}$$

Proof. The characteristic function χ_E is positive and integrable, so by Lemma 8.2.1,

$$\chi_E = g - h,$$

where g, h are positive elements of L^{inc} and $\int h \leqslant \epsilon/3$. Thus

$$\int g \leqslant \bar{\mu}(E) + \epsilon/3. \tag{2}$$

We may find an increasing sequence $\{\phi_n\}$ of positive elementary functions such that

$$\phi_n(x) \to g(x)$$

for all x outside a null set A and

$$\int \phi_n \to \int g.$$

We may also find an increasing sequence $\{\psi_n\}$ of positive elementary functions such that $\{\psi_n\}$ diverges on A and

$$\int \psi_n \leqslant \epsilon/3.$$

Let

$$\theta_n = \phi_n + \psi_n.$$

Then

$$\int \theta_n \leqslant \int g + \epsilon/3$$

and so, by (2),
$$\int \theta_n \leqslant \bar{\mu}(E) + 2\epsilon/3; \tag{3}$$
furthermore, $\chi_E \leqslant g$, whence
$$\chi_E \leqslant \lim \theta_n. \tag{4}$$
(This inequality is given the obvious interpretation at any point x for which $\theta_n(x) \to \infty$.) Now let c be defined by
$$c(\bar{\mu}(E) + \epsilon) = \bar{\mu}(E) + 2\epsilon/3 \tag{5}$$
so that
$$0 < c < 1, \tag{6}$$
and let
$$S_n^\epsilon = \{x \in X : \theta_n(x) \geqslant c\}. \tag{7}$$
Then $\{S_n^\epsilon\}$ is an increasing sequence of elementary sets whose union is S^ϵ, say. In view of (4), (6), (7) it follows that
$$E \subset S^\epsilon.$$
Finally,
$$\chi_{S_n^\epsilon} \leqslant \frac{1}{c}\theta_n$$
and so, using (3) and (5),
$$\bar{\mu}(S_n^\epsilon) \leqslant \frac{1}{c}\int \theta_n \leqslant \bar{\mu}(E) + \epsilon$$
for all n, which implies that
$$\bar{\mu}(S^\epsilon) \leqslant \bar{\mu}(E) + \epsilon.$$

Corollary. *Let $\mathscr{S}$ be the σ-ring generated by the elementary sets*
$$\{x \in X : \phi(x) \geqslant c\}$$
for all ϕ in L, c in $\mathbf{R}$. *If $\bar{\mu}(E)$ is finite, then there is a set S in $\mathscr{S}$ such that $E \subset S$ and $\bar{\mu}(E) = \bar{\mu}(S)$.*

Proof. The sets S_n^ϵ, S^ϵ of Theorem 1 all belong to $\mathscr{S}$. Let
$$S = \bigcap_{n=1}^{\infty} S^{1/n}.$$
Then $S \in \mathscr{S}$, $E \subset S$ and
$$\bar{\mu}(E) \leqslant \bar{\mu}(S) \leqslant \bar{\mu}(E) + 1/n$$
for all $n \geqslant 1$.

There is a slightly different way of interpreting the approximation theorem. Intuitively we think of two sets A, B as 'nearly the same' if their symmetric difference $A \triangle B$ is 'small' (Fig. 16).

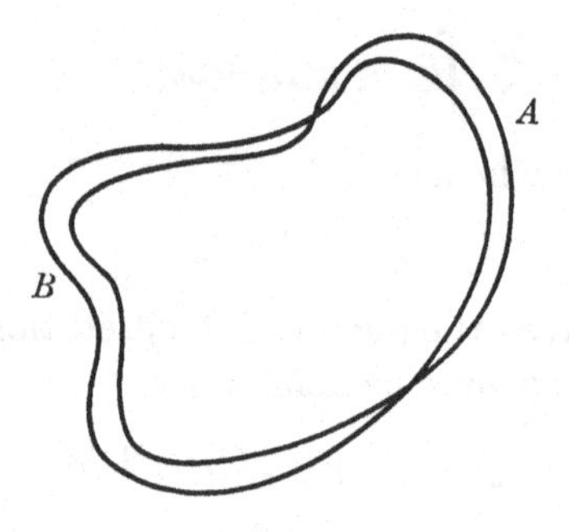

Fig. 16

More precisely, let μ be a measure on a ring $\mathscr{R}$ and for any A, B in $\mathscr{R}$ consider $\mu(A \triangle B)$ as a 'distance' between A and B. The triangle inequality

$$\mu(A \triangle C) \leqslant \mu(A \triangle B) + \mu(B \triangle C)$$

is certainly satisfied: for

$$\int |\chi_A - \chi_C| \, d\mu \leqslant \int |\chi_A - \chi_B| \, d\mu + \int |\chi_B - \chi_C| \, d\mu.$$

By taking $C = \varnothing$ we also deduce that

$$|\mu(A) - \mu(B)| \leqslant \mu(A \triangle B).$$

(Recall from Chapter 7 the important distance $\int |f-g|$ between functions f, g of L^1.) From Theorem 1 we now deduce that there is an elementary set A arbitrarily close, in the sense of the $\bar{\mu}$ distance, to any given measurable set E, provided only that $\bar{\mu}(E)$ is finite.

Theorem 2. *If $\bar{\mu}(E)$ is finite, then, to any $\epsilon > 0$, there is an elementary set A such that*

$$\bar{\mu}(E \triangle A) \leqslant \epsilon.$$

Proof. Given $\epsilon > 0$; find S_n^ϵ and S^ϵ as in the proof of Theorem 1. Since $S_n^\epsilon \uparrow S^\epsilon$ we may choose N so large that

$$\bar{\mu}(S^\epsilon \setminus S_N^\epsilon) \leqslant \epsilon.$$

Set $A = S_N^\epsilon$, then

$$\bar{\mu}(E \triangle A) \leqslant \bar{\mu}(E \triangle S^\epsilon) + \bar{\mu}(S^\epsilon \triangle A) \leqslant 2\epsilon.$$

The usual replacement of ϵ by $\epsilon/2$ leads to the required conclusion.

Exercises

1. Let μ be a measure on the ring $\mathscr{R}$ and write

$$d(A, B) = \mu(A \triangle B)$$

for any A, B in $\mathscr{R}$. How may d be modified to define a metric?

2. Let $\mathscr{S}$ be the σ-ring generated by the ring $\mathscr{R}$ and μ, ν two σ-finite measures on $\mathscr{S}$. For any set E in $\mathscr{S}$ with $\mu(E)$, $\nu(E)$ finite and any $\epsilon > 0$, there is a set A in $\mathscr{R}$ such that

$$\mu(E \triangle A) < \epsilon \quad \text{and} \quad \nu(E \triangle A) < \epsilon.$$

3. Suppose that $f \in L^1$ and $\epsilon > 0$. Adapt the proof of Theorem 1 to construct an increasing sequence $\{\theta_n\}$ of elementary functions such that

$$\int \theta_n < \int f + \epsilon$$

for all n, and

$$f \leqslant \lim \theta_n$$

(with the natural interpretation of this inequality at a point x for which $\lim \theta_n(x) = \infty$).

4. Let P, Q be elementary integrals on the linear lattice L and denote by R the elementary integral $P+Q$. These elementary integrals extend to Daniell integrals again denoted by P, Q, R on spaces $L^1(P)$, $L^1(Q)$, $L^1(R)$. Use the construction of Ex. 3 to show that

$$L^1(R) = L^1(P) \cap L^1(Q)$$

and that

$$R(f) = P(f) + Q(f)$$

for any f in $L^1(R)$. (Cf. Exx. 8.1.A–G on upper and lower Daniell integrals.)

5. In the notation of Ex. 4 show that f is R-measurable if and only if f is P-measurable and Q-measurable. In particular R satisfies the Stone Condition (1) of p. 22 if and only if P and Q also satisfy this condition.

13.4 Uniqueness of measures and Daniell integrals

The approximation theorems of the previous section give us a very strong grip on the sets of finite measure with respect to $\bar{\mu}$. First of all, we may now make precise what we meant earlier (when we proved the Extension Theorem 11.4.2) by saying that the measure $\bar{\mu}$ has a 'predisposition to give as large a measure as possible'.

Proposition 1. *Let μ be a measure on a ring $\mathscr{R}$ and μ' any measure extending μ to a ring $\mathscr{R}'$. Let $\bar{\mu}$ be the measure derived from μ by the Daniell* *construction as in* § 11.4. *Then*

$$\mu'(E) \leqslant \bar{\mu}(E)$$

for any set E for which both sides are defined.

Proof. The above inequality is obviously true if $\bar{\mu}(E) = \infty$. Assume therefore that $\bar{\mu}(E)$ is finite and $E \in \mathscr{R}'$. For any $\epsilon > 0$, consider the sets S_n^ϵ constructed in Theorem 13.3.1. In this case they all belong to $\mathscr{R}_0$ (the subring of $\mathscr{R}$ on which μ is finite) and so

$$\mu'(S_n^\epsilon) = \bar{\mu}(S_n^\epsilon).$$

Both S_n^ϵ and E belong to $\mathscr{R}'$, thus

$$\mu'(S_n^\epsilon \cap E) \leqslant \mu'(S_n^\epsilon) = \bar{\mu}(S_n^\epsilon) \leqslant \bar{\mu}(E) + \epsilon.$$

The union S^ϵ of $\{S_n^\epsilon\}$ contains E, whence

$$S_n^\epsilon \cap E \uparrow E$$

and $$\mu'(E) \leqslant \bar{\mu}(E) + \epsilon$$

by the σ-additivity of μ'. As ϵ is arbitrarily small

$$\mu'(E) \leqslant \bar{\mu}(E).$$

Proposition 1 is useful as it stands, but there is a much more precise statement.

Theorem 1. *Let μ be a measure on a ring $\mathscr{R}$ and μ' any measure extending μ to a ring $\mathscr{R}'$. Let $\bar{\mu}$ be the extension of μ given by the Daniell construction as in Theorem* 11.4.2. *Then*

$$\mu'(E) = \bar{\mu}(E)$$

for any set E in $\mathscr{R}'$ which has finite measure with respect to $\bar{\mu}$.

In rough terms, $\bar{\mu}$ can only differ from μ' by taking the value ∞.

Proof. For any $\epsilon > 0$, let A be the elementary set given by Theorem 13.3.2. Thus $\mu(A)$ is finite and

$$\bar{\mu}(E \triangle A) \leqslant \epsilon.$$

By Proposition 1 $$\mu'(E \triangle A) \leqslant \epsilon.$$

But we now have $$|\mu'(E) - \mu'(A)| \leqslant \epsilon,$$

$$|\bar{\mu}(E) - \bar{\mu}(A)| \leqslant \epsilon$$

and $$\mu'(A) = \bar{\mu}(A),$$

from which it follows that

$$|\mu'(E) - \bar{\mu}(E)| \leqslant 2\epsilon.$$

As ϵ is arbitrarily small, $$\mu'(E) = \bar{\mu}(E).$$

This result gives a second, and almost immediate proof of the fundamental uniqueness Theorem 13.2.1. In that context, let μ' be any measure on $\mathscr{A}$ extending μ, and A any element of $\mathscr{A}$. By Theorem 1

$$\mu'(A \cap X_n) = \bar{\mu}(A \cap X_n)$$

for all $n \geq 1$, and so

$$\mu'(A) = \bar{\mu}(A)$$

by the σ-additivity of μ' and $\bar{\mu}$.

Let us now consider the general situation where we are given a linear lattice L of elementary functions $\phi: X \to \mathbf{R}$. We shall say that X is *σ-finite with respect to* L if there is a sequence $\{\phi_n\}$ of positive functions ϕ_n in L such that $X = \bigcup X_n$, where

$$X_n = \{x \in X: \phi_n(x) \geq 1\}.$$

As usual, we may replace X_n by $X_1 \cup \ldots \cup X_n$ (corresponding to the function $\max\{\phi_1, \ldots, \phi_n\}$ in L) and suppose that $X_n \uparrow X$.

For any elementary integral P on L (satisfying the Stone condition (1) of p. 22) there is a Daniell integral $\int$ on a space L^1 and a complete measure $\bar{\mu}$ on a σ-algebra $\mathscr{M}$ (where L^1, $\bar{\mu}$ and $\mathscr{M}$ all depend on P). The point to note is that X is σ-finite with respect to *any* of these measures $\bar{\mu}$, for

$$\chi_{X_n} \leq \phi_n \Rightarrow \bar{\mu}(X_n) \leq \int \phi_n$$

and so $\bar{\mu}(X_n)$ is finite. In the very important case where X is a locally compact metric space and L consists of continuous functions of compact support (see § 10.1), the sets X_n are compact subsets of X, and so X is σ-compact (as described in Theorem 10.2.3). We return to this example in Chapter 15 when we discuss Borel sets. In the more familiar example where μ is a measure on a ring $\mathscr{R}$ and $\mathscr{R}_0$ is the subring on which μ is finite, let L be the linear lattice $L(\mathscr{R}_0)$. Then X is σ-finite with respect to L if and only if X is σ-finite with repect to μ, for in this case we may take for each ϕ_n a characteristic function χ_{A_n}, where $A_n \in \mathscr{R}_0$.

A fairly surprising necessary and sufficient condition for X to be σ-finite with respect to L is that, for the Daniell integral extending just *one* of the elementary integrals on L, there is a strictly positive integrable function (see Ex. 2).

The following uniqueness theorem links the Daniell integral with the classical approach to integration via measure. In the proof we shall use the Comparison Theorem of § 8.3.

Theorem 2. *Let L be a linear lattice of functions $\phi: X \to \mathbf{R}$ and let $\mathscr{A}$ be the smallest σ-algebra for which all ϕ in L are measurable. Suppose that X is σ-finite with respect to L. Then, to each elementary integral P on L (satisfying the Stone condition* (1) *of p.* 22) *there is one and only one measure λ on $\mathscr{A}$ such that*

$$P(\phi) = \int \phi \, d\lambda \tag{1}$$

for all ϕ in L; moreover, the measure $\bar{\mu}$ obtained from P by the Daniell construction is the completion of λ.

The σ-algebra $\mathscr{A}$ is the smallest σ-algebra (of subsets of X) containing the sets

$$\{x \in X: \phi(x) \geqslant c\}$$

for all ϕ in L and all c in $\mathbf{R}$. This notion of measurability with respect to $\mathscr{A}$ and the integral $\int \phi \, d\lambda$ are described in Chapter 12.

Proof. (i) The elementary integral P extends to a Daniell integral which we again denote by P. The corresponding measure $\bar{\mu}$ is defined on a σ-algebra $\mathscr{M}$ which clearly contains $\mathscr{A}$. The restriction of $\bar{\mu}$ to $\mathscr{A}$ is one measure λ satisfying (1).

(ii) Let λ be any measure on $\mathscr{A}$ satisfying equation (1) and define a Daniell integral Q by means of λ as in § 11.4, so that

$$P(\phi) = Q(\phi) = \int \phi \, d\lambda$$

for all ϕ in L. According to Theorem 8.3.1,

$$P(f) = Q(f)$$

for all f in $L^1(P)$. Now we use the σ-finiteness: let X_n be as described above and let A belong to $\mathscr{A}$; then $A \cap X_n \in \mathscr{A}$ and $\chi_{A \cap X_n} \in L^1(P)$ so that

$$P(\chi_{A \cap X_n}) = Q(\chi_{A \cap X_n}) = \lambda(A \cap X_n)$$

determines $\lambda(A \cap X_n)$ uniquely in terms of P. But $A \cap X_n \uparrow A$, so the uniqueness of λ follows by σ-additivity.

(iii) As the measure $\bar{\mu}$ is complete and $\mathscr{M} \supset \mathscr{A}$ it follows that $\mathscr{M} \supset \tilde{\mathscr{A}}$ and $\bar{\mu}$ agrees with $\tilde{\lambda}$ on $\tilde{\mathscr{A}}$ ($\tilde{\mathscr{A}}$ and $\tilde{\lambda}$ are defined in Proposition 13.2.1). It only remains to show that $\mathscr{M} \subset \tilde{\mathscr{A}}$.

Let E be an element of $\mathscr{M}$ with $\bar{\mu}(E)$ finite. By Theorem 13.3.1, Corollary, there is a set B in $\mathscr{A}$ such that $E \subset B$ and $\bar{\mu}(E) = \bar{\mu}(B)$. Thus $\bar{\mu}(B \setminus E) = 0$. By the same token, there is a set C in $\mathscr{A}$ such that $B \setminus E \subset C$ and $\bar{\mu}(C) = 0$. Without loss of generality $C \subset B$ (otherwise, replace C by $B \cap C$). Let $A = B \setminus C \in \mathscr{A}$. Then $A \subset E \subset B$ and $\lambda(B \setminus A) = 0$. In other words $E \in \tilde{\mathscr{A}}$.

For an arbitrary element E of $\mathscr{M}$ this argument shows that $E \cap X_n \in \tilde{\mathscr{A}}$ for all n, and so $E \in \tilde{\mathscr{A}}$, as required.

Exercises

1. Let X be a locally compact metric space and L the space of continuous functions $\phi: X \to \mathbf{R}$ of compact support. Show that X is σ-finite with respect to L if and only if X is σ-compact.

2. Let L be a linear lattice of functions $\phi: X \to \mathbf{R}$ and P the Daniell integral obtained by extending an elementary integral on L.

(i) If X is σ-finite with respect to L show that there exists a strictly positive function in $L^1(P)$.

(ii) If there is a sequence $\{\theta_n\}$ of positive functions in L such that

$$X = \bigcup_{n=1}^{\infty} \{x \in X: \theta_n(x) > 0\},$$

show that X is σ-finite with respect to L.

(iii) Suppose that f is a strictly positive function in $L^1(P)$. Construct (as in the proof of Theorem 13.3.1) an increasing sequence $\{\theta_n\}$ of positive functions in L such that $\lim \theta_n \geqslant f$ (with an obvious interpretation if the left hand side is infinite). Deduce from (ii) that X is σ-finite with respect to L.

(iv) Combine (i) and (iii).

3. What is the relevance to Theorem 2 of Exx. 8.4.16, 8.5.2? Work out the details in the case where $X = [0, 1]$, $\mathscr{R}$ is the ring of finite subsets of X, $L = L(\mathscr{R})$ and P is the zero functional on L.

4. Illustrate Theorem 2 by means of Ex. 8.5.1.

13.5 The Carathéodory extension

In 1892 C. Jordan formalised the classical ideas of length, area and volume by giving a precise definition of the *content* of a set. His definition was based on the principle of 'exhaustion' which goes back to Archimedes (3rd century B.C.). The measure $m(I)$ of a bounded interval I in $\mathbf{R}^k$ was given as the product of the lengths of its sides. This measure was extended naturally (and consistently) to the measure of an *elementary figure* $S = I_1 \cup \ldots \cup I_r$ ($I_1, \ldots, I_r$ disjoint) by the definition

$$m(S) = m(I_1) + \ldots + m(I_r)$$

(cf. § 4.1). For any bounded set E in $\mathbf{R}^k$ Jordan then defined the *outer content* $c^*(E)$ to be the greatest lower bound of the measures of all elementary figures containing E, and the *inner content* $c_*(E)$ to be the least upper bound of the measures of all elementary figures contained in E (see Fig. 17). In symbols:

$$c^*(E) = \inf\{m(S): S \in \mathscr{R}, S \supset E\},$$

$$c_*(E) = \sup\{m(R): R \in \mathscr{R}, R \subset E\}.$$

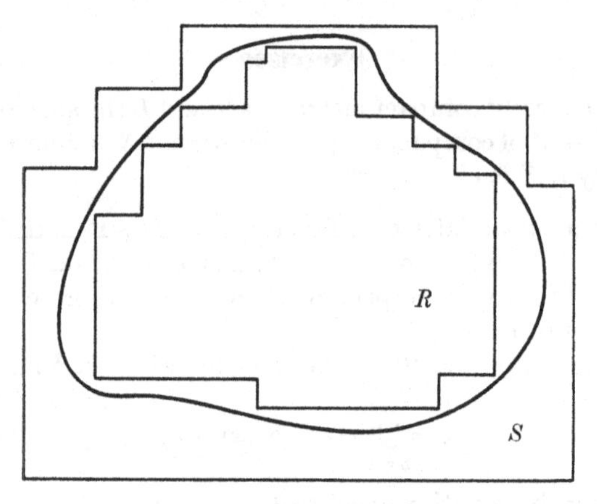

Fig. 17

It is clear from this definition that

$$c_*(E) \leqslant c^*(E).$$

If $c_*(E) = c^*(E)$ then we may say that the bounded set E is *contented*† and has *Jordan content*

$$c(E) = c_*(E) = c^*(E).$$

For any elementary figure S, $c(S) = m(S)$ (Ex. 1) and so c extends the notion of measure from the elementary figures to the (bounded) contented sets. It is not difficult to show that the (bounded) contented sets form a ring $\mathscr{C}$ and that c is an additive function on this ring (Ex. 2). Although $\mathscr{C}$ contains most of the classical geometrical figures, it does not include all the bounded open subsets or all the compact subsets of $\mathbf{R}^k$ (Ex. 3). Any finite set of points has zero content, but $\mathscr{C}$ does not contain the set of points $(x_1, \ldots, x_k)$ with rational coordinates satisfying $0 \leqslant x_i \leqslant 1$ for $i = 1, \ldots, k$ (Ex. 3). In measure theoretic terms, $\mathscr{C}$ fails to be a σ-ring and therefore is not a satisfactory domain for a measure (though, in fact, c is the restriction of Lebesgue measure to $\mathscr{C}$ and so is σ-additive: see Ex. 6).

Now suppose that E is contained in a bounded interval I. Then

$$\begin{aligned} c^*(I \setminus E) &= \inf\{m(I \setminus R) : R \in \mathscr{R}, R \subset E\} \\ &= m(I) - c_*(E). \end{aligned}$$

Thus the condition for E to be contented may be written

$$m(I) = c^*(E) + c^*(I \setminus E)$$

for all bounded intervals I containing E.

† Loomis–Sternberg [14].

If the set E is unbounded we may extend the above definitions by saying that E is *contented* if $I \cap E$ is contented for all bounded intervals I and

$$c(E) = \sup c(I \cap E)$$

for all bounded intervals I (with a natural interpretation of $c(E) = \infty$). It then follows that E is contented if and only if

$$m(I) = c^*(I \cap E) + c^*(I \cap E^c)$$

for all bounded intervals I (Ex. 4).

We may write the definition of outer content in the form

$$c^*(E) = \inf\{m(I_1) + \ldots + m(I_r) : I_1 \cup \ldots \cup I_r \supset E\}.$$

In 1902 H. Lebesgue made a very significant extension of this definition by allowing a *sequence* of intervals $\{I_n\}$ to cover the set E. He defined the *outer measure*

$$m^*(E) = \inf\left\{\sum_{n=1}^{\infty} m(I_n) : \bigcup_{n=1}^{\infty} I_n \supset E\right\}.$$

In this definition there is no need to restrict E to be bounded, as any subset E of $\mathbf{R}^k$ may be covered by a suitable sequence of bounded intervals, but we must interpret $m^*(E)$ as ∞ if there is no such sequence of bounded intervals for which the series of measures is convergent. As we shall now prove, the set E is Lebesgue measurable, and has Lebesgue measure $m^*(E)$, if and only if

$$m(A) = m^*(A \cap E) + m^*(A \cap E^c)$$

for all elementary figures A (or equivalently, for all bounded intervals A). Indeed we can prove a much more general result which links the idea of outer measure to the Daniell construction.

Let $\mathscr{R}$ be a ring of subsets of X and let μ be a measure on $\mathscr{R}$. For any subset E of X, define the *outer measure*

$$\mu^*(E) = \inf\{\Sigma\mu(A_n) : A_n \in \mathscr{R}\ (n = 1, 2, \ldots), \bigcup A_n \supset E\}.$$

The normal interpretation of this infimum depends on a treble condition: the existence of a sequence $\{A_n\}$ of sets in $\mathscr{R}$ which covers E, for which all $\mu(A_n)$ are finite and $\Sigma\mu(A_n)$ is convergent. If no such sequence exists then we interpret the above definition as saying that $\mu^*(E) = \infty$.† It makes no difference to this definition if we replace $\mathscr{R}$

† The Halmos school define μ^* on the collection $\mathscr{H}$ of all sets E which can be covered by a sequence $\{A_n\}$ of sets in $\mathscr{R}$: this collection $\mathscr{H}$ is a σ-ring and is also *hereditary* in the sense that

$$E \in \mathscr{H} \quad \text{and} \quad F \subset E \Rightarrow F \in \mathscr{H}.$$

We have extended Halmos' definition to all subsets of X by giving μ^* the value ∞ outside $\mathscr{H}$.

by the subring $\mathscr{R}_0$ on which μ is finite. Thus the outer measure μ^* has the same 'predisposition for infinity' that we attached to $\bar{\mu}$ in the proof of the Extension Theorem 11.4.2. The ring $\mathscr{R}$ may also be replaced by a collection $\mathscr{C}$ which satisfies the Disjointness Condition of Theorem 11.3.2 and which generates $\mathscr{R}$; the only trivial adjustment to the definition is that if $\varnothing \notin \mathscr{C}$, then we must allow finite sums

$$\mu(A_1) + \ldots + \mu(A_r)$$

as well as the infinite sums $\Sigma\mu(A_n)$ (see Ex. 9).

Theorem 1. *Let μ be a measure on the ring $\mathscr{R}$ and $\bar{\mu}$ the complete measure extending μ to the σ-algebra $\mathscr{M}$ obtained by means of the Daniell construction (as in Theorem 11.4.2). Let μ^* be the outer measure derived from μ as above. Then $E \in \mathscr{M}$ if and only if*

$$\mu(A) = \mu^*(A \cap E) + \mu^*(A \cap E^c) \tag{1}$$

for all A in $\mathscr{R}$ with $\mu(A)$ finite. Moreover, the restriction of μ^* to $\mathscr{M}$ coincides with $\bar{\mu}$.

Proof. (i) Let E be an element of $\mathscr{M}$. If $E \subset \bigcup A_n (A_n \in \mathscr{R})$ then

$$\bar{\mu}(E) \leqslant \Sigma\mu(A_n),$$

by the σ-additivity of $\bar{\mu}$, and so $\bar{\mu}(E) \leqslant \mu^*(E)$.

If $\bar{\mu}(E) = \infty$, then clearly $\bar{\mu}(E) = \mu^*(E)$. Suppose that $\bar{\mu}(E)$ is finite. To any $\epsilon > 0$, there is a sequence $\{A_n^\epsilon\}$ of sets in $\mathscr{R}$ such that

$$E \subset \bigcup A_n^\epsilon \quad \text{and} \quad \Sigma\mu(A_n^\epsilon) \leqslant \bar{\mu}(E) + \epsilon.$$

(Take $S_0^\epsilon = \varnothing$, $A_n^\epsilon = S_n^\epsilon \setminus S_{n-1}^\epsilon$ for $n \geqslant 1$ in the Approximation Theorem 13.3.1.) Thus

$$\mu^*(E) \leqslant \bar{\mu}(E) + \epsilon$$

for arbitrary $\epsilon > 0$. It follows that $\mu^*(E) \leqslant \bar{\mu}(E)$. Combining these inequalities we have

$$\mu^*(E) = \bar{\mu}(E)$$

for all E in $\mathscr{M}$.

(ii) For any E in $\mathscr{M}$ and any A in $\mathscr{R}$ with $\mu(A)$ finite,

$$\mu(A) = \bar{\mu}(A \cap E) + \bar{\mu}(A \cap E^c)$$

by the additivity of $\bar{\mu}$. Equation (1) now follows by part (i).

(iii) Let E be an element of finite outer measure $\mu^*(E)$. To any $\epsilon > 0$ we may find $A_n^\epsilon \in \mathscr{R}$ $(n = 1, 2, \ldots)$ so that $E \subset \bigcup A_n^\epsilon$ and

$$\Sigma\mu(A_n^\epsilon) < \mu^*(E) + \epsilon.$$

Denote the union $\bigcup A_n^\epsilon$ by S^ϵ. Then S^ϵ is an element of the σ-algebra $\mathscr{M}$ containing E and

$$\bar{\mu}(S^\epsilon) < \mu^*(E) + \epsilon.$$

The set

$$B = \bigcap_{n=1}^{\infty} S^{1/n}$$

also belongs to $\mathscr{M}$ and satisfies the conditions $E \subset B$,

$$\bar{\mu}(B) < \mu^*(E) + 1/n$$

for all $n \geqslant 1$, so that

$$\mu^*(E) = \bar{\mu}(B).$$

The set B is often called a *measurable cover* of E.

(iv) Assume that (1) holds for any element A of $\mathscr{R}$ with $\mu(A)$ finite. In view of part (iii) we may find sets B, C in $\mathscr{M}$ such that

$$A \cap E \subset B, \quad \bar{\mu}(B) = \mu^*(A \cap E),$$

$$A \cap E^c \subset C, \quad \bar{\mu}(C) = \mu^*(A \cap E^c).$$

Equation (1) therefore reads

$$\bar{\mu}(A) = \bar{\mu}(B) + \bar{\mu}(C).$$

But

$$\bar{\mu}(B) + \bar{\mu}(C) = \bar{\mu}(B \cup C) + \bar{\mu}(B \cap C)$$

and

$$A \subset B \cup C;$$

whence

$$\bar{\mu}(B \cap C) = 0.$$

Recall that $\bar{\mu}$ is a complete measure. The difference

$$(A \cap B) \setminus (A \cap E)$$

is a subset of the null set $A \cap B \cap C$ and as such belongs to $\mathscr{M}$. It therefore follows that $A \cap E \in \mathscr{M}$.

We have now shown that $A \cap E \in \mathscr{M}$ for any A in $\mathscr{R}$ with $\mu(A)$ finite. In terms of the Daniell integral,

$$\chi_{A \cap E} \in L^1 \quad \text{for any } \chi_A \text{ in } L(\mathscr{R}_0),$$

which is just Stone's condition for E to belong to $\mathscr{M}$. (Cf. the remarks on saturated measures at the end of § 12.3.)

This completes the proof.†

In his famous lectures on real functions [5] Carathéodory gave an elegant method for extending a general measure μ from a ring $\mathscr{R}$ to a σ-algebra $\mathscr{M}$ containing $\mathscr{R}$. The steps were essentially as follows.

† In Ex. 10 we suggest a shorter (and possibly more compelling) proof using upper and lower Daniell integrals.

(i) Define the outer measure $\mu^*(E)$, as above, for all subsets E of X.

(ii) Show that the subsets E satisfying condition (1) form a σ-algebra $\mathscr{M}$ containing $\mathscr{R}$.

(iii) Show that the restriction of μ^* to $\mathscr{M}$ is a complete measure extending μ.

Theorem 1 is significant in that it proves the exact equivalence of two approaches to the extension problem – the classical method of Carathéodory in terms of measure theory, and the Daniell construction in terms of integration.

In fairness to the classical approach we now include a few paragraphs to establish the steps of Carathéodory's construction from a purely measure theoretic point of view.

First of all, let us contrast the outer measure μ^* with the given measure μ. The most striking difference is that μ^* is defined on the σ-algebra $\mathscr{P}(X)$ of *all* subsets E of X. We still have the fairly obvious properties

$$\mu^*(\varnothing) = 0, \tag{2}$$

$$A \subset B \Rightarrow \mu^*(A) \leqslant \mu^*(B) \quad (\textit{monotonicity}). \tag{3}$$

But the countable additivity of μ must be replaced by the weaker condition of *countable subadditivity* for μ^*:

$$A \subset \bigcup_{n=1}^{\infty} A_n \Rightarrow \mu^*(A) \leqslant \sum_{n=1}^{\infty} \mu^*(A_n). \tag{4}$$

To establish (4) let $A \subset \bigcup A_n$. If any $\mu^*(A_n) = \infty$ there is nothing to prove. Suppose therefore that all $\mu^*(A_n)$ are finite. For any $\epsilon > 0$ and any $m \geqslant 1$, we may find a sequence $\{A_{mn}\}$ of sets in $\mathscr{R}$ such that

$$A_m \subset \bigcup_{n=1}^{\infty} A_{mn} \quad \text{and} \quad \sum_{n=1}^{\infty} \mu(A_{mn}) < \mu^*(A_m) + \epsilon/2^m.$$

The countable collection $\{A_{mn}: m, n \geqslant 1\}$ covers A and so

$$\mu^*(A) \leqslant \sum_{m,n} \mu(A_{mn}).$$

By a simple theorem on the summation of double series of positive terms

$$\sum_{m,n} \mu(A_{mn}) = \sum_{m} (\sum_{n} \mu(A_{mn}))$$

and so

$$\mu^*(A) < \sum_{m} \mu^*(A_m) + \epsilon.$$

Condition (4) follows as ϵ is arbitrary.

If $A \in \mathscr{R}$ then A covers A and so $\mu^*(A) \leqslant \mu(A)$. But the reverse inclusion follows (as in the first few lines of the proof of Theorem 1) by the σ-additivity of μ and so

$$\mu^*(A) = \mu(A)$$

for all A in $\mathscr{R}$. We may therefore express condition (1) in terms of outer measure, viz.

$$\mu^*(A) = \mu^*(A \cap E) + \mu^*(A \cap E^c) \tag{5}$$

for all A in $\mathscr{R}$ with $\mu^*(A)$ finite.

In view of the subadditivity of μ^*

$$\mu^*(A) \leqslant \mu^*(A \cap E) + \mu^*(A \cap E^c)$$

for all subsets A, E of X. To verify (5) we only need to verify that

$$\mu^*(A) \geqslant \mu^*(A \cap E) + \mu^*(A \cap E^c) \tag{6}$$

for the A's in question. In particular there is no need to restrict attention to those A's for which $\mu^*(A)$ is finite because (6) is always satisfied when $\mu^*(A) = \infty$. It is aesthetically pleasing (and technically helpful) to note that if equation (5) holds for all A in $\mathscr{R}$ then the same equation holds for all subsets A of X. To prove this, assume the weaker Carathéodory condition (5). Let A be any subset of X with finite outer measure $\mu^*(A)$. For any given $\epsilon > 0$, there exists a sequence $\{A_n\}$ of sets in $\mathscr{R}$ such that

$$A \subset \bigcup A_n \quad \text{and} \quad \Sigma\mu^*(A_n) < \mu^*(A) + \epsilon.$$

By condition (5)

$$\mu^*(A_n) = \mu^*(A_n \cap E) + \mu^*(A_n \cap E^c),$$

and so by the countable subadditivity of μ^*,

$$\mu^*(A \cap E) + \mu^*(A \cap E^c) < \mu^*(A) + \epsilon.$$

As ϵ is arbitrary this establishes (6) for all A with $\mu^*(A)$ finite and hence for all A.

Thus the Carathéodory condition for E to belong to $\mathscr{M}$ now reads

$$\mu^*(A) = \mu^*(A \cap E) + \mu^*(A \cap E^c) \tag{7}$$

for all subsets A of X. In this form the condition makes no reference to the measure μ from which μ^* was derived nor of the ring $\mathscr{R}$ on which μ was defined. With this in mind we say that a function

$$\nu\colon \mathscr{P}(X) \to [0, \infty]$$

is an *outer measure* on X if

$$\nu(\varnothing) = 0, \tag{8}$$

$$A \subset B \Rightarrow \nu(A) \leqslant \nu(B), \tag{9}$$

$$A \subset \bigcup_{n=1}^{\infty} A_n \Rightarrow \nu(A) \leqslant \sum_{n=1}^{\infty} \nu(A_n), \tag{10}$$

and we prove Carathéodory's extension theorem entirely in terms of outer measures.

***Theorem 2* (*Carathéodory*).** *Let ν be an outer measure on X and let $\mathscr{M}$ consist of all subsets E of X satisfying the condition*

$$\nu(A) = \nu(A \cap E) + \nu(A \cap E^c) \tag{11}$$

for all subsets A of X. Then $\mathscr{M}$ is a σ-algebra of subsets of X and the restriction of ν to $\mathscr{M}$ is a complete measure.

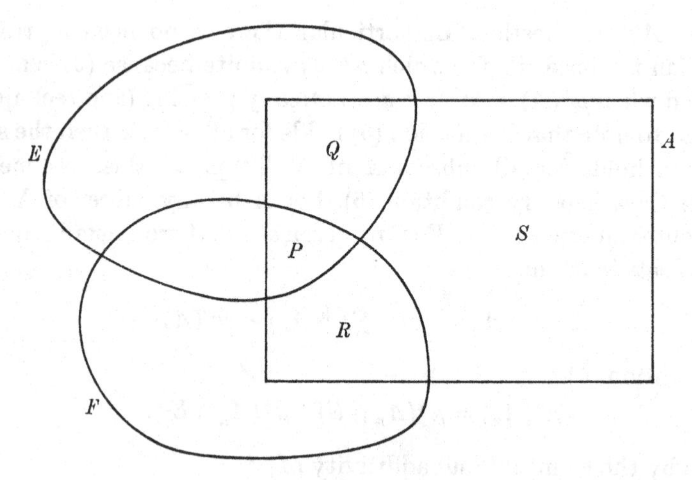

Fig. 18

Proof. (i) It is clear that $\varnothing$, X both belong to $\mathscr{M}$ and that

$$E \in \mathscr{M} \Rightarrow E^c \in \mathscr{M}.$$

Suppose that $E, F \in \mathscr{M}$. For any set A with $\nu(A)$ finite, let P, Q, R, S be the intersections of A with the disjoint sets $E \cap F$, $E \cap F^c$, $E^c \cap F$, $E^c \cap F^c$, respectively (Fig. 18).

As $E \in \mathscr{M}$, $$\nu(A) = \nu(P \cup Q) + \nu(R \cup S)$$

and $$\nu(P \cup Q \cup R) = \nu(P \cup Q) + \nu(R).$$

As $F \in \mathscr{M}$, $$\nu(R \cup S) = \nu(R) + \nu(S).$$

Combining these gives

$$\nu(A) = \nu(P \cup Q \cup R) + \nu(S)$$

for all A with $\nu(A)$ finite, and this is just another way of saying that $E \cup F \in \mathcal{M}$. At this stage we have shown that $\mathcal{M}$ is an *algebra* (Ex. 11.1.3).

(ii) For any subset A of X, define the set function ν_A by the rule

$$\nu_A(E) = \nu(A \cap E)$$

for any E in $\mathcal{M}$ (cf. § 13.2).

Let E, F be disjoint elements of $\mathcal{M}$. In the notation of part (i) above, the set P is empty and

$$\nu(Q \cup R) = \nu(Q) + \nu(R).$$

In other words
$$\nu_A(E \cup F) = \nu_A(E) + \nu_A(F),$$

and so ν_A is an *additive* set function on the algebra $\mathcal{M}$.

(iii) Let $\{E_n\}$ be a sequence of disjoint sets in $\mathcal{M}$ and let E be their union. By part (i) $S_r = E_1 \cup \ldots \cup E_r$ belongs to $\mathcal{M}$ and so

$$\nu(A) = \nu_A(S_r) + \nu_A(S_r^c) \geqslant \nu_A(S_r) + \nu_A(E^c)$$

for all A and all r. As ν_A is additive,

$$\nu(A) \geqslant \sum_{n=1}^{r} \nu_A(E_n) + \nu_A(E^c)$$

for all A and all r. Now let $r \to \infty$ and apply the countable sub-additivity of ν to deduce that

$$\nu(A) \geqslant \sum_{n=1}^{\infty} \nu_A(E_n) + \nu_A(E^c) \geqslant \nu_A(E) + \nu_A(E^c) \geqslant \nu(A)$$

for all A. This argument establishes two further facts. First of all, E belongs to $\mathcal{M}$, so that $\mathcal{M}$ is a *σ-algebra* (Ex. 11.4.5). Moreover,

$$\sum_{n=1}^{\infty} \nu_A(E_n) + \nu_A(E^c) = \nu_A(E) + \nu_A(E^c)$$

for all A, and the substitution of E for A gives

$$\nu(E) = \sum_{n=1}^{\infty} \nu(E_n),$$

so that ν is *σ-additive* on $\mathcal{M}$.

(iv) If $\nu(E) = 0$ then $\nu(A \cap E) = 0$ and

$$\nu(A) \geqslant \nu(A \cap E) + \nu(A \cap E^c)$$

for any A; thus $E \in \mathcal{M}$. From this it follows at once (by monotonicity) that the restriction of ν to $\mathcal{M}$ is *complete*.

Exercises

1. Show that $c(S) = m(S)$ for any elementary figure S.

2. Show that c is an additive function on the ring of bounded contented sets.

3. Let I be the unit cube in $\mathbf{R}^k$ defined by the inequalities $0 \leqslant x_i \leqslant 1$ for $i = 1, \ldots, k$. Let S be the set of points in I with rational coordinates. Show that S is countable but is discontented.

The points of S are written as the terms of a sequence $\{P_n\}$ (without repeats) and G_n is the open cube centre P_n and side $2^{-(n+1)}$. Show that the open set $G = \bigcup G_n$ is discontented. What can we say about $I \setminus G$?

4. Prove that a subset E of $\mathbf{R}^k$ is contented if and only if

$$m(I) = c^*(I \cap E) + c^*(I \cap E^c)$$

for all bounded intervals I.

5. Show that c is an additive function on the ring of contented sets (cf. Ex. 2).

6. Show that c is the restriction of Lebesgue measure to the ring of contented sets.

7. Let S be a subset of a bounded interval I in $\mathbf{R}^k$. Show that χ_S is Riemann integrable on I if and only if S is contented, and that in this case

$$\int_I \chi_S(x)\,dx = c(S).$$

8. Let f be a positive bounded function on the compact interval $[a, b]$ and denote by Q_f the ordinate set

$$\{(x, y) \colon a \leqslant x \leqslant b, 0 \leqslant y \leqslant f(x)\}.$$

Show that f is Riemann integrable on $[a, b]$ if and only if Q_f is contented (as a subset of $\mathbf{R}^2$) and that in this case

$$\int_a^b f(x)\,dx = c(Q_f)$$

(cf. Ex. 6.2.11).

9. Let $\mathscr{C}$ be a non-empty collection of subsets of X satisfying the Disjointness Condition of Theorem 11.3.2 and $\mathscr{R}$ the ring generated by $\mathscr{C}$. If μ is a measure on $\mathscr{R}$ show that for any subset E of X

$$\mu^*(E) = \inf\{\Sigma\mu(A_n) \colon A_n \in \mathscr{C} \cup \{\varnothing\}, \bigcup A_n \supset E\}.$$

10. Let μ be a finite measure on the ring $\mathscr{R}$ of subsets of X and P the corresponding linear functional on $L(\mathscr{R})$ which satisfies $P(\chi_A) = \mu(A)$ for all A in $\mathscr{R}$ (Theorem 11.3.1). Let P^*, P_* be the upper and lower Daniell integrals obtained from P (as in Ex. 8.1.B). Adapt the proof of the Approximation Theorem 13.3.1 to show that $P^*(\chi_E) = \mu^*(E)$ for all subsets E of X, and hence give another proof of Theorem 1.

11. In Ex. 10, can you suggest a measure theoretic definition of $\mu_*(E)$ which will ensure that $P_*(\chi_E) = \mu_*(E)$ for all subsets E of X?

12. Let X consist of just three distinct elements and define ν on $\mathscr{P}(X)$ (the collection of all subsets of X) by $\nu(X) = 2, \nu(\varnothing) = 0, \nu(E) = 1$ otherwise. Show that ν is an outer measure and find the subsets of X which are measurable with respect to ν, i.e. satisfy the Carathéodory condition (11) of Theorem 2.

Now define $\lambda(E) = \nu(X) - \nu(X \setminus E)$ and show that there exist non-measurable subsets E for which $\nu(E) = \lambda(E)$. Moreover there exist disjoint subsets A, B for which $\lambda(A) + \lambda(B) > \lambda(A \cup B)$, so we cannot regard λ as a satisfactory inner measure corresponding to the outer measure ν.

14

PRODUCT MEASURES

If A and B are any two sets, their Cartesian product $A \times B$ consists of all the ordered pairs (a, b) where $a \in A$ and $b \in B$. A familiar example is the Euclidean plane $\mathbf{R} \times \mathbf{R}$ which we have previously denoted by $\mathbf{R}^2$. If I, J are bounded intervals in $\mathbf{R}$, the product $I \times J$ is a bounded interval in $\mathbf{R} \times \mathbf{R}$ and the plane (Lebesgue) measure of $I \times J$ is $m(I)\, m(J)$. Quite generally, suppose that we are given measures μ, ν on rings $\mathscr{R}$, $\mathscr{S}$, respectively. Our purpose is to construct a *product measure* $\mu \times \nu = \lambda$, say, which satisfies

$$\lambda(A \times B) = \mu(A)\, \nu(B) \tag{1}$$

for all sets A, B for which the right hand side makes sense – and what constitutes 'sense' is to be determined by the way in which we define multiplication by ∞.

Let $\mathscr{T}$ be the smallest ring containing all the sets $A \times B$ with A in $\mathscr{R}$ and B in $\mathscr{S}$. When μ and ν are both finite we may define a (unique) measure λ on $\mathscr{T}$, by quite elementary means, so that (1) holds for all A in $\mathscr{R}$ and all B in $\mathscr{S}$. When μ and ν are σ-finite there is still a (unique) measure λ on $\mathscr{T}$ which satisfies (1) for all A in $\mathscr{R}$ and all B in $\mathscr{S}$, but here we must interpret multiplication by ∞ according to the rules

$$\begin{aligned} \infty . \infty &= \infty, \\ \infty . r = r . \infty &= \infty, \quad \text{if} \quad r > 0 \ (r \in \mathbf{R}), \\ \infty . 0 = 0 . \infty &= 0. \end{aligned}$$

These rules are dictated by the fact that the measure of a σ-finite set may be 'approached from below'.

In the general case let μ_0, ν_0 be the restrictions of μ, ν to the subrings on which they are finite, let $\lambda_0 = \mu_0 \times \nu_0$ and use the Daniell construction to extend λ_0, μ_0, ν_0 to complete measures $\bar{\lambda}$, $\bar{\mu}$, $\bar{\nu}$, respectively. By far the most significant relationship connecting these measures is Fubini's Theorem 14.2.1 which expresses any integral with respect to $\bar{\lambda}$ as a repeated integral with respect to $\bar{\mu}$ and $\bar{\nu}$. The proof of Fubini's Theorem in this general setting is very nearly the same as the one we gave in Chapter 4, but interestingly enough, the present proof is a little easier, as two of the steps in the original proof are now incorporated in the definitions of the product measure λ_0.

In § 14.3 we go a little further and show that $\bar{\lambda}$ may be regarded as a product of $\bar{\mu}$ and $\bar{\nu}$ in the sense that

$$\bar{\lambda}(A \times B) = \bar{\mu}(A)\,\bar{\nu}(B),$$

provided only that we avoid the dubious products $0.\infty$ and $\infty.0$ on the right hand side.

14.1 Products of σ-finite measures

If two rings $\mathscr{R}$, $\mathscr{S}$ are given, we shall not expect the product sets $A \times B$ $(A \in \mathscr{R}, B \in \mathscr{S})$ to form a ring, because the difference of two such sets is not in general of the same kind (Ex. 1). Bearing this in mind, we take a lead from the intervals in $\mathbf{R} \times \mathbf{R}$ and begin our construction with the following simpler and more general situation.

Let X, Y be any two sets and $\mathscr{C}$, $\mathscr{D}$ any two non-empty collections of subsets of X, Y, respectively, which satisfy the Disjointness Condition (p. 86). For $\mathscr{C}$ this reads:

> *if $A_1, A_2 \in \mathscr{C}$ then each of the sets $A_1 \cap A_2$, $A_1 \setminus A_2$ is the union of a finite collection of disjoint elements of $\mathscr{C}$;*

and similarly for $\mathscr{D}$. The bounded intervals in $\mathbf{R}$ certainly satisfy this condition; in fact they are closed with respect to intersections and so form a semi-ring as defined on p. 86.

Lemma 1. *Let $\mathscr{E} = \{A \times B : A \in \mathscr{C}, B \in \mathscr{D}\}$. If $\mathscr{C}$, $\mathscr{D}$ satisfy the Disjointness Condition, then so does $\mathscr{E}$.*

Proof. Let $A_1, A_2 \in \mathscr{C}$; $B_1, B_2 \in \mathscr{D}$. Then

$$(A_1 \times B_1) \cap (A_2 \times B_2) = (A_1 \cap A_2) \times (B_1 \cap B_2).$$

By the Disjointness Condition,

$$A_1 \cap A_2 = C_1 \cup \ldots \cup C_m,$$

where $C_1, \ldots, C_m$ are disjoint elements of $\mathscr{C}$, and

$$B_1 \cap B_2 = D_1 \cup \ldots \cup D_n,$$

where $D_1, \ldots, D_n$ are disjoint elements of $\mathscr{D}$. But this means that $(A_1 \times B_1) \cap (A_2 \times B_2)$ is the union of the disjoint sets

$$C_i \times D_j \quad (i = 1, \ldots, m;\ j = 1, \ldots, n)$$

all of which lie in $\mathscr{E}$.

If $A \in \mathscr{C}$, $B \in \mathscr{D}$ then A^c, B^c, $(A \times B)^c$ denote the complements of A, B, $A \times B$, respectively in X, Y, $X \times Y$. Thus

$$(A \times B)^c = (A^c \times B) \cup (A \times B^c) \cup (A^c \times B^c),$$

where the three sets on the right hand side are disjoint (Fig. 19). If we use this equation to expand $(A_2 \times B_2)^c$, the argument of the first paragraph shows that the difference

$$(A_1 \times B_1) \cap (A_2 \times B_2)^c$$

is a finite union of disjoint elements of $\mathscr{E}$. This completes the proof.

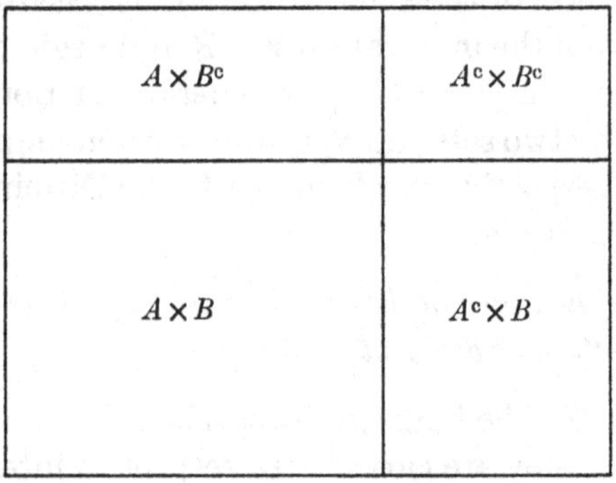

$A \times B^c$ $A^c \times B^c$

$A \times B$ $A^c \times B$

Fig. 19

Lemma 2. *Let $\mu: \mathscr{C} \to \mathbf{R}$, $\nu: \mathscr{D} \to \mathbf{R}$ be additive functions, and define $\lambda: \mathscr{E} \to \mathbf{R}$ by setting*

$$\lambda(A \times B) = \mu(A)\,\nu(B) \qquad [(1)]$$

for A in $\mathscr{C}$, B in $\mathscr{D}$. Then λ is additive.

It is possible to give a direct combinatorial proof that λ is additive (cf. Halmos [11] p. 149 (8)). It is simpler, and very much in the spirit of Fubini, to deduce Lemma 2 from the following result.

Lemma 3. *The repeated integral*

$$H(\phi) = \int\left(\int \phi(x, y)\, dy\right) dx \qquad (2)$$

defines a linear functional H on the linear space $L(\mathscr{E})$.

The notation used in equation (2) is a little ambiguous, but is clearly reminiscent of Fubini's Theorem in §4.3. If $\phi: X \times Y \to \mathbf{R}$ then $\phi_x: Y \to \mathbf{R}$ is defined by

$$\phi_x(y) = \phi(x, y).$$

The integral $$\int \phi(x, y)\, dy$$

stands for $$\int \phi_x d\nu.$$

In view of Theorem 11.3.2, this last integral is uniquely defined by the additive function ν, provided $\phi_x \in L(\mathscr{D})$. If

$$\Phi(x) = \int \phi_x d\nu$$

for all x in X, and $\Phi \in L(\mathscr{C})$, we may define

$$\int \Phi(x)\, dx = \int \Phi\, d\mu.$$

In other words $$H(\phi) = \int \Phi\, d\mu.$$

Proof of Lemma 3. In the particular case $\phi = \chi_{A\times B}$ we have

$$\phi(x, y) = \chi_A(x)\, \chi_B(y),$$

so that $$\Phi(x) = \chi_A(x)\, \nu(B)$$

and $$H(\phi) = \mu(A)\, \nu(B).$$

Any simple function in $L(\mathscr{E})$ is a linear combination of such characteristic functions $\chi_{A\times B}$. It therefore follows from the linearity of the integrals with respect to μ and ν that H is a linear functional on $L(\mathscr{E})$.

Lemma 2 now follows at once because

$$H(\chi_{A\times B}) = \lambda(A \times B),$$

and the additivity of λ is a special case of the linearity of H.

Proposition 1. *Let μ, ν be finite measures on rings $\mathscr{R}, \mathscr{S}$, respectively, and let $\mathscr{T}$ be the ring generated by $\{A \times B\colon A \in \mathscr{R},\, B \in \mathscr{S}\}$. Then there is a unique finite measure λ on $\mathscr{T}$ satisfying*

$$\lambda(A \times B) = \mu(A)\, \nu(B)$$

for all A in $\mathscr{R}$, B in $\mathscr{S}$.

We shall refer to λ as the *product measure* of the given finite measures μ, ν and write $\lambda = \mu \times \nu$.

Proof. In view of Proposition 11.2.2 the ring $\mathscr{T}$ consists of finite disjoint unions of elements of $\mathscr{E}$ and $L(\mathscr{T}) = L(\mathscr{E})$. If H denotes the linear functional of Lemma 3, define

$$\lambda(E) = H(\chi_E)$$

for any set E in $\mathscr{T}$ (cf. Theorem 11.3.2). We may now appeal to Proposition 11.4.2 which shows that the σ-additivity of λ is equivalent to the Daniell condition

$$\phi_n \downarrow 0 \Rightarrow H(\phi_n) \downarrow 0.$$

But this condition follows immediately by the same proposition and formula (2) because the Daniell condition holds for each of the integrals with respect to ν, μ separately. More specifically: if $\phi_n(x,y) \downarrow 0$ for all x, y then

$$\Phi_n(x) = \int \phi_n(x,y)\, dy \downarrow 0$$

for all x, and so $$H(\Phi_n) = \int \Phi_n(x)\, dx \downarrow 0.$$

This establishes the fact that λ is σ-additive and so λ is a finite measure on $\mathscr{T}$.

Up to this point we have only considered products of *finite* measures μ, ν. Suppose now that μ, ν are arbitrary measures on rings $\mathscr{R}$, $\mathscr{S}$. Let $\mathscr{T}$ be the ring generated by $\{A \times B: A \in \mathscr{R}, B \in \mathscr{S}\}$ and suppose that, by some means or other, we can find a measure λ on $\mathscr{T}$ which satisfies

$$\lambda(A \times B) = \mu(A)\, \nu(B),$$

whenever $\mu(A)$, $\nu(B)$ *are both finite*. In general, there are a great many different measures λ satisfying this condition for given measures μ, ν. For example, if $\mathscr{R} = \{\varnothing, X\}$ and $\mathscr{S} = \{\varnothing, Y\}$, then $\mathscr{T} = \{\varnothing, X \times Y\}$. Unless both $\mu(X)$ and $\nu(Y)$ are finite, we may give $\lambda(X \times Y)$ any value in $[0, \infty]$. On the other hand, the case that arises most frequently in practice is when both μ and ν are σ-finite and here we have a definitive result. (Recall that μ is σ-finite on $\mathscr{R}$ if and only if any element A of $\mathscr{R}$ is the union of an increasing sequence $\{A_n\}$, where $A_n \in \mathscr{R}$ and $\mu(A_n)$ is finite for each n.)

Theorem 1. *Let* μ, ν *be measures on rings* $\mathscr{R}$, $\mathscr{S}$, *respectively, and* $\mathscr{T}$ *the ring generated by* $\{A \times B: A \in \mathscr{R}, B \in \mathscr{S}\}$. *Then there exists a measure* λ *on* $\mathscr{T}$ *which satisfies*

$$\lambda(A \times B) = \mu(A)\, \nu(B), \tag{3}$$

whenever $\mu(A)$, $\nu(B)$ *are both finite.*

If μ, ν *are* σ*-finite, then there is only one such measure* λ; *moreover* λ *is* σ*-finite, and equation* (3) *extends to all sets* A *in* $\mathscr{R}$, B *in* $\mathscr{S}$ *provided multiplication by* ∞ *is interpreted according to the rules*

$$\infty . \infty = \infty, \tag{4}$$

$$\infty . r = r . \infty = \infty, \quad \text{if} \quad r > 0\ (r \in \mathbf{R}), \tag{5}$$

$$\infty . 0 = 0 . \infty = 0. \tag{6}$$

In the σ-finite case we naturally call λ the *product measure* of μ, ν, but for the moment we reserve judgment as to the definition of a product λ when μ or ν fails to be σ-finite.

Proof. To establish the existence of λ we follow the standard extension technique of Theorem 11.4.2. First of all restrict μ, ν to the subrings $\mathscr{R}_0$, $\mathscr{S}_0$ on which they are finite. The product $\lambda_0 = \mu_0 \times \nu_0$ of the restricted measures is defined on the ring $\mathscr{T}_0$ generated by $\{A \times B: A \in \mathscr{R}_0, B \in \mathscr{S}_0\}$ and, of course, satisfies

$$\lambda_0(A \times B) = \mu_0(A)\, \nu_0(B)$$

for all A in $\mathscr{R}_0$, B in $\mathscr{S}_0$. Now λ_0 extends by the Daniell construction to a (complete) measure $\bar{\lambda}$ on a σ-algebra $\mathscr{L}$ containing $\mathscr{T}_0$. In fact $\mathscr{L}$ contains $\mathscr{T}$; for if $A \in \mathscr{R}$, $B \in \mathscr{S}$, then $A \times B \in \mathscr{L}$. We verify this by showing that $\chi_{A\times B}$ is measurable or, equivalently, that

$$\min\{\chi_{A\times B}, \chi_{C\times D}\} = \chi_{(A\cap C)\times(B\cap D)}$$

is integrable for any C in $\mathscr{R}_0$, D in $\mathscr{S}_0$. This is clearly true because $A \cap C \in \mathscr{R}_0$ and $B \cap D \in \mathscr{S}_0$. The restriction λ of $\bar{\lambda}$ to $\mathscr{T}$ is a measure satisfying (3).

Suppose now that μ, ν are σ-finite. If $A \in \mathscr{R}$, $B \in \mathscr{S}$, then there are increasing sequences $\{A_n\}$, $\{B_n\}$ with terms in $\mathscr{R}_0$, $\mathscr{S}_0$, respectively and such that

$$A_n \uparrow A, \quad B_n \uparrow B.$$

It follows that

$$\lambda(A_n \times B_n) = \mu(A_n)\, \nu(B_n)$$

is finite and

$$A_n \times B_n \uparrow A \times B,$$

so that λ is σ-finite on $\mathscr{T}$ (which consists of finite disjoint unions of such sets $A \times B$). Moreover, rules (4), (5), (6) follow at once as

$$\mu(A_n) \uparrow \mu(A), \quad \nu(B_n) \uparrow \nu(B)$$

and

$$\lambda(A_n \times B_n) \uparrow \lambda(A \times B).$$

It is also clear from these observations that λ is uniquely determined by the measures μ, ν.

There is no general agreement in the literature as to the most natural candidate for the *unique product* of two given measures μ, ν. First of all, if μ, ν are defined on rings $\mathscr{R}$, $\mathscr{S}$, it seems natural to define their product λ on the ring $\mathscr{T}$ as in Theorem 1. But authors differ in their priorities here: some give μ, ν on σ-rings and define the product λ on a σ-ring (e.g. Halmos [11], Berberian [3]); some give μ, ν on σ-algebras and define λ on a σ-algebra (e.g. Rudin [19], Kingman–Taylor [12]) and some insist that λ, μ, ν should all be complete measures on σ-algebras (e.g. Royden [18]). The proof we have given of Theorem 1 caters for all these tastes! For example, if $\mathscr{R}$, $\mathscr{S}$ are σ-algebras and $\mathscr{T}$ is now the σ-algebra generated by

$$\{A \times B: A \in \mathscr{R}, B \in \mathscr{S}\},$$

then the σ-algebra $\mathscr{L}$ of the proof contains $\mathscr{T}$ just as before, and the restriction of $\bar{\lambda}$ to $\mathscr{T}$ is the required measure. The σ-finiteness of μ, ν in this case means that there are sequences

$$X_n \uparrow X, \quad Y_n \uparrow Y,$$

with $\mu(X_n)$, $\nu(Y_n)$ finite. As

$$X_n \times Y_n \uparrow X \times Y,$$

this clearly shows that λ is σ-finite, and the uniqueness of λ follows from Theorem 13.2.1. If μ, ν happen to be complete, then λ may be completed by the standard construction of Proposition 13.2.1; in the σ-finite case this completion is $\bar{\lambda}$ as constructed in the above proof (Theorem 13.4.2).

If $\mathscr{R}$, $\mathscr{S}$ are σ-rings and $\mathscr{T}$ is the σ-ring generated by

$$\{A \times B: A \in \mathscr{R}, B \in \mathscr{S}\},$$

then the σ-algebra $\mathscr{L}$ contains $\mathscr{T}$ and λ is the restriction of $\bar{\lambda}$ to $\mathscr{T}$. The σ-finiteness of μ, ν again implies the σ-finiteness and uniqueness of λ; this involves a little more argument than the above case of σ-algebras, but is left as an interesting exercise for the reader (Ex. 5). For the record:

Theorem 1 remains true if:

(i) *$\mathscr{R}$, $\mathscr{S}$ are σ-rings and $\mathscr{T}$ is the σ-ring generated by*

$$\{A \times B: A \in \mathscr{R}, B \in \mathscr{S}\};$$

(ii) *$\mathscr{R}$, $\mathscr{S}$ are σ-algebras and $\mathscr{T}$ is the σ-algebra generated by*

$$\{A \times B: A \in \mathscr{R}, B \in \mathscr{S}\}.$$

Exercises

1. Let $\mathscr{R}$, $\mathscr{S}$ be rings of sets and let $\mathscr{T} = \{A \times B: A \in \mathscr{R}, B \in \mathscr{S}\}$. If $\mathscr{R}$, $\mathscr{S}$ each contain at least three distinct sets, show that $\mathscr{T}$ is not a ring.

2. If $\mathscr{C}$, $\mathscr{D}$ are semi-rings, show that $\{A \times B: A \in \mathscr{C}, B \in \mathscr{D}\}$ is a semi-ring.

3. Let $\mathscr{R}$, $\mathscr{S}$ be σ-rings of subsets of X, Y, respectively, and $\mathscr{T}$ the σ-ring generated by $\{A \times B: A \in \mathscr{R}, B \in \mathscr{S}\}$. If $C \in \mathscr{T}$ show that $\{y \in Y: (x, y) \in C\} \in \mathscr{S}$

for all x in X and $\{x \in X: (x,y) \in C\} \in \mathscr{R}$ for all y in Y. In rough terms, all cross-sections of a measurable set are measurable.

Deduce that $A \times B \in \mathscr{T}$, $A \times B \neq \varnothing \Rightarrow A \in \mathscr{R}, B \in \mathscr{S}$.

4. Consider in relation to Ex. 3 the case where $X = Y$ is uncountable, $\mathscr{R} = \mathscr{S}$ consists of all countable subsets, and

$$D = \{(x,y) \in X \times Y: x = y\}$$

is the 'diagonal' set in $X \times Y$.

5. Show that Theorem 1 remains true if $\mathscr{R}$, $\mathscr{S}$ are σ-rings and $\mathscr{T}$ is the σ-ring generated by $\{A \times B: A \in \mathscr{R}, B \in \mathscr{S}\}$.

6. Let μ, ν be σ-finite measures on σ-algebras $\mathscr{R}$, $\mathscr{S}$, respectively, $\mathscr{T}$ the σ-algebra generated by $\{A \times B: A \in \mathscr{R}, B \in \mathscr{S}\}$ and λ the unique product measure on $\mathscr{T}$ given by Theorem 1 (variation (ii) as stated above). Give an example where μ, ν are complete but λ is incomplete.

7. Let μ_i be a σ-finite measure on the σ-algebra $\mathscr{A}_i$ of subsets of X_i $(i = 1, 2, 3)$. Show that

$$(\mu_1 \times \mu_2) \times \mu_3 = \mu_1 \times (\mu_2 \times \mu_3).$$

(As usual $(X_1 \times X_2) \times X_3$ and $X_1 \times (X_2 \times X_3)$ are identified with

$$X_1 \times X_2 \times X_3.)$$

14.2 Fubini's Theorem

Suppose that we are given rings $\mathscr{R}$, $\mathscr{S}$ of subsets of X, Y and measures μ, ν on $\mathscr{R}$, $\mathscr{S}$. Let μ_0, ν_0 be their restrictions to the subrings on which they are finite, let $\lambda_0 = \mu_0 \times \nu_0$ as in Proposition 14.1.1, and let $\bar{\lambda}$, $\bar{\mu}$, $\bar{\nu}$ be the complete measures obtained from λ_0, μ_0, ν_0 by the Daniell construction procedure of Theorem 11.4.2. If we do not assume that μ and ν are σ-finite, then the measures $\bar{\lambda}$, $\bar{\mu}$, $\bar{\nu}$ are uniquely determined by μ, ν only in the sense that they are given by this quite definite construction.

To simplify the notation and also to bring out parallels with the familiar special case of $\mathbf{R}^2$, we shall denote by $L^1(X)$, $L^1(Y)$, $L^1(X \times Y)$ the spaces of functions which are integrable with respect to $\bar{\mu}$, $\bar{\nu}$, $\bar{\lambda}$, respectively, and write

$$\int F(x)\,dx \quad \text{for} \quad \int F\,d\bar{\mu} \quad \text{if} \quad F \in L^1(X),$$

$$\int G(y)\,dy \quad \text{for} \quad \int G\,d\bar{\nu} \quad \text{if} \quad G \in L^1(Y),$$

$$\int H(x,y)\,d(x,y) \quad \text{for} \quad \int H\,d\bar{\lambda} \quad \text{if} \quad H \in L^1(X \times Y).$$

We shall even allow the notation $\int F$, $\int G$, $\int H$ for these integrals if the meaning is clear from the context.

Theorem 1 (Fubini's Theorem). *If $f \in L^1(X \times Y)$, then*

$$\int f(x,y)\,d(x,y) = \int\left(\int f(x,y)\,dy\right)dx. \tag{1}$$

This equation is to be interpreted as follows:

$$F(x) = \int f(x,y)\,dy$$

exists for almost all x in X; and then

$$\int F(x)\,dx$$

exists and equals

$$\int f(x,y)\,d(x,y).$$

As before,

$$\int f(x,y)\,dy$$

stands for

$$\int f_x\,d\bar{\nu},$$

where $f_x\colon Y \to \mathbf{R}$ is defined for each x in X by

$$f_x(y) = f(x,y) \quad (y \in Y).$$

(For definiteness, if $\int f(x,y)\,dy$ does not exist, we may set $F(x) = 0$.)

The general proof of Fubini's Theorem is virtually the same as the one given in § 4.3, but it is simpler in two significant ways. First of all, the product measure λ_0 is given in terms of equation (2) of the previous section, so that the first two steps in the proof are now effectively part of the *definition of* λ_0. Secondly, in the Daniell construction, as described in § 8.1, a null set is *defined* in terms of increasing sequences of simple functions, and so the fundamental lemma on cross-sections of null sets follows without reference to any major theorem (such as Theorem 4.2.1).

If S is any subset of $X \times Y$, the *cross-section of S at x* is

$$S_x = \{y \in Y\colon (x,y) \in S\}.$$

Note that S_x is a subset of Y (and not of X, in general).

Lemma 1. *If S is null (with respect to $\bar{\lambda}$), then S_x is null (with respect to $\bar{\nu}$) for almost all x in X (with respect to $\bar{\mu}$).*

Proof. By the definition of a null set in $X \times Y$, we can find an increasing sequence $\{\psi_n\}$ of simple functions in $L(\mathscr{T}_0)$ such that $\psi_n(x,y)$ diverges for all (x,y) in S, but $\int\psi_n$ is bounded. Define the simple function Ψ_n on X by

$$\Psi_n(x) = \int \psi_n(x,y)\,dy. \tag{2}$$

Thus

$$\int \Psi_n = \int \psi_n.$$

Now $\{\Psi_n\}$ is an increasing sequence of simple functions and $\int\Psi_n$ is bounded; the definition of null sets in X therefore shows that $\Psi_n(x)$ is bounded for almost all x in X. If x_0 is any point for which $\Psi_n(x_0)$ is bounded, then (2) shows that $\psi_n(x_0, y)$ is bounded for almost all y in Y. But $\psi_n(x_0, y)$ diverges for all y in S_{x_0} and so S_{x_0} is null. As this is true for almost all x_0 in X, the lemma is proved.

Proof of Theorem 1. (i) Suppose that $f \in L^{\text{inc}}(X \times Y)$, that $\phi_n \uparrow f$ almost everywhere in $X \times Y$ and $\int\phi_n \uparrow \int f$. Define Φ_n by

$$\Phi_n(x) = \int \phi_n(x,y)\,dy \quad (x \in X).$$

Then $\{\Phi_n\}$ is an increasing sequence of simple functions and

$$\int \Phi_n = \int \phi_n \uparrow \int f.$$

Thus $\Phi_n(x)$ is bounded for almost all x.

Let S be the set of points (x,y) for which $\phi_n(x,y)$ fails to converge to $f(x,y)$; we are given that S is null. According to Lemma 1, S_x is null for almost all x. Choose x_0 in X for which S_{x_0} is null and also $\Phi_n(x_0)$ is bounded; this will be true for almost all x_0. Then

$$\phi_n(x,y) \uparrow f(x_0, y)$$

for almost all y (in fact for y not in S_{x_0}) and

$$\int \phi_n(x_0, y)\,dy = \Phi_n(x_0)$$

is bounded. This shows that $f_{x_0} \in L^{\text{inc}}(Y)$ and

$$\Phi_n(x_0) \uparrow \int f_{x_0},$$

i.e.

$$\int \phi_n(x_0, y)\,dy \uparrow \int f(x_0, y)\,dy.$$

Let
$$F(x) = \int f(x,y)\,dy$$
when this integral exists, and $F(x) = 0$ otherwise. Then
$$\Phi_n \uparrow F$$
almost everywhere in X and
$$\int \Phi_n = \int \phi_n \uparrow \int f,$$
so that $F \in L^{\mathrm{inc}}(X)$ and
$$\int F = \int f.$$
This is equivalent to (1).

(ii) The result extends immediately from L^{inc} to L^1 as the integrals on both sides of (1) are linear operators.

Interchanging the roles of x, y throughout, we have:

Corollary. *If $f \in L^1(X \times Y)$ then*
$$\int \left(\int f(x,y)\,dx \right) dy = \int \left(\int f(x,y)\,dy \right) dx.$$

Although we have proved Fubini's Theorem for quite general measures, it still has a σ-finite flavour: this is because *any function f in $L^1(X \times Y)$ vanishes outside a σ-finite subset of $X \times Y$*. We can prove this quite easily as follows. Assume that f is integrable and define
$$S_n = \{(x,y) \colon |f(x,y)| \geqslant 1/n\}.$$
If S is the union of the sequence $\{S_n\}$ then f vanishes outside S. Also
$$\chi_{S_n} \leqslant n|f|$$
implies that
$$\bar{\lambda}(S_n) \leqslant n \int |f|\,d\bar{\lambda},$$
and so S is σ-finite.

With this point in mind we can state a partial converse of Fubini's Theorem without assuming that the spaces X, Y are σ-finite.

Theorem 2 (Tonelli's Theorem). *Let $f \colon X \times Y \to \mathbf{R}$ be a measurable function which vanishes outside a σ-finite subset of $X \times Y$. If one of the repeated integrals*
$$\int \left(\int |f(x,y)|\,dx \right) dy, \quad \int \left(\int |f(x,y)|\,dy \right) dx$$

exists, then f is integrable, and hence the repeated integrals

$$\int\left(\int f(x,y)\,dx\right)dy, \quad \int\left(\int f(x,y)\,dy\right)dx,$$

both exist and are equal.

Proof. There is an increasing sequence $\{S_n\}$ of subsets of $X \times Y$, each of finite measure, such that f vanishes outside $\bigcup S_n$. The truncated functions

$$h_n = \min\{|f|, n\chi_{S_n}\}$$

are integrable and satisfy

$$\int h_n = \int\left(\int h_n(x,y)\,dx\right)dy = \int\left(\int h_n(x,y)\,dy\right)dx,$$

by Fubini's Theorem. The increasing sequence $\{h_n\}$ converges to $|f|$ (everywhere in $X \times Y$) and $\int h_n$ is bounded above by one of the given repeated integrals; the Monotone Convergence Theorem therefore shows that $|f|$ is integrable. But f is measurable, so in fact f is integrable. This completes the proof.

Exercises

1. What do Fubini's and Tonelli's Theorems give when

$$X = Y = \{1,2,3,\ldots\} \quad \text{and} \quad \mu = \nu$$

is the counting measure (Ex. 8.4.1)?

Consider the case where

$$f(x,y) = \begin{cases} 2-2^{-x} & \text{if } x = y, \\ -2+2^{-x} & \text{if } x = y+1, \\ 0 & \text{otherwise.} \end{cases}$$

2. What does Tonelli's Theorem give when $Y = \{1,2,3,\ldots\}$ and ν is the counting measure (Ex. 8.4.1)?

3. Let $f \in L^1(X)$, $g \in L^1(Y)$ and define $h(x,y) = f(x)g(y)$ for all x in X, y in Y. Show that $h \in L^1(X \times Y)$ and

$$\int h(x,y)\,d(x,y) = \int f(x)\,dx \int g(y)\,dy.$$

4. If $f(x,y) = xy/(x^2+y^2)^2$, then

$$\int_{-1}^{1}\left(\int_{-1}^{1} f(x,y)\,dx\right)dy = \int_{-1}^{1}\left(\int_{-1}^{1} f(x,y)\,dy\right)dx,$$

but f is not integrable over the square $\{(x,y) \in \mathbf{R}^2\colon |x| \leqslant 1, |y| \leqslant 1\}$.

5. This exercise uses the notation of Chapter 12: it should also be compared with Ex. 6.2.11. Let $\mathscr{A}$ be a σ-algebra of subsets of X, $\mathscr{B}$ the σ-algebra of Lebesgue measurable subsets of $\mathbf{R}$ and $\mathscr{C}$ the σ-algebra generated by $\{A \times B: A \in \mathscr{A}, B \in \mathscr{B}\}$. To each positive function f on X define the *ordinate set*

$$P_f = \{(x, y): x \in X, y \in \mathbf{R}, 0 \leqslant y < f(x)\}.$$

(i) Show that the ordinate set P_ϕ of a positive $\mathscr{A}$-simple function ϕ is $\mathscr{C}$-measurable.

(ii) Use Proposition 12.1.1 and Ex. 14.1.3 to show that f is $\mathscr{A}$-measurable if and only if P_f is $\mathscr{C}$-measurable.

Let μ be a σ-finite measure on $\mathscr{A}$, ν Lebesgue measure on $\mathscr{B}$ and λ the unique product measure $\mu \times \nu$.

(iii) Show that f is integrable with respect to μ if and only if the ordinate set P_f has finite measure with respect to λ and that

$$\int f\, d\mu = \lambda(P_f)$$

in this case.

6. Let μ, ν be finite measures on σ-algebras $\mathscr{A}$, $\mathscr{B}$, respectively, and $\mathscr{C}$ the σ-algebra generated by $\{A \times B: A \in \mathscr{A}, B \in \mathscr{B})$. Show that the collection of sets E in $X \times Y$ for which

$$\int \left(\int \chi_E(x, y)\, dx \right) dy = \int \left(\int \chi_E(x, y)\, dy \right) dx,$$

is monotone, and deduce from Theorem 13.1.1, Corollary that this equation holds for all E in $\mathscr{C}$. If $\lambda(E)$ denotes the value of this repeated integral show that λ is the product measure $\mu \times \nu$.

Extend this result to the case where μ, ν are σ-finite.

7. Let μ, ν be σ-finite (but not necessarily complete) measures on σ-algebras $\mathscr{A}$, $\mathscr{B}$, respectively, and λ the corresponding product measure on the σ-algebra $\mathscr{C}$ given by the construction of Ex. 6. If f is integrable with respect to λ (as defined in § 12.2) show that

$$\int f\, d\lambda = \int \left(\int f(x, y)\, dx \right) dy = \int \left(\int f(x, y)\, dy \right) dx.$$

In what ways does this result differ from Fubini's Theorem as stated in the text?

8. Let $X = Y = [0, 1]$, $\mathscr{A} = \mathscr{B}$ the σ-algebra of subsets of $[0, 1]$ generated by the intervals (i.e. the Borel sets in $[0, 1]$), μ Lebesgue measure on X, ν the counting measure on Y (which gives the measure 1 to each singleton $\{y\}, y \in Y$ as in Ex. 8.4.1), D the diagonal set $\{(x, y): x = y\}$. Show that D belongs to the σ-algebra generated by $\{A \times B: A \in \mathscr{A}, B \in \mathscr{B}\}$ and

$$\int \left(\int \chi_D(x, y)\, dy \right) dx = 1, \quad \int \left(\int \chi_D(x, y)\, dx \right) dy = 0.$$

What is the relevance of this example to Tonelli's Theorem and to Ex. 6?

9. In this question integration and measure are in the sense of Lebesgue.

(i) Let E be a subset of $\mathbf{R}$. Consider the geometrical relation between the subsets

$$E' = E \times \mathbf{R}, \quad E'' = \{(x,y) : x-y \in E\}$$

of $\mathbf{R}^2$. Hence shows that if E is measurable as a subset of $\mathbf{R}$, then E'' is measurable as a subset of $\mathbf{R}^2$.

(ii) If f is a measurable function on $\mathbf{R}$, show that

$$F(x,y) = f(x-y)$$

defines a measurable function F on $\mathbf{R}^2$.

(iii) If f, g are integrable functions on $\mathbf{R}$, show that

$$h(x) = \int f(x-y)\, g(y)\, dy$$

is defined for almost all x in $\mathbf{R}$. Let $h(x) = 0$ when this integral is undefined. Show that h is integrable and

$$\int h = \int f \int g, \quad \int |h| \leqslant \int |f| \int |g|.$$

(iv) The function h defined in (iii) is written $f * g$ and called the *convolution* of f and g. Show that for any integrable functions e, f, g

$$f*g = g*f, \quad (e*f)*g = e*(f*g)$$

almost everywhere.

(v) For any integrable function f define the *Fourier transform* $\hat{f}: \mathbf{R} \to \mathbf{C}$ by the rule

$$\hat{f}(t) = \int f(x)\, e^{-ixt}\, dx \quad (t \in \mathbf{R}).$$

(As in § 10.3, express this integral in terms of its real and imaginary parts: cf. the Riemann–Lebesgue Lemma of § 5.2 or § 7.6.)

If f, g are integrable and $h = f*g$ as in (iv), show that

$$\hat{h}(t) = \hat{f}(t)\, \hat{g}(t) \quad (t \in \mathbf{R}).$$

In other words, the Fourier transform converts convolutions into pointwise products.

14.3 Products of general measures

The rest of this chapter explores the general question of product measures when no assumption is made about σ-finiteness. It may fairly be regarded as beyond the normal call of duty and we recommend that it be omitted at a first reading. Suppose that μ, ν are measures on rings $\mathscr{R}$, $\mathscr{S}$, as before, and that λ is a measure on the ring

$\mathcal{T}$ generated by $\{A \times B \colon A \in \mathcal{R}, B \in \mathcal{S}\}$. For λ to qualify as a product of μ, ν we shall certainly insist that

$$\lambda(A \times B) = \mu(A)\,\nu(B) \tag{1}$$

whenever $\mu(A)$, $\nu(B)$ are both finite. But what are we to make of this equation when ∞ occurs as one or both of the terms on the right hand side? How do we view the three rules for multiplication by ∞ as given (for σ-finite measures) in Theorem 14.1.1? We repeat them for ease of reference.

$$\infty . \infty = \infty,$$

$$\infty . r = r . \infty = \infty \quad \text{if} \quad r > 0 \ (r \in \mathbf{R}),$$

$$\infty . 0 = 0 . \infty = 0.$$

In our opinion the first two are quite reasonable, but the third appears to be a convention peculiar to the measure theorist.† In fact we are inclined to resist the third rule on the ground that it can be misleading in the non-σ-finite case. To express this in rough and ready terms, we are only prepared to accept the equations

$$\infty . 0 = 0 . \infty = 0$$

without question if the ∞ is 'not too large', viz. the measure of a σ-finite set. The above examples in which $\mathcal{R}$ and $\mathcal{S}$ each consist of two elements (p. 142) illustrate the point. A much less trivial example is given by taking $X = Y = [0, 1]$ and defining $\mu(A) = \nu(A) = 0$ if A is a countable subset of $[0, 1]$ and $\mu(A) = \nu(A) = \infty$ if A is an uncountable subset of $[0, 1]$. One measure λ on $[0, 1]^2$ which has a natural claim to be a product of μ, v, gives zero measure to all countable subsets of $[0, 1]^2$ and infinite measure to all uncountable subsets of $[0, 1]^2$. The 'horizontal' and 'vertical' lines $[0, 1] \times \{y\}$ and $\{x\} \times [0, 1]$ are uncountable and so have infinite measure with respect to λ. Thus equation (1) can only mean that

$$\infty . 0 = \infty \quad \text{or} \quad 0 . \infty = \infty$$

in the case of these lines. (Of course, the products $A \times \varnothing$ and $\varnothing \times B$ are always empty so the zero measure corresponding to $\varnothing$ is able to 'veto' any other measure and give the product measure zero.)

With this in mind we shall agree to call the measure λ a *product* of the measures μ, ν if

$$\lambda(A \times B) = \mu(A)\,\nu(B)$$

† It also clashes with natural caution in elementary calculus and with the notion of *place* in algebraic geometry as expounded by Artin and Lang: cf. S. Lang [13] p. 1.

for any sets for which the right hand side is defined, it being understood that $\infty.\infty = \infty$, that $r.\infty = \infty.r = \infty$ for $r > 0$ ($r \in \mathbf{R}$), but that $\infty.0$ *and* $0.\infty$ *are undefined.*

The reader should realise that this convention is not the one used in the best known texts on measure theory – though most of them consider only products of σ-finite measures in which case the distinction is vacuous. But the following fundamental theorem is not true as it stands if we insist that $\infty.0 = 0.\infty = 0$ (see Ex. 1).

Theorem 1. *Let μ, ν be measures on rings $\mathscr{R}$, $\mathscr{S}$, respectively, μ_0, ν_0 their restrictions to the subrings on which they are finite, and $\lambda_0 = \mu_0 \times \nu_0$. If $\bar{\lambda}$, $\bar{\mu}$, $\bar{\nu}$ are the complete measures obtained by extending λ_0, μ_0, ν_0 by the Daniell construction, then $\bar{\mu}$, $\bar{\nu}$ extend μ, ν, respectively, and*

$$\bar{\lambda}(A \times B) = \bar{\mu}(A)\,\bar{\nu}(B) \tag{2}$$

for any sets A, B for which the right hand side is defined. In other words $\bar{\lambda}$ is a product of $\bar{\mu}$, $\bar{\nu}$ as described above.

Proof. Let the measures $\bar{\lambda}$, $\bar{\mu}$, $\bar{\nu}$ be defined on σ-algebras $\mathscr{L}$, $\mathscr{M}$, $\mathscr{N}$, respectively.

(i) Suppose that $A \in \mathscr{M}$, $B \in \mathscr{N}$ and $\bar{\mu}(A)$, $\bar{\nu}(B)$ are both finite. The problem is to show that $A \times B \in \mathscr{L}$, and this is not as obvious as one might expect. For any $\epsilon > 0$, the Approximation Theorem 13.3.1 gives two increasing sequences $\{R_n{}^\epsilon\}$, $\{S_n{}^\epsilon\}$ whose terms are in $\mathscr{R}_0$, $\mathscr{S}_0$ and whose unions R^ϵ, S^ϵ satisfy

$$A \subset R^\epsilon, \quad \bar{\mu}(R^\epsilon) \leqslant \bar{\mu}(A) + \epsilon,$$

$$B \subset S^\epsilon, \quad \bar{\nu}(S^\epsilon) \leqslant \bar{\nu}(B) + \epsilon.$$

The intersections $$R = \bigcap_{n=1}^{\infty} R^{1/n}, \quad S = \bigcap_{n=1}^{\infty} S^{1/n},$$

also lie in $\mathscr{M}$, $\mathscr{N}$ and satisfy

$$A \subset R, \quad \bar{\mu}(R) \leqslant \bar{\mu}(A) + 1/n,$$

$$B \subset S, \quad \bar{\nu}(S) \leqslant \bar{\nu}(B) + 1/n,$$

for all $n \geqslant 1$, whence

$$\bar{\mu}(R) = \bar{\mu}(A), \quad \bar{\nu}(S) = \bar{\nu}(B).$$

Now, in the product space $X \times Y$,

$$R_n{}^{1/p} \times S_n{}^{1/q} \uparrow R^{1/p} \times S^{1/q}$$

and $$\bigcap_{p,q} (R^{1/p} \times S^{1/q}) = R \times S.$$

Thus $R^{1/p} \times S^{1/q}$, $R \times S$ belong to the σ-algebra $\mathscr{L}$ and by the σ-additivity of $\bar{\lambda}$, $\bar{\mu}$, $\bar{\nu}$,

$$\bar{\lambda}(R^{1/p} \times S^{1/q}) = \bar{\mu}(R^{1/p})\,\bar{\nu}(S^{1/q})$$

and
$$\bar{\lambda}(R \times S) \leqslant (\bar{\mu}(A) + 1/p)\,(\bar{\nu}(B) + 1/q)$$

for all $p, q \geqslant 1$. Therefore

$$\bar{\lambda}(R \times S) \leqslant \bar{\mu}(A)\,\bar{\nu}(B).$$

In particular, if $\bar{\mu}(A) = 0$ or if $\bar{\nu}(B) = 0$, we have $\bar{\lambda}(R \times S) = 0$ from which we deduce, by the completeness of $\bar{\lambda}$, that $A \times B \in \mathscr{L}$ and $\bar{\lambda}(A \times B) = 0$. This argument applies to

$$(R \setminus A) \times S \quad \text{and} \quad R \times (S \setminus B),$$

because $\bar{\mu}(R \setminus A) = 0$ and $\bar{\nu}(S \setminus B) = 0$. But

$$(R \times S) \setminus (A \times B) \subset ((R \setminus A) \times S) \cup (R \times (S \setminus B)),$$

and so the completeness of $\bar{\lambda}$ ensures that

$$(R \times S) \setminus (A \times B) \in \mathscr{L}$$

and
$$\bar{\lambda}((R \times S) \setminus (A \times B)) = 0.$$

Thus $(A \times B) \in \mathscr{L}$ and

$$\bar{\lambda}(A \times B) = \bar{\lambda}(R \times S) \leqslant \bar{\mu}(A)\,\bar{\nu}(B)$$

is finite. Fubini's Theorem now tidies up the proof by showing that

$$\bar{\lambda}(A \times B) = \bar{\mu}(A)\,\bar{\nu}(B).$$

(ii) Let A, B be arbitrary elements of $\mathscr{M}$, $\mathscr{N}$, respectively. If $C \in \mathscr{R}_0$, $D \in \mathscr{S}_0$, then

$$\min\{\chi_{A \times B}, \chi_{C \times D}\} = \chi_{(A \cap C) \times (B \cap D)}$$

is integrable by part (i) because $\bar{\mu}(A \cap C)$, $\bar{\nu}(B \cap D)$ are both finite. Thus $\chi_{A \times B}$ is measurable, i.e. $A \times B \in \mathscr{L}$. If $\bar{\lambda}(A \times B)$ is finite then Fubini's Theorem shows that

$$\bar{\lambda}(A \times B) = \int \left(\chi_A(x) \int \chi_B(y)\,dy\right) dx.$$

The case where $\bar{\mu}(A)$, $\bar{\nu}(B)$ are both finite has been treated in (i). If $\bar{\mu}(A) = \infty$, we must have $\bar{\nu}(B) = 0$, otherwise the integral on the right hand side could not exist. Similarly, if $\bar{\nu}(B) = \infty$ we must have $\bar{\mu}(A) = 0$, because the integrand $\chi_A(x)\,\bar{\nu}(B)$ must be finite for almost all x. This argument shows that the only products on the right hand side of (2) involving ∞ which can have a *finite* value are $\infty \,.\, 0$ and $0 \,.\, \infty$. It therefore establishes the rules $\infty.\infty = \infty$, $\infty.r = r.\infty = \infty$ for $r > 0$ (but says nothing to exclude the possibility of products like $\infty.0 = \infty$ or $0.\infty = \infty$).

Theorem 1 shows that the Daniell extension procedure is 'compatible' with the formation of product measures. This means, first of all, that equation (2) for $\bar{\lambda}, \bar{\mu}, \bar{\nu}$ follows from the corresponding equation for λ_0, μ_0, ν_0. But the idea goes further than this: let $\bar{\mu}, \bar{\nu}$ be defined on σ-algebras $\mathscr{M}$, $\mathscr{N}$, respectively. Suppose that $\mathscr{R}_1$, $\mathscr{S}_1$ are any two rings satisfying

$$\mathscr{R}_0 \subset \mathscr{R}_1 \subset \mathscr{M}, \quad \mathscr{S}_0 \subset \mathscr{S}_1 \subset \mathscr{N}$$

and μ_1, ν_1 are the restrictions of $\bar{\mu}, \bar{\nu}$ to $\mathscr{R}_1, \mathscr{S}_1$, respectively. In view of the fundamental comparison Theorem 8.3.1 and Ex. 3, the Daniell construction extends μ_1, ν_1 to exactly the same measures $\bar{\mu}, \bar{\nu}$ as before; moreover, equation (2), for finite values on the right hand side, ensures that the same product measure $\bar{\lambda}$ is obtained from μ_1, ν_1. As extreme cases we may, of course, take

$$\mu_1 = \mu_0, \quad \nu_1 = \nu_0,$$

or

$$\mu_1 = \bar{\mu}, \quad \nu_1 = \bar{\nu}.$$

This freedom of choice for μ_1, ν_1 may be convenient in several different settings. For example, if we wish to construct Lebesgue measure on $\mathbf{R}^k$ as the product of Lebesgue measure on $\mathbf{R}^l$ with Lebesgue measure on $\mathbf{R}^m$ $(k = l+m)$ we may as well construct the product at the level of elementary figures (or even more simply, as we did for $k = 2$, at the level of bounded intervals, which form a semi-ring). The complete measures obtained by extending these finite measures by the Daniell construction are the required Lebesgue measures, and uniqueness in this case follows because all the spaces $\mathbf{R}^k$, $\mathbf{R}^l$, $\mathbf{R}^m$ are σ-finite. This is in some ways more natural than the construction which begins with the complete Lebesgue measures $\bar{\mu}$, $\bar{\nu}$ on $\mathbf{R}^l$, $\mathbf{R}^m$ and yields the same Lebesgue measure $\bar{\lambda}$ on $\mathbf{R}^k$ as their (completed) product.

To provide a second example of this 'compatibility', consider an entirely different situation. We are given two Daniell integrals on spaces X, Y. Can we define a 'product' Daniell integral on the product space $X \times Y$? And, if so, is Fubini's Theorem true for these integrals on the spaces X, Y, $X \times Y$? We answer both of these questions in the affirmative, and quite simply, by throwing them back on the idea of product measure. Corresponding to the integrals on X, Y there are complete measures μ, ν, respectively. Let λ be the complete measure on the σ-algebra $\mathscr{L}$ defined as in the proof of Theorem 14.1.1 by the Daniell construction. (We have dropped the bars over λ, μ, ν for simplicity of notation.) The steps are as follows. First of all, μ, ν are restricted to the subrings $\mathscr{R}_0, \mathscr{S}_0$ on which they are finite so that the corresponding linear lattices $L(\mathscr{R}_0)$, $L(\mathscr{S}_0)$ consist of the integrable simple functions for the given integrals. (These are just the integrable functions which assume only a finite set of values.) Let

$$\mathscr{E} = \{A \times B: A \in \mathscr{R}_0, B \in \mathscr{S}_0\}.$$

Then $L(\mathscr{E})$ is a linear lattice and the linear functional H is defined on $L(\mathscr{E})$ as a repeated integral

$$H(\phi) = \int \left(\int \phi(x, y)\, dy \right) dx. \tag{3}$$

This functional H is an elementary integral on $L(\mathscr{E})$, i.e. satisfies the Daniell condition, and so extends to a Daniell integral which yields the corresponding measure λ. From this construction in terms of measure it is clear that Fubini's Theorem holds for the integrals with respect to λ, μ, ν. Stone's Theorem 8.5.1 identifies $\int d\mu$, $\int d\nu$ with the given Daniell integrals on X, Y and $\int d\lambda$ is the required *product integral* on $X \times Y$. The question of compatibility crops up in the following way. Suppose that the Daniell integrals on X, Y arise from elementary integrals on linear lattices M, N, respectively. We may define a linear space L of 'elementary functions' on $X \times Y$ which are linear combinations of 'product functions' ϕ, where

$$\phi(x, y) = \xi(x)\, \eta(y),$$

with ξ in M, η in N. Unfortunately, there is a major snag here: the linear space L may not be a lattice! (See Ex. 4.) But if L *is* a lattice, and if the Stone Condition can be verified for L, we may define H on L, by means of (3), verify the Daniell condition as before, and apply the same extension procedure in each of the three spaces. The three measures λ, μ, ν that we obtain will be exactly the same as those obtained above by the construction which starts with the integrable simple functions on X and Y. Although Fubini's Theorem is expressed entirely in terms of integrals, the above difficulty about the space L strengthens our conviction that this famous result has its natural place in the setting of product measure. (On this point it may be helpful to consult Stone's original paper [21], Note III; also Asplund–Bungart [1], §4.1.)

Exercises

1. Let $X = Y = [0,1]$, $\mathscr{R} = \mathscr{S}$ consist of the finite subsets of $[0,1]$ and $\mu = \nu = 0$. Illustrate Theorem 1 by means of this example, showing in particular that there are sets A, B in $[0,1]$ such that $\bar{\mu}(A) = 0$, $\bar{\nu}(B) = \infty$ and $\bar{\lambda}(A \times B) = \infty$.

2. In the setting of Theorem 1, if $\bar{\mu}(A) = 0$, $\bar{\nu}(B) = \infty$, what possible values can be taken by $\bar{\lambda}(A \times B)$?

3. Let Q denote a Daniell integral and let P be the restriction of Q to the space L of integrable simple functions defined in §8.4 (L consists of the functions in $L^1(Q)$ that attain only finitely many values). Show that P is an elementary integral on L and that P extends to Q by the Daniell construction. In other words the Daniell integral Q may be constructed starting with the simple functions rather than the given elementary functions (cf. Ex. 8.3.2).

4. (Asplund–Bungart [1], Ex. 4.1.1.) A piecewise linear function on $[0,1]$ is a continuous real valued function on $[0,1]$ whose graph consists of a finite number of straight line segments. Show that the piecewise linear functions on $[0,1]$ form a linear lattice M. Show that the functions h on $[0,1]^2$ given by

$$h(x,y) = \sum_{i=1}^{n} f_i(x)\, g_i(y)$$

($n \geqslant 1, f_i, g_i \in M, i = 1, 2, \ldots, n$) form a linear space L and that $h(x,y)$ is of the form $a + bx + cy + dxy$ in each of a finite collection of disjoint intervals whose union is the unit square $[0,1]^2$.

If $n = 2, f_1(x) = g_1(x) = x, f_2(x) = -g_2(x) = \frac{1}{2}$ for x in $[0,1]$, show that the corresponding function $|h|$ is not in L.

15

BOREL MEASURES

The bounded intervals in $\mathbf{R}^k$ played a central role in the construction of the Lebesgue integral and measure. In § 15.1 we consider the σ-algebra $\mathscr{B}$ generated by these intervals; $\mathscr{B}$ may also be described as the smallest σ-algebra containing the open subsets of $\mathbf{R}^k$. The elements of $\mathscr{B}$ are the *Borel sets* in $\mathbf{R}^k$, and any measure on $\mathscr{B}$ which is finite on compact subsets is a *Borel measure* on $\mathbf{R}^k$. In this context the measures μ provided by the Riesz Representation Theorem are Borel measures with the further property of *regularity*: to each Borel set for which $\mu(E)$ is finite and each $\epsilon > 0$, there is a compact set F and an open set G such that $F \subset E \subset G$ and $\mu(G \setminus F) < \epsilon$.

In § 15.2 we prove that the only Borel measures on $\mathbf{R}^k$ which are invariant under translations are the constant multiples of Lebesgue measure (restricted to $\mathscr{B}$). This has immediate applications to the affine and Euclidean geometry of measure. In § 15.3 this idea is also applied to homeomorphisms in $\mathbf{R}^k$ with density functions.

The functions given by Baire in his famous classification in terms of iterated limits of sequences [2] are identified in § 15.5 with the Borel measurable functions on $\mathbf{R}^k$.

In the final section we consider the much more general situation where X is a locally compact Hausdorff space. This leads to a distinction between Baire and Borel measures on X and a general form of the Riesz Representation Theorem.

15.1 Borel measures

Let $\mathscr{B}$ denote the σ-algebra generated by the bounded intervals in $\mathbf{R}^k$. As we saw in § 6.3 (and again in § 9.3) any open set A in $\mathbf{R}^k$ is the union of a sequence of disjoint bounded intervals; thus $\mathscr{B}$ contains all the open subsets of $\mathbf{R}^k$. Conversely, if $\mathscr{C}$ is a σ-algebra which contains all the open sets in $\mathbf{R}^k$, then $\mathscr{C}$ contains all the bounded intervals, because any such interval is the difference of two open sets. (For example, in $\mathbf{R}^1$ the interval $(a, b]$ may be expressed as $(a, c) \setminus (b, c)$ for any $c > b$.) Thus $\mathscr{B}$ may also be described as the σ-algebra generated by the open sets in $\mathbf{R}^k$. This leads to the following general definition.

Let X be an arbitrary topological space and denote by $\mathscr{B}(X)$ the

σ-algebra generated by the open sets in X; the elements of $\mathscr{B}(X)$ are called the *Borel sets* in X. According to this definition the Borel sets are topological objects, viz.

Proposition 1. *Let X, Y be topological spaces and T a homeomorphism of X onto Y. Then $TA \in \mathscr{B}(Y)$ if and only if $A \in \mathscr{B}(X)$.*

Proof. The proof follows immediately as T is one–one and TA is an open subset of Y if and only if A is an open subset of X.

The σ-algebra of Lebesgue measurable sets in $\mathbf{R}^k$ includes all bounded intervals and so includes all Borel sets. But we saw at the end of § 6.3 (Ex. 6.3.8) that homeomorphisms do not respect Lebesgue measurability. In view of Proposition 1 this means that there are Lebesgue measurable sets which are not Borel sets, i.e. the σ-algebra of Lebesgue measurable sets is strictly greater than $\mathscr{B}$.†

Any measure μ on $\mathscr{B}$ which is finite on the bounded intervals will be finite on the compact sets, because any compact set in $\mathbf{R}^k$ is bounded and so contained in a bounded interval. This leads to another general definition.

Let X be a Hausdorff topological space; then any measure on $\mathscr{B}(X)$ which is finite on compact sets is called a *Borel measure* on X. (Recall that a *Hausdorff* topological space is one in which, to each pair of distinct points x, y there are disjoint open sets U, V such that $x \in U$, $y \in V$; in particular, any metric space satisfies this condition. Any compact subset of a Hausdorff space X is closed and therefore lies in $\mathscr{B}(X)$: cf. Ex. 10.1.1 or Ex. 23 of the Appendix to Volume 1.)

It is important to realise that each space X has only one collection of Borel sets though there are, in general, many Borel measures on X. The fundamental Uniqueness Theorem 13.4.2 has a great deal to say about these Borel measures. First of all, if L is the linear lattice of step functions on $\mathbf{R}^k$ then the corresponding σ-algebra generated by the elementary sets $\{x \in X : \phi(x) \geqslant c\}$ $(\phi \in L, c \in \mathbf{R})$ is $\mathscr{B}$ itself. Each elementary integral P on L yields a unique Borel measure μ which satisfies

$$P(\phi) = \int \phi \, d\mu$$

for all ϕ in L; the completion of μ is the measure $\bar{\mu}$ obtained by applying the Daniell extension procedure to P and, of course, μ is the restriction of $\bar{\mu}$ to $\mathscr{B}$. These measures $\bar{\mu}$ are exactly the Lebesgue–

† Let $\mathbf{c}$ denote the cardinal number of the 'continuum' $\mathbf{R}$. It may be shown that the cardinal number of $\mathscr{B}$ is $\mathbf{c}$ and the cardinal number of the collection of Lebesgue measurable sets is $2^{\mathbf{c}}$ (see [12], p. 94).

Stieltjes measures of Theorem 9.2.1; their description in terms of Stieltjes functions guarantees a rich selection of Borel measures μ. We note in passing that if $\bar{\mu}$ is Lebesgue measure on $\mathbf{R}^k$, then the corresponding Borel measure μ cannot be complete, otherwise all Lebesgue measurable sets would belong to $\mathscr{B}$, contrary to our remark after Proposition 1.

More generally, if X is a locally compact metric space, the Riesz Representation Theorem 10.1.4 gives a Borel measure for every positive linear functional on $C_c(X)$. In order to apply the Uniqueness Theorem 13.4.2 we require X to be σ-finite with respect to $C_c(X)$, i.e. we look for elementary sets

$$X_n = \{x \in X: \phi_n(x) \geqslant 1\} \quad (\phi_n \in C_c(X))$$

whose union is X. Each X_n is a closed subset of the support of the corresponding ϕ_n and so is compact. Thus we require X to be σ-compact, as described in the introduction to Theorem 10.2.3. For the remainder of this section we shall make the simplifying assumption that X is a σ-compact, locally compact metric space: all results will therefore apply in particular to $\mathbf{R}^k$. In this case we may identify the σ-algebra $\mathscr{A}$ of the Uniqueness Theorem:

Proposition 2. *If X is a σ-compact, locally compact metric space, then $\mathscr{B}(X)$ is the smallest σ-algebra for which all the elements of $C_c(X)$ are measurable.*

Proof. Let $\mathscr{A}$ be the σ-algebra in question, i.e. $\mathscr{A}$ is the σ-algebra generated by the sets

$$\{x \in X: \phi(x) \geqslant c\}$$

$(\phi \in C_c(X),\ c \in \mathbf{R})$. Each of these sets is closed and therefore belongs to $\mathscr{B}$. Thus $\mathscr{A} \subset \mathscr{B}$.

On the other hand, every compact subset K of X belongs to $\mathscr{A}$, because K may be expressed as

$$\{x \in X: \psi(x) \geqslant 1\}$$

by means of Proposition 10.1.1. Let F be an arbitrary closed set in X and $\{X_n\}$ a sequence of compact sets whose union is X. Then each $F \cap X_n$ is compact and so belongs to $\mathscr{A}$. But F is the union of these sets $F \cap X_n$ so that $F \in \mathscr{A}$. As $\mathscr{A}$ includes all closed sets, $\mathscr{A}$ includes all open sets and hence $\mathscr{A} \supset \mathscr{B}$.

Theorem 1. *Let X be a σ-compact, locally compact metric space. Then to each positive linear functional P on $C_c(X)$ there corresponds a unique*

Borel measure μ which satisfies

$$P(\phi) = \int \phi \, d\mu \tag{1}$$

for all ϕ in $C_c(X)$, and every Borel measure μ on X arises in this way. The measure $\bar{\mu}$ obtained from P by the Daniell construction is the completion of μ.

Proof. As in the proof of Theorem 10.1.4, Dini's Theorem shows that any positive linear functional P on $C_c(X)$ is an elementary integral, i.e. satisfies the Daniell condition. The existence and uniqueness of the measure μ now follow immediately by Theorem 13.4.2.

Conversely, let μ be a Borel measure on X and ϕ a continuous function on X of compact support K. The sets $\{x \in X : \phi(x) \geqslant c\}$ are closed and so ϕ is measurable with respect to $\mathscr{B}(X)$. Also ϕ is bounded on the compact set K and so ϕ is integrable with respect to μ (Proposition 12.3.2). Thus equation (1) defines a positive linear functional P on $C_c(X)$. This completes the proof.

In view of Theorem 10.2.3 there is a corresponding statement for signed measures.

Theorem 2. *Let X be a σ-compact, locally compact metric space. Then to each bounded linear functional F on $C_c(X)$ there corresponds a unique finite signed Borel measure μ such that*

$$F(\phi) = \int \phi \, d\mu$$

for all ϕ in $C_c(X)$, and every finite signed Borel measure arises in this way.

As we should expect, a *finite signed Borel measure* is the difference of two finite Borel measures.

Proof. All that requires proof is the uniqueness. Suppose that $\mu_1 - \mu_2$, $\nu_1 - \nu_2$ are two finite signed Borel measures corresponding to F. Then

$$\int \phi \, d\mu_1 - \int \phi \, d\mu_2 = \int \phi \, d\nu_1 - \int \phi \, d\nu_2$$

for all ϕ in $C_c(X)$. In other words

$$\int \phi \, d\mu_1 + \int \phi \, d\nu_2 = \int \phi \, d\nu_1 + \int \phi \, d\mu_2$$

for all ϕ in $C_c(X)$. But now the uniqueness part of Theorem 1 shows that

$$\mu_1 + \nu_2 = \nu_1 + \mu_2,$$

and so

$$\mu_1 - \mu_2 = \nu_1 - \nu_2.$$

In the present topological setting the approximation theorems of § 13.3 may be sharpened significantly.

Theorem 3. *Let X be a σ-compact, locally compact metric space, μ a Borel measure on X and let ε > 0 be given. To each set E in $\mathscr{B}(X)$ there is a closed set F and an open set G such that*

$$F \subset E \subset G \quad \text{and} \quad \mu(G \setminus F) < \epsilon.$$

A measure which satisfies the conclusion of Theorem 3 is said to be *regular*.

Proof. (i) First of all consider the case where $\mu(E)$ is finite and appeal to the Approximation Theorem 13.3.1. In the proof we constructed the sets

$$S_n = \{x \in X: \theta_n(x) \geqslant c\}.$$

(For simplicity of notation we now omit the exponent ϵ.) The proof would have been unaltered if we had chosen to consider, in place of S_n, the sets

$$T_n = \{x \in X: \theta_n(x) > c\}.$$

In the present context the functions θ_n belong to $C_c(X)$ and so each T_n is an open subset of X. The union T of the sequence $\{T_n\}$ is again open and satisfies the conditions

$$E \subset T, \quad \mu(T \setminus E) \leqslant \epsilon.$$

(ii) Suppose now that X is the union of the sequence $\{X_n\}$ of compact sets and let

$$E_n = E \cap X_n.$$

For each n apply part (i) to find an open set G_n such that

$$E_n \subset G_n, \quad \mu(G_n \setminus E_n) < \epsilon/2^{n+1}.$$

Let G be the open set $\bigcup G_n$. Then $E \subset G$ and the inclusion

$$(G \setminus E) \subset \bigcup (G_n \setminus E_n)$$

implies that

$$\mu(G \setminus E) < \epsilon/2.$$

The same argument applies to E^c and yields an open set F^c, say, for which

$$E^c \subset F^c, \quad \mu(F^c \setminus E^c) < \epsilon/2.$$

Now F is closed (as F^c is open) and

$$F^c \setminus E^c = E \setminus F,$$

whence

$$F \subset E, \quad \mu(E \setminus F) < \epsilon/2.$$

Thus

$$F \subset E \subset G \quad \text{and} \quad \mu(G \setminus F) < \epsilon.$$

Corollary. *Under the conditions of Theorem 3, if $\mu(E)$ is finite, we may also insist that F is compact.*

Proof. Let K_n be the compact set $X_1 \cup \ldots \cup X_n$ and let $F_n = F \cap K_n$. Then F_n is compact and $F_n \uparrow F$. It follows that $\mu(F_n) \uparrow \mu(F)$. As $\mu(E)$ is finite we may therefore choose n large enough to ensure that

$$\mu(G \setminus F_n) < \epsilon.$$

This completes the proof.

Following the pioneer Hausdorff, the letters G, F are frequently used to denote open and closed sets, respectively. He also introduced a useful notation for some special Borel sets: if B is the intersection of a sequence of open sets, then B is called a G_δ (δ for *Durchschnitt*–intersection); and if B is the union of a sequence of closed sets, then B is called an F_σ (σ for *Summe*–union). In terms of these special sets there is an even stronger approximation property.

Theorem 4. *Let X be a σ-compact, locally compact metric space and μ a Borel measure on X. To each set E in $\mathscr{B}(X)$ there are sets A, B in $\mathscr{B}(X)$ such that A is an F_σ, B is a G_δ,*

$$A \subset E \subset B \quad \textit{and} \quad \mu(B \setminus A) = 0.$$

Proof. For each n find F_n (closed), G_n (open) satisfying

$$F_n \subset E \subset G_n, \quad \mu(G_n \setminus F_n) < 1/n$$

and let $$A = \bigcup F_n, \quad B = \bigcap G_n.$$

Then A is an F_σ, B is a G_δ,

$$A \subset E \subset B, \quad \mu(B \setminus A) < 1/n$$

for all $n \geqslant 1$.

Exercises

1. Show that $\mathscr{B}(\mathbf{R}^k)$ is the σ-ring generated by the half-open intervals of type $[\ ,\ )^k$.

2. Let X be a discrete topological space (i.e. all subsets of X are open). Describe the Borel measures on X.

3. Show that any half-open interval of type $[\ ,\ )^k$ in $\mathbf{R}^k$ is an F_σ and a G_δ.

4. Consider a bounded function $f: \mathbf{R} \to \mathbf{R}$. For any point p in $\mathbf{R}$ let

$$g(p) = \sup\{\inf\{f(x): |x-p| < r\}: r > 0\}$$
$$h(p) = \inf\{\sup\{f(x): |x-p| < r\}: r > 0\}.$$

(These functions g, h are the lower and upper approximations to f which were used in Exx. 3.3.E, F to link Riemann integration with Lebesgue integration.) Show that (i) $g \leqslant f \leqslant h$, (ii) f is continuous at p if and only if $g(p) = h(p)$, (iii) g is lower semicontinuous and h is upper semicontinuous (Ex. 9.1.3).

Deduce that the points of continuity of f form a G_δ. Extend this to an arbitrary function $f: \mathbf{R} \to \mathbf{R}$.

5. Extend Ex. 4 to a real valued function on a topological space X.

15.2 Translation invariance

The Borel sets in $\mathbf{R}^k$ have been defined in terms which show their independence of the standard frame of reference in $\mathbf{R}^k$. We may expect the Borel sets to help in freeing the notion of Lebesgue measure from its apparent bondage to the coordinate axes and the rectangular intervals from which we originally constructed it. First of all, we need a criterion which picks out Lebesgue measure from all the other Borel measures on $\mathbf{R}^k$. The theorem which follows requires not only the topological properties of $\mathbf{R}^k$ but also refers to the group theoretic operation of addition in $\mathbf{R}^k$. It is a special case of a deeper uniqueness theorem for the Haar measure on a (locally compact Hausdorff) topological group.

Theorem 1. *Let μ be a Borel measure on $\mathbf{R}^k$. If μ is translation invariant, i.e.*

$$\mu(a+B) = \mu(B)$$

for all a in $\mathbf{R}^k$ and all B in $\mathscr{B}$, then μ is a constant multiple of Lebesgue measure (restricted to $\mathscr{B}$).

Recall that $$a+B = \{a+b : b \in B\}.$$

Proof. Let m denote Lebesgue measure, let I denote the unit cube $\{(x_1, \ldots, x_k): 0 \leqslant x_i < 1 \text{ for } i = 1, \ldots, k\}$, and let $\mu(I) = c$. Consider the collection $\mathscr{C}$ of all subsets B of $\mathscr{B}$ for which

$$\mu(B) = c\, m(B),$$

including ∞ as a possible value on either side. (For the purpose of this proof, if c happens to be zero, we interpret the right hand side as zero, even if $m(B) = \infty$.) Clearly $\mathscr{C}$ contains I, because $m(I) = 1$. As in the proof of Proposition 6.3.1 we may subdivide I into 2^{kn} disjoint cubes of type $[\,,\,)^k$, each of side 2^{-n}, and see at once, by the translation invariance of μ and m, that these cubes all belong to $\mathscr{C}$. In fact, the whole class $\mathscr{S}_n$ of cubes of side 2^{-n} described in that proof is contained in $\mathscr{C}$, and as μ and m are σ-additive, $\mathscr{C}$ includes every open subset of $\mathbf{R}^k$. (This is where we have to check the above conventions about ∞.) Any bounded interval is the difference $A \setminus B$ of two bounded open sets $A \supset B$ and hence $\mathscr{C}$ includes all bounded intervals; thus $\mathscr{C}$ contains the ring $\mathscr{R}$ of elementary figures which are finite disjoint unions of bounded intervals. The first uniqueness theorem (Theorem 13.2.1), applied to μ and m, now shows that $\mathscr{C} = \mathscr{B}$, and completes the proof.

A variation of this proof, using monotone collections of sets, is given in Ex. 1. A quite different proof in terms of Stieltjes functions is given in Ex. 9.2.4.

Theorem 1 has a delightfully simple application to the affine mappings in $\mathbf{R}^k$ (cf. Theorem 6.3.2).

Theorem 2. *Let T be an invertible affine mapping on $\mathbf{R}^k$. Then TB is a Borel set if and only if B is a Borel set. Moreover, there is a non-zero constant Δ_T (depending only on T) such that*

$$m(TB) = \Delta_T m(B) \tag{1}$$

for all Borel sets B.

As usual m denotes Lebesgue measure.

Proof. We remark first of all that any affine mapping is continuous (see the paragraph immediately after the proof of Proposition 6.3.5). Thus an invertible affine mapping is a homeomorphism and so preserves Borel sets (Proposition 1) and compactness. We may therefore define a Borel measure μ by the equation

$$\mu(B) = m(TB)$$

for all B in $\mathscr{B}$. The σ-additivity of μ follows from that of m as T is one–one, and $\mu(B)$ is finite for compact B. Now μ is translation invariant; this is almost obvious from the geometry, but is verified strictly as follows:

$$Tx = a + Lx$$

for all x in $\mathbf{R}^k$, where $a \in \mathbf{R}^k$ and L is linear; then

$$\begin{aligned}\mu(c+B) &= m(a+Lc+LB) \\ &= m(a+LB) \qquad \text{(as } m \text{ is translation invariant)} \\ &= \mu(B).\end{aligned}$$

Theorem 1 now shows that there is a constant Δ_T such that

$$\mu(B) = \Delta_T m(B)$$

for all B in $\mathscr{B}$. Clearly Δ_T cannot be zero as T is invertible.

If T' is another invertible affine mapping, it follows at once from (1) that

$$\Delta_{TT'} = \Delta_T \Delta_{T'}.$$

This is reminiscent of a formula for determinants – in fact Δ_T is the absolute value of $\det T$, as we saw in § 6.3 – but it is gratifying to have a proof of Theorem 2 that makes no reference to matrices, shears, reflections or diagonal transformations, let alone determinants!

The restriction of Theorem 2 to the case of *invertible* affine mappings is not very significant, as a non-invertible (i.e. singular) affine mapping sends the whole of $\mathbf{R}^k$ onto a translated linear subspace of lower dimension which therefore has Lebesgue measure zero (see Ex. 2). Thus (1) holds in this case with $\Delta_T = 0$.

The special case of a Euclidean transformation, i.e. an isometry, T of $\mathbf{R}^k$ is now so obvious that it hardly needs comment. By considering the effect of T on a unit ball we verify at once that $\Delta_T = 1$ and so we have

Corollary. *An isometry of* $\mathbf{R}^k$ *preserves Borel sets and also preserves their Lebesgue measure.*

We may take Theorem 2 one stage further; we have seen that Lebesgue measure is obtained from its restriction to $\mathscr{B}$ by the simple construction of Proposition 13.2.1. If E is Lebesgue measurable, then there are Borel sets A, B for which

$$A \subset E \subset B \quad \text{and} \quad m(B \setminus A) = 0.$$

It follows from Theorem 2 that

$$TA \subset TE \subset TB \quad \text{and} \quad m(TB \setminus TA) = 0.$$

Thus TE is Lebesgue measurable and

$$m(TE) = \Delta_T m(E). \tag{2}$$

This gives a large part of Theorem 6.3.2 (bearing in mind the above remark that a singular affine mapping carries the whole of $\mathbf{R}^k$ onto a null set). For convenience we restate the result.

Theorem 3. *Let* $T: \mathbf{R}^k \to \mathbf{R}^k$ *be an affine mapping and let* E *be an arbitrary subset of* $\mathbf{R}^k$ *of finite (Lebesgue) measure; then* TE *has finite (Lebesgue) measure*

$$m(TE) = \Delta_T m(E),$$

where Δ_T *is a constant (depending only on* T*).*

More generally,

$$\int g = \Delta_T \int f$$

for any f in L^1, where f, g are related by the equation

$$g(Tx) = f(x)$$

for all x in $\mathbf{R}^k$ and g vanishes outside $T\mathbf{R}^k$.

To complete the proof, let f be an element of L^1 and define g by

$$g(Tx) = f(x)$$

for all x in $\mathbf{R}^k$. (Without loss of generality we assume that T is invertible so that $T(\mathbf{R}^k) = \mathbf{R}^k$.) Equation (2) shows that

$$\int g = \Delta_T \int f$$

in the special case where $f = \chi_E$ and E has finite Lebesgue measure. This now extends to integrable simple functions f by linearity, then to positive integrable functions f by Proposition 6.2.2 (or Proposition 8.4.3), and finally to all integrable functions f by means of the decomposition

$$f = f^+ - f^-.$$

Exercises

1. In the notation used in the proof of Theorem 1 let $\mathscr{C}$ be the collection of all B in $\mathscr{B}$ for which $\mu(B) = \mu(I)\, m(B)$. Show that $\mathscr{C}$ is monotone and contains all bounded intervals, and hence show that $\mathscr{C} = \mathscr{B}$.

2. Let T be a non-invertible (i.e. singular) affine mapping on $\mathbf{R}^k$ and let $B = T(\mathbf{R}^k)$. Show that B is a closed subset of $\mathbf{R}^k$ and that $m(B) = 0$.

3. Let m denote Lebesgue measure restricted to $\mathscr{B}(\mathbf{R}^k)$. Each cm $(c \geqslant 0)$ is a Borel measure which is invariant under rotations about the origin: are these the only ones? How is this affected if we ask for invariance under rotations about every point of $\mathbf{R}^k$?

15.3 Density functions

Having sampled the power of these topological methods applied to questions of affine and Euclidean geometry, we now have some confidence that they may also apply to the much harder topological problems discussed in § 6.4. Recall that U is an *open* subset of $\mathbf{R}^k$ and T is a *one–one continuous* mapping of U onto a subset V of $\mathbf{R}^k$. (As we remarked before, in the concluding paragraphs of Chapter 6, a famous theorem of Brouwer ensures that V is an *open* set in $\mathbf{R}^k$ and that T is a *homeomorphism*; but we expressly wish to be independent of this deep result, and so we shall only impose the apparently weaker conditions on T. With the aid of Brouwer's Theorem, condition (2) below is a special case of Proposition 15.1.1.)

We shall refer several times to the quite elementary Proposition 6.3.2 which states that any open set A in $\mathbf{R}^k$ may be expressed as the union of a sequence of disjoint cubes whose closures are contained in A.

If $\mathscr{C}$ is an arbitrary collection of subsets of X and $E \subset X$, then we write

$$\mathscr{C} \cap E = \{C \cap E \colon C \in \mathscr{C}\}.$$

Lemma 1. *Let $\mathscr{S}$ be the σ-ring generated by $\mathscr{C}$. Then $\mathscr{S} \cap E$ is the σ-ring generated by $\mathscr{C} \cap E$.*

Proof. (Cf. Halmos [11] § 5.) It is immediately verified that $\mathscr{S} \cap E$ is a σ-ring containing $\mathscr{C} \cap E$. Let $\mathscr{S}'$ be the σ-ring generated by $\mathscr{C} \cap E$; then $\mathscr{S}' \subset \mathscr{S} \cap E$. If

$$\mathscr{T} = \{A \cup B \colon A \in \mathscr{S}', B \in \mathscr{S} \cap E^c\},$$

then these sets A, B are disjoint and we again verify that $\mathscr{T}$ is a σ-ring. Moreover, any C in $\mathscr{C}$ is of the form

$$C = (C \cap E) \cup (C \cap E^c);$$

thus $\mathscr{T}$ contains $\mathscr{C}$ and so contains $\mathscr{S}$. Intersecting with E now shows that $\mathscr{T} \cap E$ contains $\mathscr{S} \cap E$, i.e. $\mathscr{S}' \supset \mathscr{S} \cap E$. This completes the proof.

There is the obvious counterpart of this lemma for the σ-algebra $\mathscr{A}$ generated by $\mathscr{C}$, where of course, $\mathscr{A} \cap E$ will be a σ-algebra of subsets of E (not of X in general).

The σ-algebra $\mathscr{B}(U)$ is generated by the open subsets of U. As U is itself open in $\mathbf{R}^k$, the open subsets of U are exactly the open subsets of $\mathbf{R}^k$ which are contained in U. Similarly, the σ-algebra $\mathscr{B}(V)$ is generated by the open subsets of V; but as we have not assumed that V is an open subset of $\mathbf{R}^k$, an 'open subset of V' is to be interpreted as an open set in the topology on V inherited from $\mathbf{R}^k$, i.e. an open subset of V is of the form $G \cap V$, where G is open in $\mathbf{R}^k$ (Appendix to Volume 1). In view of Lemma 1 this means that

$$\mathscr{B}(V) = \mathscr{B} \cap V,$$

as we should reasonably expect.

As a substantial part of Theorem 6.4.1 we now have

Theorem 1. *Let U be an open subset of $\mathbf{R}^k$ and T a one–one continuous mapping of U into $\mathbf{R}^k$. Then TB is a Borel set in TU if and only if B is a Borel set in U.*

Moreover, if T has density function h, then

$$m(TB) = \int_B h \tag{1}$$

for any Borel set B in U.

As usual, (1) means that $m(TB) = \infty$ if and only if $h \notin L^1(B)$.

Proof. Write $V = TU$. As T is one–one,

$$\mathscr{C} = \{TB: B \in \mathscr{B}(U)\}$$

is a σ-algebra. As T is continuous, $\mathscr{C}$ includes all open sets in V and so $\mathscr{C} \supset \mathscr{B}(V)$.

On the other hand, the σ-algebra

$$\mathscr{A} = \{B: TB \in \mathscr{B}(V)\}$$

contains all compact sets in U. Any bounded interval in $\mathbf{R}^k$ may be expressed as the union of an increasing sequence of compact intervals (e.g. in $\mathbf{R}^1$, $[a+1/n, b] \uparrow (a, b]$, and the idea clearly generalises to $\mathbf{R}^k$). Thus $\mathscr{A}$ includes all bounded intervals whose closures are contained in U. By Proposition 6.3.2, $\mathscr{A}$ includes all open sets in U and therefore $\mathscr{A} \supset \mathscr{B}(U)$. These two inclusions combine to show that

$$TB \in \mathscr{B}(V) \quad \text{if and only if} \quad B \in \mathscr{B}(U). \tag{2}$$

Recall that h is a positive function defined on U which satisfies

$$m(TI) = \int_I h$$

for any bounded interval I with $\bar{I} \subset U$. Each of the equations

$$\mu(B) = m(TB), \quad \nu(B) = \int_B h,$$

defines a Borel measure on $\mathscr{B}(U)$. The measures μ, ν agree on the bounded intervals I with $\bar{I} \subset U$ and hence agree on the ring of elementary figures which are finite disjoint unions of these intervals. But the fundamental Uniqueness Theorem 13.2.1 now shows that μ, ν agree on the σ-algebra $\mathscr{B}(U)$. In other words, equation (1) holds for all B in $\mathscr{B}(U)$. This completes the proof.

For ease of reference we restate Theorem 6.4.1.

Theorem 2. *Let U be an open subset of $\mathbf{R}^k$ and $T: U \to \mathbf{R}^k$ a one–one continuous mapping with density function h. If $E \subset U$ and E is Lebesgue measurable, then TE is Lebesgue measurable,*

$$m(TE) = \int_E h, \tag{3}$$

and

$$\int_{TU} g(y)\,dy = \int_U g(Tx)\,h(x)\,dx. \tag{4}$$

Equations (3), (4) are interpreted in the sense that if one side is defined, then so is the other and they are equal (infinite values are allowed in (3)).

Proof. We now give an alternative proof using Theorem 1. The first part (as far as (3)) only needs the idea of completion as described in Proposition 13.2.1. If E is Lebesgue measurable, then we can find Borel sets A, B such that

$$A \subset E \subset B \quad \text{and} \quad m(B \setminus A) = 0.$$

It follows that

$$TA \subset TE \subset TB \quad \text{and} \quad m(TB \setminus TA) = 0$$

(the latter using equation (1)), so TE is Lebesgue measurable and (3) holds.

Conversely, if TE is Lebesgue measurable, then there exist Borel sets TA, TB such that

$$TA \subset TE \subset TB \quad \text{and} \quad m(TB \setminus TA) = 0.$$

From this we deduce that $A \subset E \subset B$ and

$$\int_{B \setminus A} h = 0. \tag{5}$$

It follows from (5) that $\quad h\chi_A = h\chi_B$
almost everywhere. Thus

$$\int_A h = \int_E h = \int_B h$$

(possibly all infinite) and (3) holds once again. This establishes (3) under the assumption that either E or TE is Lebesgue measurable.

Now let the functions $f: U \to \mathbf{R}$, $g: V \to \mathbf{R}$ be related by the equation

$$g(Tx) = f(x)$$

for all x in U. Equation (4) may be rewritten

$$\int_V g = \int_U fh. \tag{6}$$

We have proved that (6) holds for $f = \chi_E$, $g = \chi_{TE}$ when either E or TE is Lebesgue measurable. Thus (6) holds if either f or g is a simple function. A familiar argument, applying Proposition 6.2.2 (or Proposition 8.4.3) to g^+, g^- shows that (6) holds if $g \in L^1(V)$.

On the other hand, if $fh \in L^1(U)$, let

$$f_0 = (fh)\frac{1}{h},$$

i.e.

$$\begin{aligned} f_0(x) &= f(x) \quad \text{if} \quad h(x) \neq 0 \\ &= 0 \qquad\ \text{if} \quad h(x) = 0. \end{aligned}$$

Now h is measurable, because $h\chi_I$ is integrable for every bounded interval I with $\bar{I} \subset U$ and we may apply Proposition 6.3.2. Thus f_0 is measurable (as the product of the measurable functions fh, $1/h$). Let g_0 be defined by

$$g_0(Tx) = f_0(x)$$

for all x in U. Then the same argument as above, applied to f_0^+, f_0^- shows that

$$\int_V g_0 = \int_U f_0 h.$$

Let

$$S = \{x \in U : h(x) \neq 0\}.$$

For any bounded interval I with $\bar{I} \subset U$ we may take $f = \chi_I$, $f_0 = \chi_{I \cap S}$ and deduce that

$$m(TI \cap TS) = \int_{I \cap S} h = \int_I h = m(TI).$$

Thus $TI \setminus TS$ is null. In view of Proposition 6.3.2, $V \setminus TS$ is null. This shows that $g_0 = g$ almost everywhere in V and, of course, $f_0 h = fh$; so finally (6) holds in this case also.

It is interesting to note that we did not deduce from (5) that $m(B \setminus A) = 0$, nor did we deduce that E is Lebesgue measurable. If S is the measurable set defined above, we can deduce from (5) that

$$m((B \setminus A) \cap S) = 0$$

and hence that $E \cap S$ is Lebesgue measurable and

$$m(A \cap S) = m(E \cap S) = m(B \cap S).$$

In practice, the set on which h vanishes, viz. $U \setminus S$ is usually null; in this case TE is Lebesgue measurable if and only if E is Lebesgue measurable.

We have now given the promised, rather more streamlined, treatment of the Fundamental Theorems 6.3.2, 6.4.1, i.e. Theorem 15.2.3

and Theorem 2 above. But this discussion has achieved far more: it has established the fact that Borel sets are well adapted to answer topological questions, and that significant results may be deduced for Lebesgue measure from the corresponding results for Borel measure by a simple completion argument.

Exercise

1. Prove the counterpart of Lemma 1 for the σ-algebra $\mathscr{A}$ generated by $\mathscr{C}$.

15.4 Measurable transformations

The affine mapping T of Theorem 15.2.2 and the continuous mapping T of Theorem 15.3.1 are important examples of measurable transformations. Quite generally, suppose that $\mathscr{A}$ is a σ-algebra of subsets of X and $\mathscr{B}$ is a σ-algebra of subsets of Y; a transformation (or mapping) $T: X \to Y$ will be called *$(\mathscr{A}, \mathscr{B})$-measurable* if $T^{-1}(B) \in \mathscr{A}$ for every set B in $\mathscr{B}$. If there is no possibility of confusion about the σ-algebras $\mathscr{A}$, $\mathscr{B}$, then we may say that T is *measurable*.

It follows at once from the definition

$$T^{-1}(\mathscr{B}) = \{T^{-1}(B) : B \in \mathscr{B}\}$$

that $T^{-1}(\mathscr{B})$ is a σ-algebra of subsets of X; furthermore, if $\mathscr{B}$ is generated by $\mathscr{C}$ then $T^{-1}(\mathscr{B})$ is generated by $T^{-1}(\mathscr{C})$ (cf. Ex. 11.1.6). Thus T is $(\mathscr{A}, \mathscr{B})$-measurable if and only if $T^{-1}(C) \in \mathscr{A}$ for all C in $\mathscr{C}$. This sheds light on our previous definition of an $\mathscr{A}$-measurable function $f: X \to \mathbf{R}$ (p. 102). Let $\mathscr{C}$ consist of the intervals $[c, \infty)$ in $\mathbf{R}$. Then the corresponding σ-algebra $\mathscr{B}$ generated by $\mathscr{C}$ consists of the Borel sets in $\mathbf{R}$. Recall that f is $\mathscr{A}$-measurable if and only if

$$\{x \in X : f(x) \geqslant c\}$$

belongs to $\mathscr{A}$ for all c in $\mathbf{R}$. But this is just another way of saying that $f^{-1}(C) \in \mathscr{A}$ for all C in $\mathscr{C}$. *Thus an $\mathscr{A}$-measurable function $f: X \to \mathbf{R}$ is the same as an $(\mathscr{A}, \mathscr{B})$-measurable transformation $f: X \to \mathbf{R}$ when $\mathscr{B}$ consists of the Borel sets in $\mathbf{R}$.*

There is a simple abstract theorem which is reminiscent of the integration theorems in the previous two sections.

Theorem 1. *Let $T: X \to Y$ be an $(\mathscr{A}, \mathscr{B})$-measurable transformation and let μ be a measure on $\mathscr{A}$. We may define a measure ν on $\mathscr{B}$ by the rule*

$$\nu(B) = \mu(T^{-1}(B))$$

for all B in $\mathscr{B}$. Let $g: Y \to \mathbf{R}$ be a $\mathscr{B}$-measurable function and define the composite function $f: X \to \mathbf{R}$ by the rule

$$f(x) = g(Tx)$$

for all x in X. Then f is $\mathscr{A}$-measurable and

$$\int f d\mu = \int g d\nu$$

in the sense that if one integral exists, so does the other and they are equal.

Proof. This is set as Ex. 2 below.

Exercises

1. Let $\mathscr{A}$ be a σ-algebra of subsets of X. If $f: \mathbf{R} \to \mathbf{R}$ is Borel measurable and $g: X \to \mathbf{R}$ is $\mathscr{A}$-measurable, show that the composite function $f \circ g$ is $\mathscr{A}$-measurable. Would this statement remain true if Borel were replaced by Lebesgue?

2. Prove Theorem 1 by considering the successive cases where g is a characteristic function, a $\mathscr{B}$-simple function, a positive $\mathscr{B}$-measurable function, a $\mathscr{B}$-measurable function.

3. Let $\mathscr{A}$ be a σ-algebra of subsets of X, and T an arbitrary mapping of X onto Y. Define

$$\mathscr{B} = \{B \subset Y: T^{-1}(B) \in \mathscr{A}\}.$$

Show that $\mathscr{B}$ is the largest σ-algebra of subsets of Y for which T is $(\mathscr{A}, \mathscr{B})$-measurable.

If the functions $f: X \to \mathbf{R}$, $g: Y \to \mathbf{R}$ are related by the equation $f(x) = g(Tx)$ for all x in X, show that f is $\mathscr{A}$-measurable if and only if g is $\mathscr{B}$-measurable. What becomes of Theorem 1 in this case?

4. Give an example of a one–one $(\mathscr{A}, \mathscr{B})$-measurable transformation whose inverse is not $(\mathscr{B}, \mathscr{A})$-measurable.

15.5 Baire functions

In § 6.1 we mentioned the Baire classes of real valued functions on $\mathbf{R}^k$. Recall that the Baire class 0 consists of the real valued continuous functions on $\mathbf{R}^k$, and for each ordinal α the Baire class α consists of the real valued functions which are limits of convergent sequences of functions in the Baire classes β with $\beta < \alpha$. To obtain all the functions in Baire's classification this inductive definition has to be interpreted transfinitely. A simpler, though in some ways less illuminating, way of obtaining these functions is as follows: let L be the class of all continuous real valued functions on $\mathbf{R}^k$ and let M be the smallest class of functions which contains L and is closed with respect to pointwise convergence of sequences (everywhere); then M is the class of *Baire functions* on $\mathbf{R}^k$. From now on we accept this as our 'official' definition of the Baire functions on $\mathbf{R}^k$. The existence of M is established by taking the intersection of all classes of real valued functions on $\mathbf{R}^k$ which contain L and which are closed with respect to pointwise convergence of sequences – there is at least one such class, viz. the class of *all* $f: \mathbf{R}^k \to \mathbf{R}$. This is a good example of a construction which guarantees the existence of a certain algebraic object M but gives very little idea of the complexity of the way in which M depends on the 'generating' set L. To gain some idea of the complexity in this case the reader may care to consult [8], Ch. 11 or [15], Ch. 15.

In a much more general setting we have the following simple result.

Proposition 1. *Let L be a linear lattice of real valued functions on a set X and let M be the smallest class of real valued functions on X which contains L and is closed with respect to pointwise convergence of sequences. Then M is a linear lattice.*

Proof. (Cf. the proof of Theorem 13.1.1.) Suppose that $f \in L$ and consider

$$M_f = \{g \in M : f+g \in M\}.$$

We verify at once that M_f is closed with respect to pointwise convergence, and certainly $M_f \supset L$, so that $M_f = M$ by the minimal property of M. This means that if $f \in L$, $g \in M$ then $f+g \in M$. Now, for any g in M,

$$M_g = \{f \in M : f+g \in M\}$$

contains L and is closed with respect to pointwise convergence; thus $M_g = M$. In other words, if $f, g \in M$, then $f+g \in M$.

In almost exactly the same way we may prove that

$$\{f \in M : cf \in M\}, \quad \{f \in M : |f| \in M\}$$

$(c \in \mathbf{R})$ are both equal to M, and so M is a linear lattice.

If L is the linear lattice of step functions on $\mathbf{R}^k$ or the linear lattice of continuous functions on $\mathbf{R}^k$ of compact support, it is easy to check that the class M provided by Proposition 1 coincides with the class of Baire functions on $\mathbf{R}^k$ (Ex. 1). Quite generally, we may say that M is the class of *Baire functions relative to the given linear lattice L.*

The Baire functions were first mentioned in § 6.1 when we introduced the Lebesgue measurable functions on $\mathbf{R}^k$. This was to reinforce the claim that the Lebesgue measurable functions form a very comprehensive class. For an arbitrary topological space X let $\mathscr{B}(X)$ denote the

σ-algebra of Borel sets in X; then the *Borel measurable functions* on X are simply the $\mathscr{B}(X)$-measurable functions in the sense of § 12.1, i.e. the functions $f: X \to \mathbf{R}$ for which all sets

$$A_c = \{x \in X : f(x) \geqslant c\}$$

($c \in \mathbf{R}$) belong to $\mathscr{B}(X)$. According to this definition, every continuous function f on X is Borel measurable (because all the corresponding sets A_c are closed). Thus the Borel measurable functions form a linear lattice which contains all continuous functions and is closed with respect to pointwise convergence of sequences (Theorem 12.1.1). Under the rather stringent assumptions of § 15.1 about the topological space X, we may now identify the Baire functions relative to $C(X)$ (or $C_c(X)$) with the Borel measurable functions on X.

Theorem 1. *Let X be a σ-compact, locally compact metric space. Then the Borel measurable functions on X form the smallest class of functions on X which contains the continuous functions and is closed with respect to pointwise convergence of sequences.*

As a particular case of this theorem we see that the Baire functions on $\mathbf{R}^k$ coincide with the Borel measurable functions.

Proof. Let M be the 'smallest class' mentioned in the statement of the theorem. According to Proposition 1, M is a linear lattice; thus

$$\mathscr{M} = \{A : \chi_A \in M\}$$

is a ring of subsets of X (cf. the proof of Proposition 11.2.2). As M is closed with respect to (increasing) pointwise convergence, $\mathscr{M}$ is in fact a σ-ring. If K is any compact subset of X, then by Proposition 10.1.1 there is a continuous function ψ of compact support for which $\psi^n \downarrow \chi_K$. This implies that $\chi_K \in M$ and so $K \in \mathscr{M}$. As X is σ-compact there is an increasing sequence $\{X_n\}$ of compact subsets of X whose union is X. This ensures that $X \in \mathscr{M}$, i.e. $\mathscr{M}$ is a σ-algebra; more generally, if F is an arbitrary closed set in X then $F \cap X_n$ is compact and $F \cap X_n \uparrow F$ so that $F \in \mathscr{M}$. Thus $\mathscr{M}$ contains the whole σ-algebra $\mathscr{B}(X)$ of Borel sets. It follows that M contains all simple Borel functions (i.e. functions in $L(\mathscr{B}(X))$) and hence contains all Borel measurable functions by Proposition 12.1.1. The reverse inclusion follows from the general remarks before the statement of Theorem 1, and this completes the proof.

Exercise

1. Find the linear lattice M of Proposition 1 in the cases where (i) L consists of the step functions on $\mathbf{R}^k$, (ii) $L = C_c(\mathbf{R}^k)$.

15.6 Baire measures

Up to this point almost all topological theorems of any substance have been limited to metric spaces. In this section we shall merely assume that X *is a locally compact Hausdorff space*. First of all we prove a variation of Urysohn's Lemma which is adapted to this kind of topological space.

Theorem 1. *Let X be a locally compact Hausdorff space, A and B non-empty disjoint subsets of X, A closed and B compact. Then there is a continuous function $\psi: X \to [0,1]$ (of compact support) such that $\psi(x) = 0$ for all x in A and $\psi(x) = 1$ for all x in B.*

In preparation for the proof of Theorem 1 we shall make more explicit the separation properties of the space X.

Lemma 1. *Let X be a Hausdorff space, K a compact subset and $p \in K^c$. Then there exist disjoint open subsets G, H such that $p \in G$ and $K \subset H$.*

Proof. To any point x of K there exist disjoint open sets A_x, B_x such that $p \in A_x$, $x \in B_x$. From the covering $\{B_x\}$ of K there is a finite sub-covering $B_{x_1}, \ldots, B_{x_n}$ and the sets

$$G = \bigcap_{i=1}^{n} A_{x_i}, \quad H = \bigcup_{i=1}^{n} B_{x_i},$$

satisfy the required conditions.

Lemma 2. *Let X be a locally compact Hausdorff space, K a compact subset, U an open subset and $K \subset U$. Then there exists an open subset V with compact closure $\overline{V}$ such that*

$$K \subset V \subset \overline{V} \subset U.$$

Proof. Let G be the open set with compact closure $\overline{G}$ provided by Lemma 10.1.1. If $U = X$ we simply take $V = G$. In general G is too large: the open set $G \cap U$ is a more promising candidate for V as its closure is a subset of $\overline{G}$ and so is compact, but its closure may still contain points outside U.

This is where we use the Hausdorff property in the form of Lemma 1. We may as well assume that the complement F of U is not empty. To any point p of F there are disjoint open sets G_p, H_p such that $p \in G_p$, $K \subset H_p$. As $F \cap \overline{G}$ is compact there are points $p_1, \ldots, p_n$ in F such that

$G_{p_1}, \ldots, G_{p_n}$ cover $F \cap \bar{G}$. We now verify at once that the open set

$$V = G \cap H_{p_1} \cap \ldots \cap H_{p_n}$$

satisfies the conditions of the lemma.

To motivate the proof of Theorem 1 consider the following simple picture. There is an island consisting of a single mountain with a flat top. On a map (Fig. 20) the sea surrounding the island is denoted by A and the flat top is denoted by B. The height of B above sea level is, say, 800 feet and there are contour lines at intervals of 100 feet above sea level.

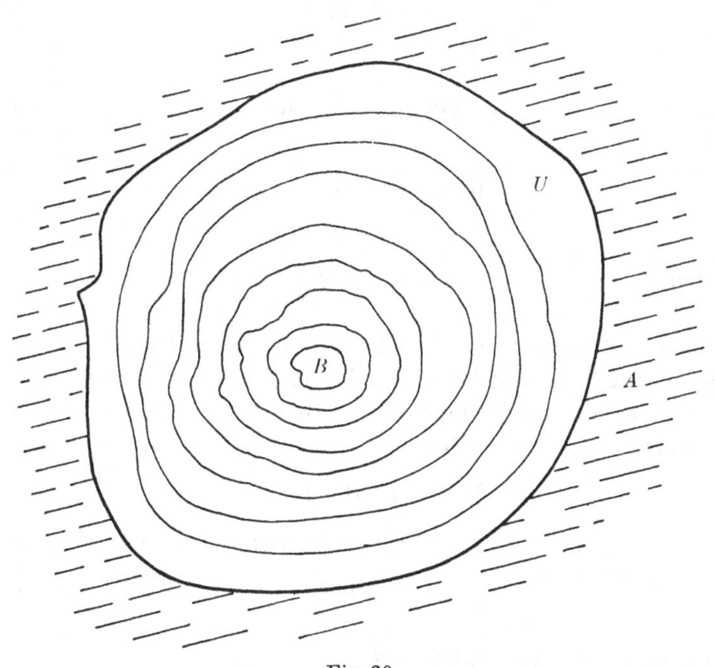

Fig. 20

To make a three-dimensional model of the island we may cut out of thick cardboard an outline U of the shore and an outline V_r of each contour, then paste these together in the appropriate positions indicated by the map. As there are only 8 contours, this of course gives a terraced effect to the model. Intuitively it is clear that we should obtain a 'smoother' model if there were more contours – say 32 contours at 25 feet intervals – with suitably thinner cardboard. A 'smooth' three-dimensional model may be regarded as the graph of a continuous height function which takes the minimum value 0 on A and the maximum value 800 on B.

Proof of Theorem 1. Let U be the complement of A. According to Lemma 2 there is an open set $V_{\frac{1}{2}}$ with compact closure such that

$$B \subset V_{\frac{1}{2}} \subset \overline{V}_{\frac{1}{2}} \subset U;$$

and then there are open sets $V_{\frac{1}{4}}$, $V_{\frac{3}{4}}$, with compact closures such that

$$B \subset V_{\frac{3}{4}} \subset \overline{V}_{\frac{3}{4}} \subset V_{\frac{1}{2}} \subset \overline{V}_{\frac{1}{2}} \subset V_{\frac{1}{4}} \subset \overline{V}_{\frac{1}{4}} \subset U.$$

Continuing in this way we obtain open sets V_r for each dyadic rational number $r = p/2^m$ in $(0,1)$ such that

$$B \subset V_r \subset \overline{V}_r \subset U$$

and

$$\overline{V}_r \subset V_s \quad \text{for} \quad r > s. \tag{1}$$

Now we must construct a continuous function $\psi: X \to [0,1]$. To this end define, for each $r = p/2^m$ in $(0,1)$

$$\psi_r(x) = r \quad \text{if} \quad x \in V_r,$$
$$= 0 \quad \text{otherwise;}$$

and then

$$\psi = \sup_r \psi_r.$$

It follows at once that $0 \leqslant \psi \leqslant 1$, that $\psi = 0$ on A and $\psi = 1$ on B. It also follows by Exx. 9.1.3, 4 that ψ_r and ψ are lower semicontinuous. To prove that ψ is continuous we introduce the upper semicontinuous functions θ_r and θ defined by

$$\theta_r(x) = 1 \quad \text{if} \quad x \in \overline{V}_r,$$
$$= r \quad \text{otherwise;}$$

and

$$\theta = \inf_r \theta_r.$$

It is sufficient to show that $\psi = \theta$.

We can only have $\psi_r(x) > \theta_s(x)$ if $r > s$, $x \in V_r$ and $x \notin \overline{V}_s$. But this is impossible by (1), whence $\psi_r \leqslant \theta_s$ for all r, s and so $\psi \leqslant \theta$.

On the other hand suppose that $\psi(x) < \theta(x)$. Then there are dyadic rationals r, s in $(0,1)$ such that

$$\psi(x) < r < s < \theta(x).$$

As $\psi(x) < r$ we have $x \notin V_r$ and as $\theta(x) > s$ we have $x \in \overline{V}_s$, which again contradicts (1). Thus $\psi \geqslant \theta$. Combining these inequalities gives $\psi = \theta$ and establishes the continuity of ψ.

The proof may be modified very simply to ensure that ψ has compact support. We only need to find an open set V with compact closure satisfying $K \subset V \subset \bar{V} \subset U$ and to replace U by V at the beginning of the proof.

Using Theorem 1, and virtually the same proof as before, Proposition 10.1.1 now takes the form:

Proposition 1. *Let X be a locally compact Hausdorff space and K a compact subset of X. Then there exists a continuous function $\psi: X \to [0, 1]$ of compact support such that $\psi(x) = 1$ for all x in K.*

We can no longer deduce from this that $\psi^n \downarrow \chi_K$: in fact $\psi^n \downarrow \chi_Q$, where

$$Q = \{x \in X: \psi(x) = 1\}$$

and, of course, $Q \supset K$. This set Q is by no means arbitrary. First of all, Q is a closed subset of the support of ψ and as such is compact. Moreover, Q may be expressed as the intersection of the open sets

$$\{x \in X: \psi(x) > 1 - 1/n\}$$

for $n = 1, 2, \ldots$, and so is a G_δ (as defined at the end of § 15.1).

Proposition 2. *Let X be a locally compact Hausdorff space and Q a compact subset of X. A necessary and sufficient condition for the existence of a continuous function $\psi: X \to [0, 1]$ of compact support with*

$$Q = \{x \in X: \psi(x) = 1\}$$

is that Q should be a G_δ.

Proof. One half has already been proved. Suppose that Q is the intersection of the sequence $\{G_n\}$ of open sets. There is an open set G containing Q with compact closure $\bar{G}$ (Lemma 10.1.1). By replacing each G_n by $G_n \cap G$, if necessary, we may arrange that every $\bar{G}_n$ is contained in the compact set $\bar{G}$. For each n there is a continuous function $\psi_n: X \to [0, 1]$ such that $\psi_n(x) = 1$ if $x \in Q$ and $\psi_n(x) = 0$ if $x \notin G_n$. Thus ψ_n vanishes outside the compact set $\bar{G}$. Let

$$\psi(x) = \Sigma \psi_n(x)/2^n.$$

This series is dominated by $\Sigma \frac{1}{2}^n$ and so is absolutely and uniformly convergent by Weierstrass' M-test. Thus ψ is continuous and has compact support (contained in $\bar{G}$). Clearly $0 \leqslant \psi \leqslant 1$ and $\psi(x) = 1$ for all x in Q. Finally, if $x \notin Q$ then $x \notin G_N$ for some N so that $\psi_N(x) = 0$ and $\psi(x) \leqslant 1 - \frac{1}{2}^N < 1$.

Let $\mathscr{A}(X)$ denote the smallest σ-algebra for which all functions ϕ of $C_c(X)$ are measurable. Then $\mathscr{A}(X)$ is generated by the sets

$$(x \in X: \phi(x) \geqslant c\}$$

($\phi \in C_c(X)$, $c \in \mathbf{R}$). We shall refer to the elements of $\mathscr{A}(X)$ as the *Baire sets* in X.

Theorem 2. *Let X be a locally compact Hausdorff space. The σ-algebra of Baire sets in X is generated by the compact G_δ's.*

Proof. By Proposition 2 any compact G_δ is a Baire set. Let $\mathscr{A}'$ be the σ-algebra generated by the compact G_δ's. Each set

$$A_c = \{x \in X: \phi(x) \geqslant c\}$$

is closed and is the intersection of the open sets

$$\{x \in X: \phi(x) > c + 1/n\}.$$

Moreover, if $c > 0$, A_c is contained in the support of ϕ and hence is compact. Thus $A_c \in \mathscr{A}'$ for all $c > 0$ (and all ϕ). Also

$$B_c = \{x \in X: \phi(x) > c\} = \bigcup_n A_{c+1/n} \in \mathscr{A}'$$

for $c > 0$. If we now replace ϕ by $-\phi$ and take complements in X we see that $A_c \in \mathscr{A}'$ for all $c < 0$. Finally

$$A_0 = \bigcap_n A_{-1/n}.$$

Thus $\mathscr{A}'$ contains all A_c and so contains all Baire sets.

Let $\mathscr{A}(X)$ denote the σ-algebra of Baire sets in X; then the *Baire measurable functions* on X are the $\mathscr{A}(X)$-measurable functions.

Theorem 3. *Let X be a locally compact Hausdorff space. Then the Baire measurable functions on X form the smallest linear lattice M of functions on X which contains the continuous functions of compact support and the constant functions, and is closed with respect to pointwise convergence of sequences.*

In other words, the Baire measurable functions on X are the Baire functions relative to the smallest linear lattice L containing 1 *and* $C_c(X)$.

Proof. (Cf. Theorem 15.5.1.) The linear lattice M is the intersection of *all* linear lattices which contain 1 and $C_c(X)$ and are closed with respect to pointwise convergence of sequences. First of all, the Baire measurable functions form a linear lattice containing M (Theorem 12.1.1).

Consider the collection of sets

$$\mathscr{M} = \{A: \chi_A \in M\}.$$

As M is a linear lattice closed with respect to pointwise convergence of sequences, $\mathscr{M}$ is a σ-ring. As M contains the constant function 1, $\mathscr{M}$ is a σ-algebra. If Q is a compact G_δ, then by Proposition 2 there is a function ψ in $C_c(X)$ such that $\psi^n \downarrow \chi_Q$. Thus $\chi_Q \in M$ and $Q \in \mathcal{M}$. It now follows from Theorem 2 that $\mathcal{M}$ contains all Baire sets. Thus M contains all simple Baire functions (the elements of $L(\mathcal{A}(X))$) and hence contains all Baire measurable functions by Proposition 12.1.1. This completes the proof.

These two theorems help to compare the Baire sets and the Baire measurable functions with the rather more comprehensive Borel sets and Borel measurable functions. $\mathcal{A}(X)$ is the σ-algebra of Baire sets on X; any measure on $\mathcal{A}(X)$ which is finite on compact Baire sets is called a *Baire measure* on X. The promised general form of the Riesz Representation Theorem now reads:

Theorem 4. *Let X be a locally compact Hausdorff space and $C_c(X)$ the linear space of continuous real valued functions on X of compact support. Any positive linear functional P on $C_c(X)$ extends to a Daniell integral $\int$ on a linear space L^1 containing $C_c(X)$. Thus*

$$P(\phi) = \int \phi$$

for all ϕ in $C_c(X)$. The corresponding measure μ on $\mathcal{A}(X)$ defined by

$$\begin{aligned} \mu(A) &= \int \chi_A \quad \textit{if} \quad \chi_A \in L^1, \\ &= \infty \quad \textit{otherwise}, \end{aligned}$$

is a Baire measure.

Proof. This is essentially the same as the proof already given for Theorem 10.1.4, due allowance being made for the more general form of Urysohn's Lemma, and the compact set K of the original proof being replaced by a compact G_δ.

If we are prepared to assume in Theorem 4 that X is *σ-compact*, then the Baire measure μ is uniquely determined by P (using Theorem 13.4.2 with $L = C_c(X)$); also, in this case the open sets

$$\{x \in X : \theta_n(x) > c\}$$

are Baire sets and so the proof of Theorem 15.1.3 shows that the Baire measure μ is *regular*.

To any regular Baire measure μ on X there corresponds a unique regular Borel measure μ' extending μ which may be constructed as follows:

for each open set G, $\mu'(G) = \sup\{\mu(Q) : Q \subset G, Q \text{ a compact } G_\delta\}$;

for each Borel set E, $\mu'(E) = \inf\{\mu'(G) : E \subset G, G \text{ an open set}\}$.

The verification that the function μ' so defined is indeed a regular Borel measure may be found in Royden [18] who gives a fairly condensed proof. For a more detailed treatment of the extension of Baire measures to Borel measures, the classical books of Halmos [11] and Berberian [3] are strongly recommended. It should be mentioned that for Halmos and Berberian the Borel sets are the elements of the *σ-ring* generated by the *compact* sets and the Baire sets are the elements of the *σ-ring* generated by the compact G_δ's. This means in particular that any Baire or Borel set in the sense of Halmos is contained in a σ-compact subset of X, and any Baire measure is regular (Ex. 11). Further, if we use Halmos's definition of a measurable function (given on p. 102) then the Baire measurable functions form the smallest class containing $C_c(X)$ and closed with respect to pointwise convergence of sequences (the constant function 1 may fail to be Baire measurable: cf. Theorem 3).

An alternative method of obtaining Borel measures is to modify the Daniell condition by expressing it in terms of 'directed limits' rather than sequential limits. For this approach see Stone's Note IV [21] and his references there to Bourbaki.

Exercises

1. A Hausdorff space X is said to be *normal* if to any disjoint closed subsets A, B of X there are disjoint open subsets V, W such that $A \subset V$, $B \subset W$.

Adapt the proof of Theorem 1 to prove the general form of ***Urysohn's Lemma.*** *A Hausdorff space X is normal if and only if to each pair of disjoint closed subsets A, B there is a continuous function $f: X \to [0,1]$ such that $f(x) = 0$ for all x in A and $f(x) = 1$ for all x in B.*

2. Show that a compact Hausdorff space is normal (Ex.1).

3. Let X be a locally compact Hausdorff space and ω a point not in X; write $X^* = X \cup \{\omega\}$. We shall say that a subset of X^* is open if it is either an open subset of X or the complement in X^* of a compact subset of X. Show that these open sets form a topology $\mathscr{T}^*$ on X^* (Appendix to Volume 1) and that $(X^*, \mathscr{T}^*)$ is a compact Hausdorff space – the so-called *one-point compactification* of X.

4. Use Exx. 1, 2, 3 to give another proof of Theorem 1. (In this context what becomes of the island and the 'infinite' sea surrounding it?)

5. Let X be a topological space and $\{f_n\}$ a sequence of continuous real valued functions on X such that $f_n \downarrow \chi_A$. Show that A is a closed G_δ.

6. Let the compact set Q be the intersection of a sequence $\{G_n\}$ of open sets. According to Theorem 1 there are continuous functions $\psi_n: X \to [0,1]$

of compact support such that $\psi_n = 1$ on Q and $\psi_n = 0$ outside G_n. If $\phi_n = \min\{\psi_1, \ldots, \psi_n\}$ show that $\phi_n \in C_c(X)$ and $\phi_n \downarrow \chi_Q$.

7. (The Baire Sandwich Theorem.) Let X be a locally compact Hausdorff space, K a compact subset, U an open subset and $K \subset U$. Then there exist Baire sets V, Q such that V is a σ-compact open set, Q is a compact G_δ and

$$K \subset V \subset Q \subset U.$$

8. A topological space is *separable* if there is a sequence $\{G_n\}$ of open subsets with the property that to each x in X and each open set U containing x there is an integer N such that $x \in G_N \subset U$. (Briefly, X has a countable base.) Prove that in a separable Hausdorff space every compact subset is a G_δ.

9. Let X be an uncountable set with the discrete topology (i.e. every subset of X is open) and let X^* be the one-point compactification of X (Ex. 3).

(i) What are $C_c(X)$ and $C(X^*)$?

(ii) What are the Baire and Borel sets in X?

(iii) What are the Baire and Borel sets in X^*?

(iv) Show that X^* has a compact subset which is not a Baire set.

10. Let X be a locally compact Hausdorff space and let $\mathscr{S}$ be the σ-ring generated by the compact G_δ's in X. (According to Halmos the elements of $\mathscr{S}$ are the Baire sets in X.) For any $f: X \to \mathbf{R}$ let

$$N(f) = \{x \in X: f(x) \neq 0\}, \quad A_c = \{x \in X: f(x) \geqslant c\}$$

($c \in \mathbf{R}$); then f is *measurable* with respect to the σ-ring $\mathscr{S}$ if and only if $N(f) \cap A_c \in \mathscr{S}$ for every c in $\mathbf{R}$. Show that this agrees with our definition of measurability if $\mathscr{S}$ is a σ-algebra (cf. the discussion of measurability on p. 102).

Adapt the proof of Theorem 3 to show that the class of $\mathscr{S}$-measurable functions is the smallest class of functions on X which contains $C_c(X)$ and is closed with respect to pointwise convergence of sequences.

11. Let $\mathscr{S}$ be the σ-ring of Ex. 10 and let μ be a Baire measure on X. Show that the restriction ν of μ to $\mathscr{S}$ is regular. If μ is the Baire measure of Theorem 4 show that ν is the one and only measure on $\mathscr{S}$ satisfying $P(\phi) = \int \phi \, d\nu$ for all ϕ in $C_c(X)$.

Give an example where there is more than one Baire measure μ on $\mathscr{A}$ satisfying $P(\phi) = \int \phi \, d\mu$ for all ϕ in $C_c(X)$.

12. Let X be a locally compact Hausdorff space. The purpose of this exercise is to prove that a compact Baire set in X is a G_δ (Halmos [11] §51).

(i) Let $\mathscr{A}$ be the σ-algebra generated by an arbitrary collection $\mathscr{C}$ of subsets of X; then $\mathscr{A}$ is the union of the σ-algebras generated by the countable subsets of $\mathscr{C}$.

(ii) Let K be a compact Baire set in X. According to (i) there is a sequence $\{Q_n\}$ of compact G_δ's such that K is in the σ-algebra generated by $Q_1, Q_2, \ldots$.

For each n there is a continuous function $\psi_n: X \to [0, 1]$ (of compact support) such that $Q_n = \{x \in X: \psi_n(x) = 1\}$. If $x, y \in X$ define the 'distance'

$$d(x,y) = \Sigma\,|\psi_n(x) - \psi_n(y)|/2^n.$$

(iii) Use (the pseudo-metric) d to define a metric on the quotient space with respect to the equivalence relation $\sim$ defined by

$$x \sim y \Leftrightarrow d(x,y) = 0.$$

(iv) Any closed subset of a metric space is a G_δ. Deduce that K is a G_δ.

16

REAL AND COMPLEX MEASURES

We have already met real and complex measures in the context of Riesz's Representation Theorem (Chapter 10). They arose there quite naturally as real or complex linear combinations of finite measures. Let $\mathscr{C}$ be a non-empty collection of subsets of X. A σ-additive function $\mu: \mathscr{C} \to \mathbf{R}$ will be called a *real measure* on $\mathscr{C}$, and a σ-additive function $\mu: \mathscr{C} \to \mathbf{C}$ will be called a *complex measure* on $\mathscr{C}$. As we have only allowed these measures to take finite values, the real measures on $\mathscr{C}$ form a linear space over $\mathbf{R}$, and the complex measures on $\mathscr{C}$ form a linear space over $\mathbf{C}$. It is convenient also to use the term *positive measure* for a finite measure as previously defined, viz. a σ-additive function $\mu: \mathscr{C} \to [0, \infty)$.

In § 16.1 we prove the Jordan Decomposition Theorem which shows that a real measure μ on a σ-ring $\mathscr{S}$ is the difference of two positive measures and so is a *signed measure* as defined in § 10.2 (this fails to be true, in general, if $\mathscr{S}$ is replaced by a ring $\mathscr{R}$). The real measure μ has unique positive and negative parts μ^+, μ^- and an absolute value $|\mu| = \mu^+ + \mu^-$, where $|\mu|$ is the smallest positive measure satisfying

$$|\mu(E)| \leqslant |\mu|(E)$$

for all E in $\mathscr{S}$. More generally, each complex measure λ on a σ-ring $\mathscr{S}$ possesses a unique absolute value $|\lambda|$ which is the smallest positive measure satisfying

$$|\lambda(E)| \leqslant |\lambda|(E)$$

for all E in $\mathscr{S}$ (Theorem 16.2.1).

In the last section we give alternative proofs of some of these results in terms of linear functionals.

16.1 Real measures

We begin where we broke off at the end of § 11.3. Let $\mathscr{C}$ be a non-empty collection of subsets of X satisfying the Disjointness Condition and let μ be an additive real valued function on $\mathscr{C}$. According to Theorem 11.3.2, μ extends uniquely to an additive function (again denoted by μ) on the ring $\mathscr{R}$ generated by $\mathscr{C}$. From the equation $\varnothing \cup \varnothing = \varnothing$ it follows that $\mu(\varnothing) + \mu(\varnothing) = \mu(\varnothing)$ and so $\mu(\varnothing) = 0$.

If $\mu = \mu_1 - \mu_2$, where μ_1, μ_2 are positive additive functions on $\mathscr{R}$, then μ is *relatively bounded* in the sense that

$$\{\mu(A) : A \in \mathscr{R}, A \subset E\}$$

is bounded for each set E in $\mathscr{R}$. This follows from the inequalities

$$\mu(A) \leqslant \mu_1(A) \leqslant \mu_1(E), \quad \mu(A) \geqslant -\mu_2(A) \geqslant -\mu_2(E).$$

The converse of this simple statement is already quite significant.

Proposition 1. *Let μ be an additive real valued function on a ring $\mathscr{R}$. If μ is relatively bounded, then μ may be expressed as*

$$\mu = \mu^+ - \mu^-,$$

where μ^+, μ^- are the positive additive functions on $\mathscr{R}$ defined by

$$\mu^+(E) = \sup\{\mu(A) : A \in \mathscr{R}, A \subset E\}, \quad (1)$$

$$\mu^-(E) = \sup\{-\mu(A) : A \in \mathscr{R}, A \subset E\}. \quad (2)$$

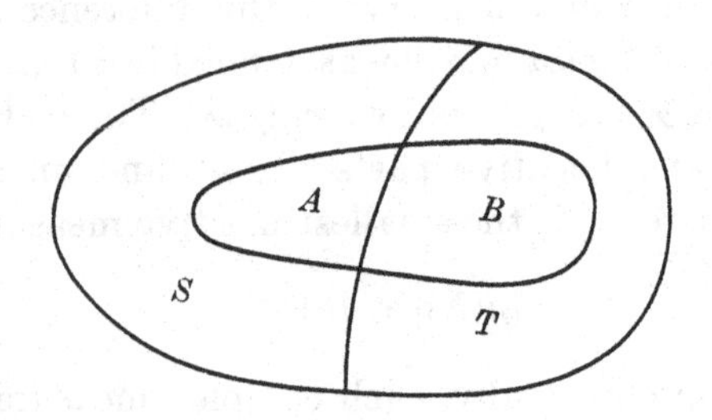

Fig. 21

Proof. First of all, define μ^+ by the given formula (1). As $\varnothing \in \mathscr{R}$, μ^+ is clearly positive. We shall show that μ^+ is additive. Let S, T be disjoint elements of $\mathscr{R}$; then any subset C of $S \cup T$ is uniquely of the form $C = A \cup B$ where $A \subset S$, $B \subset T$ (viz. $A = C \cap S$, $B = C \cap T$), and conversely, any pair of such subsets A, B gives a subset $C = A \cup B$ of $S \cup T$ (Fig. 21). Moreover, C belongs to $\mathscr{R}$ if and only if A and B both belong to $\mathscr{R}$. The equation

$$\mu(C) = \mu(A) + \mu(B)$$

therefore gives $\quad \mu^+(S \cup T) = \mu^+(S) + \mu^+(T)$
(cf. Ex. 1.2.2).

If we define μ^- by the equation

$$\mu^- = \mu^+ - \mu,$$

then
$$\mu^-(E) = \sup\{\mu(A) - \mu(E) : A \in \mathscr{R}, A \subset E\}$$
$$= \sup\{-\mu(B) : B \in \mathscr{R}, B \subset E\},$$
where we have written $B = E \setminus A$ and note that $B \in \mathscr{R}$ and $B \subset E$ if and only if $A \in \mathscr{R}$ and $A \subset E$. This completes the proof.

The pair of positive additive functions μ^+, μ^- is a 'best possible' solution of the equation
$$\mu = \mu^+ - \mu^-$$
in the following sense. If
$$\mu = \mu_1 - \mu_2,$$
where μ_1, μ_2 are also positive additive functions, then
$$\mu^+(E) \leqslant \mu_1(E)$$
and
$$\mu^-(E) \leqslant \mu_2(E)$$
for all E in $\mathscr{R}$. We express this succinctly by writing
$$\mu^+ \leqslant \mu_1, \quad \mu^- \leqslant \mu_2.$$
In this sense, then, μ^+ and μ^- are the unique smallest positive additive functions satisfying $\mu = \mu^+ - \mu^-$. These inequalities follow at once from the definition of μ^+, μ^- for, as we saw above,
$$\mu(A) \leqslant \mu_1(E)$$
and
$$-\mu(A) \leqslant \mu_2(E)$$
for all $A \subset E$ $(A \in \mathscr{R})$.

Now let
$$|\mu| = \mu^+ + \mu^-.$$
It follows from (1) and (2) that
$$|\mu|(E) = \sup\{\mu(A) - \mu(B) : A, B \in \mathscr{R}, A, B \subset E\}. \tag{3}$$
If $D = A \cap B$ then
$$\mu(A) - \mu(B) = \mu(A \setminus D) - \mu(B \setminus D)$$
and so we may as well assume in (3) that A, B are disjoint.

Let ν be any positive additive function on $\mathscr{R}$ which *dominates* μ, i.e.
$$|\mu(E)| \leqslant \nu(E)$$
for all E in $\mathscr{R}$. Then, for any disjoint A, B contained in E $(A, B \in \mathscr{R})$, it follows that
$$\mu(A) - \mu(B) \leqslant \nu(A) + \nu(B) \leqslant \nu(E).$$
Thus
$$|\mu|(E) \leqslant \nu(E)$$
for all E in $\mathscr{R}$. On the other hand
$$|\mu(E)| = |\mu^+(E) - \mu^-(E)| \leqslant \mu^+(E) + \mu^-(E) = |\mu|(E).$$

This proves that $|\mu|$ *is the unique smallest positive additive function on* $\mathscr{R}$ *which dominates* μ.

If $X \in \mathscr{R}$, i.e. if $\mathscr{R}$ is an *algebra*, then relative boundedness is equivalent to boundedness: we need only take $E = X$ in the definition of relative boundedness. Any positive additive function on a ring $\mathscr{R}$ is relatively bounded, but certainly need not be bounded: Lebesgue measure on the ring of elementary figures in $\mathbf{R}^k$ is an obvious example. At first sight the following result may be little surprising.

Lemma 1. *Let* $\mathscr{S}$ *be a* σ*-ring. Then any positive additive function* $\mu: \mathscr{S} \to \mathbf{R}$ *is bounded.*

Proof. Assume, on the contrary, that μ is unbounded. We may therefore find an increasing sequence $\{A_n\}$ of sets in $\mathscr{S}$ for which $\mu(A_n) \to \infty$. But the union A of the sets A_n also belongs to $\mathscr{S}$ and so

$$\mu(A_n) \leqslant \mu(A),$$

which provides a contradiction.

Now suppose that μ is a *real measure* on $\mathscr{C}$, i.e. μ is a σ-additive function on $\mathscr{C}$ which takes its values in $\mathbf{R}$. To say that μ is σ*-additive* means that μ is additive† and

$$\mu(\bigcup A_n) = \Sigma \mu(A_n) \tag{4}$$

for any sequence $\{A_n\}$ of disjoint sets A_n in $\mathscr{C}$, whose union is also in $\mathscr{C}$. As the union $\bigcup A_n$ does not depend on the order of the terms A_n, equation (4) shows that the sum on the right hand side is independent of the order of the terms; by a well-known elementary theorem of Riemann this implies that the series $\Sigma \mu(A_n)$ is absolutely convergent (Ex. 5).

Once again we assume that $\mathscr{C}$ satisfies the Disjointness Condition. The proof of Proposition 11.4.1 shows that the real measure μ on $\mathscr{C}$ may be extended uniquely to a real measure on the ring $\mathscr{R}$ generated by $\mathscr{C}$. The only change in the proof is in the last sentence, where the gathering of terms is rather simpler in the present case as there are no infinite values to contend with. From now on we shall therefore assume that any real measure is defined on a ring. As we shall see this is not a sufficiently sweeping assumption – it is much more natural to consider the real measures which are defined on σ-rings.

Lemma 2. *Let* μ *be a real measure on a* σ*-ring* $\mathscr{S}$. *Then* μ *is relatively bounded.*

† If we assume that the empty set belongs to $\mathscr{C}$ then we may deduce the additivity of μ from equation (4) by taking $A_{m+1}, A_{m+2}, \ldots$ equal to the empty set.

Proof. For any E in $\mathscr{S}$, denote by μ_E the *contraction* of μ by E, viz. the real measure on $\mathscr{S}$ defined by

$$\mu_E(A) = \mu(A \cap E)$$

for all A in $\mathscr{S}$. The sets $A \cap \mathrm{E}$ give all the elements of $\mathscr{S}$ which are contained in E; thus μ is relatively bounded if and only if μ_E is bounded for all E in $\mathscr{S}$.

Assume that μ_E is unbounded for some E in $\mathscr{S}$. We shall construct a decreasing sequence $\{A_n\}$ of subsets of E such that $A_n \in \mathscr{S}$, μ_{A_n} is unbounded, and $|\mu(A_n)| \geqslant n-1$. To begin the construction let $A_1 = E$. Suppose that $A_1, \ldots, A_n$ have been found to satisfy these conditions. As μ_{A_n} is unbounded, there is a set B in $\mathscr{S}$, contained in A_n, such that

$$|\mu(B)| \geqslant |\mu(A_n)| + n.$$

If μ_B is unbounded, take $A_{n+1} = B$ and note that $|\mu(B)| \geqslant n$. If μ_B is bounded, consider $C = A_n \setminus B$: as

$$\mu_{A_n} = \mu_B + \mu_C,$$

it follows that μ_C is unbounded. Also

$$|\mu(C)| = |\mu(A_n) - \mu(B)| \geqslant |\mu(B)| - |\mu(A_n)| \geqslant n$$

by our choice of B. In this case we take $A_{n+1} = C$. The sequence $\{A_n\}$ may now be constructed by induction.

The intersection of this decreasing sequence $\{A_n\}$ is a set A belonging to the σ-ring $\mathscr{S}$. It follows from the σ-additivity of μ that

$$\mu(A_n) \to \mu(A),$$

which is contrary to the inequality

$$|\mu(A_n)| \geqslant n-1.$$

Thus μ_E is bounded for all E in $\mathscr{S}$, and this establishes the result.

Theorem 1. *Let μ be a real measure on a σ-ring $\mathscr{S}$. Then μ may be expressed as*

$$\mu = \mu^+ - \mu^-, \tag{5}$$

where μ^+, μ^- are positive measures on $\mathscr{S}$ and

$$|\mu| = \mu^+ + \mu^- \tag{6}$$

is the smallest positive measure dominating μ. The positive measures μ^+, μ^- are uniquely determined by these conditions.

This expression of μ as the difference $\mu^+ - \mu^-$ is usually called the *Jordan Decomposition* of μ. The positive measures μ^+, μ^-, $|\mu|$ are often called the *positive, negative* and *total* variations of μ.

Proof. By Lemma 2, μ is relatively bounded. Hence, by Proposition 1, $\mu = \mu^+ - \mu^-$, where μ^+, μ^- are given by (1), (2). As we saw above, $|\mu|$ is the smallest positive additive function on $\mathscr{S}$ dominating μ. This determines $|\mu|$ uniquely, and

$$\mu^+ = \tfrac{1}{2}(|\mu| + \mu) \quad \text{and} \quad \mu^- = \tfrac{1}{2}(|\mu| - \mu).$$

It only remains to prove that μ^+, μ^- and $|\mu|$ are σ-additive. In the light of (5) and (6), it is enough to show that μ^+ is σ-additive.

Suppose that $E_n \uparrow E$ (all sets in $\mathscr{S}$). As μ^+ is positive and additive,

$$\mu^+(E_n) \leqslant \mu^+(E);$$

thus the increasing sequence $\{\mu^+(E_n)\}$ converges to a limit l, say, where

$$l \leqslant \mu^+(E).$$

For any subset A of E ($A \in \mathscr{S}$),

$$\mu(A \cap E_n) \leqslant \mu^+(A \cap E_n) \leqslant \mu^+(E_n);$$

also, from the σ-additivity of μ,

$$\mu(A \cap E_n) \to \mu(A);$$

hence

$$\mu(A) \leqslant l.$$

By (1), as A ranges over the subsets of E, this gives

$$\mu^+(E) \leqslant l.$$

It follows that $\mu^+(E) = l$, i.e.

$$\mu^+(E_n) \to \mu^+(E).$$

Combined with the additivity of μ^+, this shows that μ^+ is σ-additive (Proposition 11.4.2). Incidentally, by Lemma 1, μ is actually bounded!

It is of interest to show that the conclusion of Theorem 1 is not true, in general, for a real measure μ on a ring $\mathscr{R}$. If we examine the proof of Lemma 2 it suggests that we find a ring $\mathscr{R}$ in which there is a decreasing sequence $\{A_n\}$ whose intersection A does not lie in $\mathscr{R}$. With this in mind, let $\{A_n\}$ be a strictly decreasing sequence of sets (in almost any universal set you care to mention) whose intersection A is not empty, and let $\mathscr{R}$ be the ring generated by the sets A_n. There is an additive function $\mu: \mathscr{R} \to \mathbf{R}$ which satisfies $\mu(A_n) = n$. The only kind of decreasing sequence of elements of $\mathscr{R}$ which has empty intersection is one in which all but a finite number of terms are empty; thus σ-additivity means no more than additivity in this case. Now $\mathscr{R}$ is an *algebra* of subsets of A_1, and μ is an *unbounded* real measure on $\mathscr{R}$: therefore μ

cannot be expressed as a difference of positive measures. (More details are worked out in Ex. 2.)

The following example is a particularly important source of real measures. Let $\int$ be a Daniell integral on a space L^1 of integrable functions and $\mathscr{M}$ the corresponding σ-algebra of measurable sets. If h is any element of L^1 then the function ν defined by

$$\nu(E) = \int_E h \tag{7}$$

for all E in $\mathscr{M}$, is a real measure on $\mathscr{M}$. Recall that

$$\int_E h = \int h\chi_E.$$

The additivity of ν follows immediately from the definition (7) and then the σ-additivity of ν is a simple application of the Dominated Convergence Theorem: for if $E_n \uparrow E$, then

$$h\chi_{E_n} \to h\chi_E,$$

and $h\chi_{E_n}$ is dominated by $|h|$. This real measure ν is called the *indefinite integral* of h: it is the analogue in the present general situation of the function F defined by

$$F(x) = \int_a^x f,$$

in the case of the Lebesgue integral on $\mathbf{R}$ – or, rather, of the Lebesgue–Stieltjes measure that F defines (cf. § 3.4 and Theorem 9.1.1). The function h is also reminiscent of the density functions studied in § 6.4 and § 15.3.

As the positive functions $|h|$, h^+, h^- all belong to L^1, their indefinite integrals are positive measures. In fact, as we shall prove, ν is relatively bounded, and

$$|\nu|(E) = \int_E |h|,$$

$$\nu^+(E) = \int_E h^+,$$

$$\nu^-(E) = \int_E h^-,$$

for all E in $\mathscr{M}$. It is only necessary to establish one of these equations as the other two will follow from the relations

$$h = h^+ - h^-, \quad |h| = h^+ + h^-.$$

For the moment let ν_1, ν_2 denote the indefinite integrals of h^+, h^-, respectively. As

$$\nu = \nu_1 - \nu_2,$$

ν is relatively bounded, and we certainly have

$$\nu^+ \leqslant \nu_1, \quad \nu^- \leqslant \nu_2.$$

Now let
$$S = \{x \in X: h(x) \geqslant 0\},$$
$$T = \{x \in X: h(x) < 0\};$$

then S, T are disjoint elements of $\mathscr{M}$ and

$$\int_E h^+ = \int_{E \cap S} h,$$

i.e.
$$\nu_1(E) = \nu(E \cap S)$$

for all E in $\mathscr{M}$. Thus
$$\nu_1(E) \leqslant \nu^+(E)$$

by the definition of ν^+. This identifies ν_1 with ν^+. As

$$\nu_1 - \nu_2 = \nu^+ - \nu^-,$$

it also follows that $\nu_2 = \nu^-$.

The positive measures ν^+, ν^- are *contractions* ν_S, $-\nu_T$, viz.

$$\nu^+(E) = \nu(E \cap S), \quad \nu^-(E) = -\nu(E \cap T),$$

for all E in $\mathscr{M}$. As S, T are disjoint this is equivalent to saying that ν^+, ν^- are *concentrated* on S, T, respectively, i.e.

$$\nu^+(E) = \nu^+(E \cap S), \quad \nu^-(E) = \nu^-(E \cap T),$$

for all E in $\mathscr{M}$ (see Ex. 3). Any two real measures which are concentrated on disjoint sets in this way are said to be *mutually singular*.

We shall now see that this property of the indefinite integrals ν^+, ν^- is true in a very general setting. (An attractive alternative proof of Theorem 2 will be available in the final chapter as the Radon–Nikodym Theorem shows that *any* real measure on a σ-algebra is an indefinite integral.)

Theorem 2. *Let μ be a real measure on a σ-algebra $\mathscr{A}$ of subsets of X. Then there exist elements S, T of $\mathscr{A}$ such that*

$$S \cup T = X, \quad S \cap T = \varnothing$$

and μ^+, μ^- are concentrated on S, T, respectively.

Proof. By the definition (1) of μ^+ we may find a sequence $\{A_n\}$ of sets in $\mathscr{A}$ which satisfy

$$\mu(A_n) > \mu^+(X) - \tfrac{1}{2}^n \tag{8}$$

for all n. Now define

$$S = \bigcap_m \Big(\bigcup_{n>m} A_n \Big)$$

(which is the exact counterpart of $\limsup a_n$ for sequences of real numbers: see Ex. 4) and let $T = X \setminus S$. It is clear that $S, T \in \mathscr{A}$. We shall show that

$$\mu^-(S) = 0, \quad \mu^+(T) = 0,$$

from which the theorem follows at once.

From (8) we deduce that

$$\mu^+(A_n) - \mu^-(A_n) > \mu^+(A_n) - \tfrac{1}{2}^n,$$

whence

$$\mu^-(A_n) < \tfrac{1}{2}^n.$$

The σ-additivity of μ^- then gives

$$\mu^-\Big(\bigcup_{n>m} A_n \Big) < \sum_{n>m} \tfrac{1}{2}^n = \tfrac{1}{2}^m,$$

from which

$$\mu^-(S) < \tfrac{1}{2}^m$$

for all $m \geqslant 1$. Thus

$$\mu^-(S) = 0.$$

Similarly,

$$\mu^+(X \setminus A_n) = \mu^+(X) - \mu^+(A_n) \leqslant \mu^+(X) - \mu(A_n) < \tfrac{1}{2}^n.$$

But

$$X \setminus S = \bigcup_m \Big(\bigcap_{n>m} (X \setminus A_n) \Big),$$

where

$$\mu^+\Big(\bigcap_{n>m} (X \setminus A_n) \Big) < \tfrac{1}{2}^n$$

for each $n > m$. Hence

$$\mu^+\Big(\bigcap_{n>m} (X \setminus A_n) \Big) = 0$$

for all $m \geqslant 1$, and

$$\mu^+(X \setminus S) = 0.$$

The decomposition of X described in Theorem 2 is usually attributed to Hahn. In general it is not uniquely determined by μ: in the case of the indefinite integral of h, we took

$$S = \{x \in X : h(x) \geqslant 0\}, \quad T = \{x \in X : h(x) < 0\};$$

the set

$$\{x \in X : h(x) = 0\}$$

may be transferred from S to T without altering their essential properties.

Exercises

1. Let $\mathscr{C}$ consist of the non-empty subintervals of the compact interval $[a, b]$. For any continuous function $F: [a, b] \to \mathbf{R}$, define μ_F on $\mathscr{C}$ by

$$\mu_F(I) = F(d) - F(c),$$

where c, d are, respectively, the left hand and right hand end points of I. Extend μ_F to the ring $\mathscr{R}$ generated by $\mathscr{C}$ (using Theorem 11.3.2). Show that μ_F is bounded on $\mathscr{R}$ if and only if F has bounded variation on $[a, b]$. Identify the set functions μ_F^+, μ_F^-, $|\mu_F|$ in this case.

2. Let $\{A_n\}$ be a strictly decreasing sequence of sets whose intersection A is not empty. Describe the ring $\mathscr{R}$ generated by the sets A_n. Show that there is an additive function $\mu: \mathscr{R} \to \mathbf{R}$ such that $\mu(A_n) = n$. By referring to Proposition 11.4.2 (iii) show that μ is a real measure. Show that μ cannot be expressed as the difference of two positive measures.

3. Let μ, ν be real measures on a ring $\mathscr{R}$ which are concentrated respectively on the sets S, T of $\mathscr{R}$, i.e.

$$\mu(E) = \mu(E \cap S), \quad \nu(E) = \nu(E \cap T)$$

for all E in $\mathscr{R}$. Let $\lambda = \mu + \nu$. If S, T are disjoint show that $\mu = \lambda_S$ and $\nu = \lambda_T$.

4. Let $\{A_n\}$ be a sequence of subsets of X and let

$$S = \bigcap_m \Big(\bigcup_{n > m} A_n \Big).$$

Show that $\chi_S = \limsup \chi_{A_n}$. Find T so that $\chi_T = \liminf \chi_{A_n}$.

5. For any permutation τ of the set $\{1, 2, 3, \ldots\}$, the series (of real numbers) $\Sigma a_{\tau(n)}$ is convergent. Show that Σa_n is absolutely convergent.

6. (Cf. Ex. 11.4.11.) Let X be an infinite set and $\mathscr{P}$ the collection of all subsets of X. Let $\{x_n\}$ be a sequence of distinct points of X and Σp_n an absolutely convergent series of real numbers. For any set A in $\mathscr{P}$ let

$$\mu(A) = \sum_{x_n \in A} p_n,$$

where the sum extends over all integers n for which $x_n \in A$. Show that μ is a real measure on $\mathscr{P}$. Would it be sufficient to assume that Σp_n is convergent?

7. Let μ be a real measure on $\mathscr{C}$. A set E in $\mathscr{C}$ is an *atom* of μ if $\mu(E) \neq 0$ and, for any A in $\mathscr{C}$ contained in E, either $\mu(A) = 0$ or $\mu(A) = \mu(E)$. What are the atoms of the real measure μ in Ex. 6?

8. Let μ be a real measure on a σ-algebra of subsets of X and suppose that μ has no atoms (Ex. 7). Show that μ takes every real value between $-\mu^-(X)$ and $\mu^+(X)$. (See Ex. 11.4.19.)

9. Let $\mathscr{A}$ be a σ-algebra of subsets of X. Show that the real measures on $\mathscr{A}$ form a linear space M over $\mathbf{R}$ and that M is a Banach space (i.e. a complete normed linear space as in §7.7, p. 221) with respect to the norm defined by

$$\|\mu\| = |\mu|\,(X).$$

16.2 Complex measures

A *complex measure* on a ring $\mathscr{R}$ is a σ-additive function

$$\lambda : \mathscr{R} \to \mathbf{C};$$

such a complex measure λ may be written uniquely in terms of its 'real' and 'imaginary' parts:

$$\lambda = \mu + i\nu,$$

where μ, ν are real measures on $\mathscr{R}$. Many results for λ may be deduced quite simply from the corresponding results for μ, ν. For example, if $\mathscr{R}$ is a σ-ring, then, by Theorem 16.1.1 $|\mu|$, $|\nu|$ are positive measures on $\mathscr{R}$ and λ is *dominated* by the positive measure $|\mu| + |\nu|$, i.e.

$$|\lambda(E)| \leqslant (|\mu| + |\nu|)(E)$$

for all E in $\mathscr{R}$; and in view of Lemma 16.1.1, λ is actually bounded. The main purpose of this section is to show that there is a smallest positive measure $|\lambda|$ dominating λ. It is of interest to give a proof that uses only Lemma 2 of the previous section, as this now provides an alternative proof of Theorem 16.1.1 in terms of the measure $|\mu|$, rather than the positive and negative parts μ^+, μ^-.

Theorem 1. *Let λ be a complex measure on a σ-ring $\mathscr{S}$. Then there exists a unique smallest positive measure $|\lambda|$ which dominates λ, defined by*

$$|\lambda|(E) = \sup\left\{\sum_{j=1}^{m} |\lambda(A_j)| : \textit{disjoint } A_1, \ldots, A_m \in \mathscr{S} \textit{ contained in } E\right\} \tag{1}$$

for all E in $\mathscr{S}$.

Proof. The sum $\qquad |\lambda(A_1)| + \ldots + |\lambda(A_m)|$

which appears in (1) is not greater than

$$|\mu(A_1)| + \ldots + |\mu(A_m)| + |\nu(A_1)| + \ldots + |\nu(A_m)|.$$

But the terms in this sum may be gathered according to the signs of $\mu(A_i)$, $\nu(A_i)$ so that it equals

$$\mu(A) - \mu(B) + \nu(C) - \nu(D),$$

where $A, B, C, D \in \mathscr{S}$ are subsets of E. The relative boundedness of μ, ν (Lemma 16.1.2) now shows that the supremum in (1) exists (i.e. is finite) and so defines $|\lambda|$ on $\mathscr{S}$.

If we take $m = 1$, $A_1 = E$ in (1) it follows that

$$|\lambda(E)| \leqslant |\lambda|(E).$$

Thus $|\lambda|$ dominates λ. On the other hand, if τ is any positive measure dominating λ (as far as we know at this stage there may be none), then

$$|\lambda(A_1)| + \ldots + |\lambda(A_m)| \leqslant \tau(A_1) + \ldots + \tau(A_m) \leqslant \tau(E)$$

and so

$$|\lambda|(E) \leqslant \tau(E).$$

The proof now consists in showing that $|\lambda|$ is σ-additive. Suppose that E is the union of the sequence $E_1, E_2, \ldots$ of disjoint sets in $\mathscr{S}$. We have to prove that

$$|\lambda|(E) = \sum_{i=1}^{\infty} |\lambda|(E_i). \tag{2}$$

If $A_1, \ldots, A_m \in \mathscr{S}$ are disjoint subsets of E, then, for each j, A_j is the union of the disjoint sets $E_i \cap A_j$ ($i = 1, 2, \ldots$). By the σ-additivity of λ,

$$\lambda(A_j) = \sum_i \lambda(E_i \cap A_j)$$

and so

$$|\lambda(A_j)| \leqslant \sum_i |\lambda(E_i \cap A_j)|.$$

For each i, the disjoint sets $E_i \cap A_j$ ($j = 1, \ldots, m$) are contained in E_i; thus

$$\sum_j |\lambda(E_i \cap A_j)| \leqslant |\lambda|(E_i).$$

By the elementary Proposition 1.1.1, extended to m sequences,

$$\sum_j |\lambda(A_j)| \leqslant \sum_i |\lambda|(E_i).$$

The definition of $|\lambda|(E)$ as a supremum now gives

$$|\lambda|(E) \leqslant \sum_i |\lambda|(E_i). \tag{3}$$

On the other hand, to any $\epsilon > 0$, we can find, for each i, a finite collection of disjoint sets A_{ij} in $\mathscr{S}$ contained in E_i, for which

$$|\lambda|(E_i) - \epsilon/2^i < \sum_j |\lambda(A_{ij})|.$$

The sets A_{ij} may be written as an array

$$\begin{matrix} A_{11} & A_{12} & \ldots \\ A_{21} & A_{22} & \ldots \\ \ldots & & \end{matrix}$$

in which every row is finite. As the disjoint sets A_{ij} are all contained in E, it follows that

$$\sum_{i=1}^{n} |\lambda|(E_i) - \epsilon < |\lambda|(E)$$

for any $n \geqslant 1$. Thus $\displaystyle\sum_{i=1}^{\infty} |\lambda|(E_i) - \epsilon \leqslant |\lambda|(E).$

As ϵ is arbitrarily small we deduce that

$$\sum_{i} |\lambda|(E_i) \leqslant |\lambda|(E). \tag{4}$$

Finally, (3) and (4) combine to give (2) and this completes the proof.

Suppose now that $\int$ is a Daniell integral on L^1 with σ-algebra $\mathscr{M}$. We may define a *complex Daniell integral* in the obvious way: if $f, g \in L^1$ and $h = f + ig$, then

$$\int h = \int f + i \int g. \tag{5}$$

Denote by $L_{\mathbf{C}}^1$ the collection of all functions $h = f + ig$, where $f, g \in L^1$, and make $L_{\mathbf{C}}^1$ into a linear space over $\mathbf{C}$ by the obvious multiplication by complex scalars:

$$(a+ib)(f+ig) = af - bg + i(ag + bf).$$

Lemma 1. *$\int$ is a linear functional on $L_{\mathbf{C}}^1$.*

Proof. The linearity over $\mathbf{R}$ follows immediately from the definition (5) of $\int h$ in terms of the real and imaginary parts f, g. It is also a trivial matter to check from (5) that

$$\int ih = i \int h,$$

and so the lemma follows.

As we should expect from the real case, the *indefinite integral* of h yields a complex measure λ on $\mathscr{M}$, viz.

$$\lambda(E) = \int_E h = \int_E f + i \int_E g \tag{6}$$

for any E in $\mathscr{M}$. The positive function $|h| = (f^2 + g^2)^{\frac{1}{2}}$ is integrable (Ex. 6.1.5) and it turns out, happily, that $|\lambda|$ is the indefinite integral of $|h|$. To prove this we need a simple result which is probably familiar from the theory of complex functions.

Lemma 2. $|\int h| \leqslant \int |h|$.

Proof. Let $$\int h = r(\cos\theta + i\sin\theta),$$
where $r \geqslant 0$, $0 \leqslant \theta < 2\pi$. Then

$$\int (\cos\theta - i\sin\theta)\, h = (\cos\theta - i\sin\theta)\int h = r,$$

by virtue of Lemma 1. As r is real, the integral of the imaginary part of $(\cos\theta - i\sin\theta)\, h$ must be zero. Thus r equals the integral of the real part, and as such is not greater than

$$\int |(\cos\theta - i\sin\theta)\, h| = \int |h|.$$

In other words, $$\left|\int h\right| \leqslant \int |h|.$$

Proposition 1. *Let $h \in L_{\mathbf{C}}^1$ and define λ by the equation*

$$\lambda(E) = \int_E h$$

for all E in $\mathscr{M}$. Then $|\lambda|$ is the indefinite integral of $|h|$, i.e.

$$|\lambda|\,(E) = \int_E |h|$$

for all E in $\mathscr{M}$.

Proof. Let τ denote the indefinite integral of $|h|$, i.e.

$$\tau(E) = \int_E |h|$$

for all E in $\mathscr{M}$. The inequality of Lemma 2 now gives

$$|\lambda(E)| \leqslant \tau(E).$$

Thus $$|\lambda| \leqslant \tau$$

as $|\lambda|$ is the smallest positive measure dominating λ.

On the other hand, as $f, g \in L^1$, there exist sequences $\{\xi_n\}$, $\{\eta_n\}$ of $\mathscr{M}$-simple functions converging to f, g (everywhere). If $\zeta_n = \xi_n + i\eta_n$ then ζ_n is a simple complex valued function. Let

$$\operatorname{sgn}\zeta_n(x) = \frac{\zeta_n(x)}{|\zeta_n(x)|}$$

for all x in X, with the usual understanding that the quotient is zero if the denominator vanishes. Then

$$\epsilon_n = \overline{\operatorname{sgn}}\,\zeta_n$$

(complex conjugate) is a simple function whose non-zero values all have modulus 1, and

$$\epsilon_n \zeta_n = |\zeta_n|.$$

The integral

$$\int_E \epsilon_n h$$

may therefore be written as a finite sum

$$a_1 \lambda(A_1) + \ldots + a_m \lambda(A_m),$$

where $A_1, \ldots, A_m \subset E$ are disjoint subsets of $\mathscr{M}$ and each $|a_i| = 1$. Thus

$$\left| \int_E \epsilon_n h \right| \leqslant |\lambda| (E).$$

But

$$\epsilon_n \zeta_n \to |h|$$

and

$$\epsilon_n(\zeta_n - h) \to 0,$$

so that

$$\epsilon_n h \to |h|.$$

The Dominated Convergence Theorem therefore gives

$$\int_E |h| \leqslant |\lambda| (E)$$

for all E in $\mathscr{M}$, i.e.

$$\tau \leqslant |\lambda|,$$

and this completes the proof that $\tau = |\lambda|$.

Exercises

1. Under the conditions of Theorem 1 show that

$$|\lambda| (E) = \sup \sum_{n=1}^{\infty} |\lambda(A_n)|,$$

where the supremum is taken over all sequences $\{A_n\}$ of disjoint sets in $\mathscr{S}$ contained in E. Will this be altered if we also insist that

$$\bigcup_{n=1}^{\infty} A_n = E?$$

2. Prove that if μ, ν are real measures on a σ-ring $\mathscr{S}$ and $\lambda = \mu + i\nu$, $\sigma = |\mu| + i|\nu|$, then $|\lambda| = |\sigma|$.

3. (Cf. Ex. 16.1.9.) Let $\mathscr{A}$ be a σ-algebra of subsets of X. Show that the complex measures on $\mathscr{A}$ form a linear space $M_{\mathbf{C}}$ over $\mathbf{C}$ and that

$$\|\lambda\| = |\lambda| (X)$$

defines a norm on $M_{\mathbf{C}}$.

Does

$$\|\lambda\|' = (|\mu| + |\nu|)(X)$$

define a norm on $M_{\mathbf{C}}$?

Show that $M_{\mathbf{C}}$ is a complex Banach space with respect to the norm $\| \;\|$.

16.3 Real and complex linear functionals

The previous two sections are closely related to the linear functionals of § 10.2 and § 10.3. For example, suppose that μ is an additive real valued function on the ring $\mathscr{R}$ and define a linear functional F on $L(\mathscr{R})$, as in Theorem 11.3.1, which satisfies $F(\chi_A) = \mu(A)$ for all A in $\mathscr{R}$. Then the relative boundedness of the set function μ as defined in § 16.1 is equivalent to the relative boundedness of the linear functional F as defined in § 10.2.

Proposition 1. *The linear functional F is relatively bounded if and only if the additive function μ is relatively bounded.*

Proof. (i) If F is relatively bounded

$$\{F(\phi): \phi \in L(\mathscr{R}), |\phi| \leqslant \psi\}$$

is bounded for each positive ψ in $L(\mathscr{R})$. Set $\phi = \chi_A$, $\psi = \chi_E$ and deduce that

$$\{\mu(A): A \in \mathscr{R}, A \subset E\}$$

is bounded for all E in $\mathscr{R}$.

(ii) Conversely, suppose that μ is relatively bounded. Let

$$\psi = e_1\chi_{E_1} + \ldots + e_r\chi_{E_r},$$

where $E_1, \ldots, E_r$ are disjoint and $e_1, \ldots, e_r$ are positive; let

$$E = E_1 \cup \ldots \cup E_r,$$

$$e = \max\{e_1, \ldots, e_r\}.$$

If

$$\phi = a_1\chi_{A_1} + \ldots + a_m\chi_{A_m}$$

with disjoint $A_1, \ldots, A_m$ satisfies $|\phi| \leqslant \psi$, then $A_i \subset E$ and $|a_i| \leqslant e$ for $i = 1, \ldots, m$. Thus

$$|a_1\mu(A_1) + \ldots + a_m\mu(A_m)| \leqslant e(|\mu(A_1)| + \ldots + |\mu(A_m)|).$$

We may gather the $\mu(A_i)$ according to their signs and write this inequality as

$$|F(\phi)| \leqslant e(\mu(A) - \mu(B)),$$

where A, B are (disjoint) elements of $\mathscr{R}$ contained in E. It follows that F is relatively bounded.

In view of this result we may deduce Proposition 16.1.1 from Proposition 10.2.1 and the formulae

$$F^+(\psi) = \sup\{F(\phi): 0 \leqslant \phi \leqslant \psi\}, \tag{1}$$

$$F^-(\psi) = \sup\{-F(\phi): 0 \leqslant \phi \leqslant \psi\}, \tag{2}$$

numbered (3), (4) in § 10.2.

It seems likely that we can also deduce the Jordan Decomposition (Theorem 16.1.1) in a similar way. Let us examine this question in detail. First of all we prove a simple result that goes back to Daniell's original paper [7].

Proposition 2. *Let L be a linear lattice of functions $\phi: X \to \mathbf{R}$ and F a relatively bounded linear functional on L satisfying the Daniell condition*

$$\phi_n \downarrow 0 \Rightarrow F(\phi_n) \to 0.$$

(Note that the sequence $\{F(\phi_n)\}$ will no longer be monotone, in general.) Then the positive linear functionals $|F|$, F^+, F^- all satisfy the Daniell condition, and so are elementary integrals on L.

Proof. Suppose that $\psi_n \downarrow 0$; we want to show that $F^+(\psi_n) \to 0$ from which the rest follows easily using the relations $F = F^+ - F^-$, $|F| = F^+ + F^-$. For any $\epsilon > 0$, we define a sequence $\{\phi_n\}$ so that $0 \leqslant \phi_n \leqslant \psi_n$ and

$$F(\phi_n) > F^+(\psi_n) - \epsilon/2^n$$

for $n \geqslant 1$. This uses equation (1). Unfortunately, the sequence $\{\phi_n\}$ may not be decreasing. Let

$$\theta_n = \min\{\phi_1, \ldots, \phi_n\}.$$

Then $0 \leqslant \theta_n \leqslant \phi_n \leqslant \psi_n$ and $\theta_n \downarrow 0$. We may verify by a simple induction argument, given in Lemma 1 below, that

$$F(\theta_n) > F^+(\psi_n) - \epsilon$$

for $n \geqslant 1$. By the Daniell condition for F there is an integer N such that

$$F(\theta_n) < \epsilon$$

for $n \geqslant N$. Finally

$$F^+(\psi_n) < 2\epsilon$$

for $n \geqslant N$, which shows that $F^+(\psi_n) \to 0$.

Lemma 1. $F(\theta_n) > F^+(\psi_n) - \epsilon(\frac{1}{2} + \ldots + \frac{1}{2}^n)$.

Proof. This inequality is true for $n = 1$ by our choice of $\theta_1 = \phi_1$. Assume that it holds for a particular value of $n \geqslant 1$ and argue inductively. By our construction

$$\theta_{n+1} = \min\{\theta_n, \phi_{n+1}\}:$$

by considering the two cases $\theta_{n+1}(x) = \theta_n(x)$, $\theta_{n+1}(x) = \phi_{n+1}(x)$, we may verify the inequality

$$0 \leqslant \phi_{n+1}(x) + \theta_n(x) - \theta_{n+1}(x) \leqslant \psi_n(x)$$

for all x. Using equation (1) we deduce that

$$F^+(\psi_n) \geqslant F(\phi_{n+1}) + F(\theta_n) - F(\theta_{n+1}).$$

This combines with the assumed inequality and the constructed inequality

$$F(\phi_{n+1}) > F^+(\psi_{n+1}) - \epsilon/2^{n+1}$$

to give

$$F(\theta_{n+1}) > F^+(\psi_{n+1}) - \epsilon(\tfrac{1}{2} + \ldots \tfrac{1}{2}^{n+1}),$$

and this completes the proof by induction.

According to Proposition 2, a relatively bounded linear functional F on L satisfying the Daniell condition may be expressed as the difference of two elementary integrals F^+, F^-. Moreover, there is a unique smallest elementary integral $|F| = T$ which *dominates* F in the sense that

$$|F(\phi)| \leqslant |F|\,(\phi)$$

for all positive ϕ in L.

By the standard Daniell construction, $|F|$ may be extended to a Daniell integral, again denoted by $|F|$, on a space $L^1(|F|)$ of integrable functions; and similarly for F^+, F^-. As

$$F^+(\phi) \leqslant |F|\,(\phi) \quad \text{and} \quad F^-(\phi) \leqslant |F|\,(\phi)$$

for all positive ϕ in L, it follows immediately from the definitions that a null set for $|F|$ is null for both F^+, F^- and then that $L^1(|F|)$ is contained in both $L^1(F^+)$, $L^1(F^-)$. In fact it is not very difficult to prove that

$$L^1(|F|) = L^1(F^+) \cap L^1(F^-)$$

(Ex. 1). In line with our previous use of signed Daniell integrals in § 10.2 we regard F^+, F^- as integrals on the space $L^1(|F|)$ and define

$$F = F^+ - F^-$$

on this space of integrable functions. (In the topological setting of Chapter 10 it was not necessary to assume any Daniell condition for F: this followed, not from any result like Proposition 2, but because we could apply Dini's Theorem to the positive linear functionals F^+, F^- as in Theorem 10.1.4.)

Now suppose that μ satisfies the conditions of Theorem 16.1.1, i.e. μ is a real measure on a σ-ring $\mathscr{S}$. According to Lemma 16.1.2, μ is relatively bounded and so the linear functional F of Proposition 1 is relatively bounded. To prove Theorem 16.1.1 it only remains to verify the Daniell condition for F, as the Daniell condition for $|F|$, F^+, F^- would then follow by Proposition 2 and the σ-additivity of $|\mu|$, μ^+, μ^- by Proposition 11.4.2. Unfortunately the Daniell condition for F does *not* follow from Proposition 11.4.2! This result was vitally dependent on the *positiveness* of μ, as demonstrated by Ex. 11.4.13. We shall not pursue this line of argument as the proof of the σ-additivity given for Theorem 16.1.1 is already easier than the proof of Proposition 2.

The use of linear functionals in the real case is a little disappointing. But in the complex case we may illustrate the methods of § 10.3 and give an alternative proof of the central Theorem 16.2.1.

Let λ be a complex measure on a σ-ring $\mathscr{S}$. There is a 'complexification' of $L_{\mathbf{R}} = L(\mathscr{S})$ which we denote by $L_{\mathbf{C}}$, viz. the linear space over $\mathbf{C}$ consisting of all functions

$$\zeta = a_1\chi_{A_1} + \dots + a_m\chi_{A_m},$$

where $A_1, \dots, A_m \in \mathscr{S}$ and $a_1, \dots, a_m \in \mathbf{C}$. Just as in § 10.3 there is a unique linear functional H on $L_{\mathbf{C}}$ such that

$$H(\chi_A) = \lambda(A)$$

for all A in $\mathscr{S}$. If $\lambda = \mu + i\nu$ and if F, G are the corresponding linear functionals on $L_{\mathbf{R}}$, then $H = F + iG$. Define

$$\|\zeta\| = \sup_{x \in X} |\zeta(x)|$$

as before. (In this case $\|\zeta\|$ is the largest of the finitely many values of $|\zeta|$.) It follows at once from the boundedness of μ, ν that F, G are bounded linear functionals on $L_{\mathbf{R}}$ and H is a bounded linear functional on $L_{\mathbf{C}}$. We now appeal to Proposition 10.3.1 which gives a smallest positive linear functional $|H|$ dominating H and satisfying

$$|H|(\psi) = \sup_{|\zeta| \leqslant \psi} |H(\zeta)|$$

for all positive ψ in $L_{\mathbf{R}}$. Admittedly, the spaces $L_{\mathbf{R}}$, $L_{\mathbf{C}}$ in Proposition 10.3.1 were different from our present ones, but the same proof applies. Even the property:

> *if ψ_1, ψ_2 are positive elements of $L_{\mathbf{R}}$, then any ζ in $L_{\mathbf{C}}$ which is dominated by $\psi_1 + \psi_2$ is the sum of two elements ζ_1, ζ_2 in $L_{\mathbf{C}}$ which are dominated by ψ_1, ψ_2, respectively,*

may be deduced by the same two arguments. The simpler of these is to set

$$\zeta_1 = \frac{\zeta\psi_1}{\psi_1 + \psi_2}, \quad \zeta_2 = \frac{\zeta\psi_2}{\psi_1 + \psi_2},$$

which are easily seen to be elements of $L_{\mathbf{C}}$.

Now if

$$\zeta = a_1\chi_{A_1} + \dots + a_m\chi_{A_m}$$

with disjoint $A_1, \ldots, A_m$ in $\mathscr{S}$ and $|\zeta| \leqslant \chi_E$ $(E \in \mathscr{S})$, then each $A_i \subset E$, and each $|a_i| \leqslant 1$. Hence

$$|a_1\lambda(A_1) + \ldots + a_m\lambda(A_m)| \leqslant |\lambda(A_1)| + \ldots + |\lambda(A_m)|.$$

Moreover, we can achieve equality here by choosing each a_i to be a suitable complex number of modulus 1 so that $a_i\lambda(A_i) = |\lambda(A_i)|$. It follows that

$$|H|(\chi_E) = \sup \Sigma |\lambda(A_i)|,$$

which identifies $|\lambda|(E)$ with $|H|(\chi_E)$.

We now appeal to Theorem 16.1.1. The Daniell condition for $|F|$, $|G|$ follows from the σ-additivity of $|\mu|$, $|\nu|$ and $|H|$ is dominated by $|F| + |G|$, so the Daniell condition holds for $|H|$; thus, finally, $|\lambda|$ is σ-additive and hence is a positive measure on $\mathscr{S}$. This completes the alternative proof of Theorem 16.2.1.

Exercises

1. In the notation of p. 202 show that

$$L^1(|F|) = L^1(F^+) \cap L^1(F^-).$$

[Cf. Exx. 8.1.F, G and Ex. 13.3.4.]

2. Let μ, ν be real measures on a σ-algebra $\mathscr{A}$ of subsets of X and let F, G be the unique linear functionals on $L(\mathscr{A})$ which satisfy $F(\chi_A) = \mu(A)$, $G(\chi_A) = \nu(A)$ for all A in $\mathscr{A}$. Show that μ, ν are mutually singular (as defined on p. 192) if and only if F, G are mutually singular (as defined in Ex. 10.2.4).

3. Deduce the Hahn Decomposition Theorem 16.1.2 from Theorem 16.1.1 and Ex. 2.

4. In the notation of p. 202 show that F^+, F^-, regarded as linear functionals on the space $L^1(|F|)$, are the positive and negative parts of F.

5. Let $L_{\mathbf{R}}$ and $L_{\mathbf{C}}$ be the linear spaces described in the above section. Use the two arguments given in the proof of Proposition 10.3.1 to show that if ψ_1, ψ_2 are positive elements of $L_{\mathbf{R}}$, then any ζ in $L_{\mathbf{C}}$ which satisfies

$$|\zeta| \leqslant \psi_1 + \psi_2$$

may be expressed as $\zeta = \zeta_1 + \zeta_2$ where $\zeta_1, \zeta_2 \in L_{\mathbf{C}}$ and $|\zeta_1| \leqslant \psi_1$, $|\zeta_2| \leqslant \psi_2$.

17

THE RADON–NIKODYM THEOREM

We have already referred to this famous theorem in several different contexts. At the end of Chapter 3 we discussed the functions H which satisfied

$$H(x) - H(a) = \int_a^x h \quad (x \in [a, b]),$$

for some h in $L^1[a, b]$. For the present purpose write $I = [a, x]$, $\nu(I) = H(x) - H(a)$ (as for the Lebesgue–Stieltjes measures of § 9.2), and let μ denote Lebesgue measure. Then the above relationship between H and h becomes

$$\nu(I) = \int_I h\,d\mu.$$

In § 6.4 and § 15.3 we introduced the idea of a *density function* h for a mapping T, defined by the condition

$$\mu(TI) = \int_I h\,d\mu,$$

where μ denotes Lebesgue measure on $\mathbf{R}^k$. Write $\nu(I) = \mu(TI)$. Then

$$\nu(I) = \int_I h\,d\mu.$$

Again, in § 16.1 we discussed the *indefinite integral* ν of h:

$$\nu(E) = \int_E h\,d\mu.$$

The Radon–Nikodym Theorem gives a strikingly simple condition for the existence of such 'density functions' h, viz. that

$$\mu(E) = 0 \Rightarrow \nu(E) = 0.$$

For ease of reference we prove in § 17.2 a measure theoretic form of the theorem as this is the one most commonly quoted in the literature. A more fundamental, though slightly harder, result will be proved independently in § 17.3. Von Neumann's beautiful proof of the Radon–Nikodym Theorem is based on Riesz's Theorem (7.4.2 or 17.1.4) which states that the space $\mathscr{L}^2$ is self-dual. For this reason we introduce the abstract spaces $\mathscr{L}^p$ in § 17.1 and return to the question of the dual spaces $\mathscr{L}^p$, $\mathscr{L}^q$ $(1/p + 1/q = 1)$ in § 17.4 as a final application of the Radon–Nikodym Theorem.

17.1 The spaces $\mathscr{L}^p$

Let P be a Daniell integral with corresponding space $L^1(P)$ of integrable functions $f\colon X \to \mathbf{R}$. For $p \geqslant 1$ we denote by $L^p(P)$ the set of all measurable functions f for which $|f|^p \in L^1(P)$. In this case 'measurable' means 'P-measurable' in the sense of Stone, i.e. f is measurable if and only if $\operatorname{mid}\{-g, f, g\} \in L^1(P)$ for every positive function g in $L^1(P)$.

Likewise, if μ is a measure on a σ-algebra $\mathscr{A}$ we may define integration with respect to μ either by the Daniell extension procedure of Chapter 11 or by the classical approach of Chapter 12. The space of integrable functions is denoted by $L^1(\mu)$, and $L^p(\mu)$ is the set of all measurable functions f for which $|f|^p \in L^1(\mu)$. In this case 'measurable' means '$\mathscr{A}$-measurable' as defined in § 12.1.

Throughout this chapter, L^p will be shorthand for either $L^p(P)$ or $L^p(\mu)$. The spaces L^p have almost all the properties that we proved in Chapter 7 for the spaces $L^p(\mathbf{R})$. It is convenient to repeat some of these results and to prove their extensions to the present abstract setting.

First of all, *L^p is a linear space over* $\mathbf{R}$. For, if $f, g \in L^p$ and $a, b \in \mathbf{R}$, then $af + bg$ is measurable and

$$|af+bg|^p \leqslant (|a|+|b|)^p \max\{|f|^p, |g|^p\},$$

whence $|af+bg|^p \in L^1$ and $af+bg \in L^p$.

If $f \in L^p$, let

$$\|f\| = \left\{\int |f|^p\right\}^{1/p}.$$

Then, for any f, g in L^p and a in $\mathbf{R}$,

N 1. $\|f\| \geqslant 0$,

N 2. $\|f\| = 0$ if and only if $f = 0$ almost everywhere,

N 3. $\|af\| = |a|\,\|f\|$,

N 4. $\|f+g\| \leqslant \|f\| + \|g\|$.

The first three of these rules follow immediately from the definition of $\|f\|$ and the fourth will be proved in Proposition 2 below.

The function $\|\ \|$ is not quite a norm, but rule N 2 suggests that we define a 'quotient space' $\mathscr{L}^p$ (i.e. $\mathscr{L}^p(P)$ or $\mathscr{L}^p(\mu)$ as the case may be) whose elements are the equivalence classes $\tilde{f}$ of the equivalence relation $\sim$ defined by

$$f \sim g \quad \text{if and only if} \quad f = g \quad \text{almost everywhere.}$$

In particular, the class $\tilde{0}$ consists of the *null functions*, viz. the functions f: $X \to \mathbf{R}$ which are zero almost everywhere. The definitions

$$\tilde{f}+\tilde{g} = (f+g)^{\sim},$$
$$a\tilde{f} = (af)^{\sim},$$
$$\|\tilde{f}\| = \|f\|,$$

are easily shown to be independent of the representatives f, g chosen for the equivalence classes $\tilde{f}$, $\tilde{g}$. Moreover $\|\tilde{f}\| = 0$ if and only if $\tilde{f} = \tilde{0}$. Thus $\mathscr{L}^p$ is a *normed linear space* over $\mathbf{R}$.

In the concrete case of Lebesgue measure on $\mathbf{R}$ the spaces L^p and $\mathscr{L}^p$ are usually 'identified' though this is not an entirely defensible action. In the abstract case, the space $\mathscr{L}^p$ may be drastically different from L^p – for example, if P is the Dirac δ-function defined in Example 3 of § 8.1, p. 3, then $L^p(P)$ consists of *all* functions f: $\mathbf{R}^k \to \mathbf{R}$, the null functions are those for which $f(0) = 0$ (a vast collection!) and $\mathscr{L}^p(P)$ is effectively the space of functions f: $\{0\} \to \mathbf{R}$. For the rest of this chapter we shall make the distinction between L^p and $\mathscr{L}^p$, and state our results with a precision that was occasionally lacking in Chapter 7. (It would be a good exercise, after reading the present section, to go quickly through §§ 7.2–7.5 and see whether or not our 'gentlemen's agreement' on p. 166 of Volume 1 was really justified!)

Suppose that $p, q > 1$ satisfy the symmetric relation

$$\frac{1}{p}+\frac{1}{q} = 1.$$

We may prove the elementary inequality

$$ab \leqslant \frac{a^p}{p}+\frac{b^q}{q} \tag{1}$$

for any $a, b \geqslant 0$. This is obviously true if $a, b = 0$ and may be established quite simply when $a, b > 0$ by considering the sign of $f'(x)$, where

$$f(x) = 1-c+cx-x^c$$

for $x > 0$ $(0 < c < 1)$ and substituting $c = 1/p$, $x = a^p/b^q$ (Ex. 7.2.2).

***Proposition 1* (Hölder's Inequality).** *Let $p, q > 1$ satisfy the relation $1/p+1/q = 1$. If $f \in L^p$ and $g \in L^q$ then $fg \in L^1$ and*

$$\int |fg| \leqslant \|f\|_p \, \|g\|_q.$$

(The suffixes refer to the spaces L^p, L^q, respectively.)

Proof. Hölder's Inequality is obviously true if f or g is null. By homogeneity we may therefore assume without loss of generality that

$$\|f\|_p = \|g\|_q = 1,$$

i.e.

$$\int |f|^p = \int |g|^q = 1.$$

By (1),

$$|fg| \leqslant \frac{|f|^p}{p} + \frac{|g|^q}{q}.$$

Thus the measurable function fg is integrable and

$$\int |fg| \leqslant \frac{1}{p} + \frac{1}{q} = 1.$$

Proposition 2 (Minkowski's Inequality). *If* $f, g \in L^p$ $(p \geqslant 1)$, *then*

$$\|f+g\| \leqslant \|f\| + \|g\|.$$

Proof. The case $p = 1$ is already known, so assume that $p > 1$. First of all,

$$|f+g|^p \leqslant |f|\,|f+g|^{p-1} + |g|\,|f+g|^{p-1}. \tag{2}$$

As $(p-1)q = p$ it follows that $|f+g|^{p-1} \in L^q$; write A for the 'norm' of $|f+g|^{p-1}$ in L^q, i.e.

$$A = \left\{\int |f+g|^p\right\}^{1/q}.$$

By Hölder's Inequality applied to (2) we have

$$\int |f+g|^p \leqslant (\|f\| + \|g\|)\,A. \tag{3}$$

If $A = 0$ then $\|f+g\| = 0$ and the result is obvious; if $A \neq 0$ then we divide both sides of (3) by A and use the relation $1 - 1/q = 1/p$ to deduce that

$$\left\{\int |f+g|^p\right\}^{1/p} \leqslant \|f\| + \|g\|.$$

Now suppose that we have a sequence $\{f_n\}$ whose terms f_n are in L^p and a function f in L^p. We shall say that $\{f_n\}$ *converges strongly* to f in L^p if $\|f_n - f\| \to 0$. In this case $\{f_n\}$ converges strongly to any other function g in L^p which is equal to f almost everywhere. Thus the strong limit of a sequence in L^p (if it exists at all) is only determined to within a null function. Here again the space $\mathscr{L}^p$ is more satisfactory: we say that $\{\tilde{f}_n\}$ *converges strongly* to $\tilde{f}$ in $\mathscr{L}^p$ if $\|\tilde{f}_n - \tilde{f}\| \to 0$. This is equivalent to saying that $\{f_n\}$ converges strongly to f in L^p, but the strong limit $\tilde{f}$ in $\mathscr{L}^p$ is now uniquely determined by the sequence $\{\tilde{f}_n\}$: for if

$\|\tilde{f}_n - \tilde{f}\| \to 0$ and $\|\tilde{f}_n - \tilde{g}\| \to 0$, it follows from Minkowski's Inequality that

$$\|\tilde{f} - \tilde{g}\| \leqslant \|\tilde{f}_n - \tilde{f}\| + \|\tilde{f}_n - \tilde{g}\| \to 0$$

and so $\|\tilde{f} - \tilde{g}\| = 0$.

Strong convergence does not necessarily imply pointwise convergence. For example, if P is the Dirac δ-function mentioned above, $\{f_n\}$ converges strongly to f in $L^p(P)$ if and only if $f_n(0) \to f(0)$, and this makes no mention of points in $\mathbf{R}^k$ other than the origin. (In this particular case, strong convergence is the same as pointwise convergence almost everywhere, but see Exx. 2, 3.)

Let (X, d) be a metric space (as defined in the Appendix to Volume 1) and $\{s_n\}$ a sequence whose terms s_n lie in X. We say that $\{s_n\}$ is a *Cauchy sequence* if

$$d(s_m, s_n) \to 0 \quad \text{as} \quad m, n \to \infty.$$

Then (X, d) is a *complete* metric space if, to any Cauchy sequence $\{s_n\}$ with terms in X, there is an element s in X such that $d(s_n, s) \to 0$. We are now able to establish the centrally important result that the space $\mathscr{L}^p (p \geqslant 1)$ is complete with respect to the metric d defined by

$$d(\tilde{f}, \tilde{g}) = \|\tilde{f} - \tilde{g}\|$$

(Theorem 2). Note that the space $\mathscr{L}^p(\mu)$ may be defined for a measure μ on a σ-algebra $\mathscr{A}$ by the classical approach of § 12.2 without the assumption that μ is a complete measure (cf. the remarks on p. 107). The completeness of $\mathscr{L}^p(\mu)$ as a metric space is not dependent on any such assumption about the measure μ.

As there is a notion of addition (of finitely many terms) in L^p and $\mathscr{L}^p$, we may make use of series and in particular of absolutely convergent series. Let $\{a_n\}$ be a sequence whose terms are in L^p and define the partial sums

$$s_n = a_1 + a_2 + \ldots + a_n$$

for $n \geqslant 1$. We say that Σa_n *converges strongly* to s in L^p if $s \in L^p$ and $\|s_n - s\| \to 0$. In the same circumstances, $\Sigma \tilde{a}_n$ *converges strongly* to $\tilde{s}$ in $\mathscr{L}^p$. We also say that Σa_n is *absolutely convergent* if $\Sigma \|a_n\|$ is convergent as a series of positive real numbers. The following result may be regarded as a generalisation of the Absolute Convergence Theorem 8.2.3 on p. 14.

Theorem 1. *Let $a_n \in L^p$ for $n = 1, 2, \ldots$. If Σa_n is absolutely convergent, then Σa_n converges strongly to a function s in L^p and $\Sigma a_n(x)$ converges to $s(x)$ for almost all x.*

Proof. Let $\Sigma \|a_n\| = M$ and let

$$g_n = |a_1| + \ldots + |a_n|$$

for $n \geqslant 1$. Then $\{g_n\}$ is an increasing sequence of positive functions in L^p and

$$\|g_n\| \leqslant \|a_1\| + \ldots + \|a_n\|$$

by the triangle inequality (extended to n terms). Thus

$$\|g_n\| \leqslant M,$$

i.e.

$$\int g_n{}^p \leqslant M^p.$$

By the Monotone Convergence Theorem the increasing sequence $\{g_n{}^p\}$ converges almost everywhere to a positive function h in L^1 and

$$\int h \leqslant M^p.$$

Thus $\{g_n\}$ converges almost everywhere to a positive measurable function g, where $g^p = h$, so that $g \in L^p$ and $\|g\| \leqslant M$.

By an elementary theorem on real series (Proposition 1.1.2), Σa_n converges almost everywhere to a function s where $|s| \leqslant g$. Also s is measurable, by Theorem 8.4.1 (iii), whence $s \in L^p$ and $\|s\| \leqslant M$. In terms of the partial sums

$$s_n = a_1 + \ldots + a_n$$

this means that $s_n - s \to 0$ almost everywhere. But

$$|s_n - s| \leqslant 2g$$

and so, by the Dominated Convergence Theorem,

$$\int |s_n - s|^p \to 0.$$

Thus $\|s_n - s\| \to 0$, i.e. Σa_n converges strongly to s in L^p.

As a direct consequence of Theorem 1 we have our main result.

Theorem 2. *The metric space $\mathscr{L}^p$ $(p \geqslant 1)$ is complete.*

Proof. (Cf. Theorem 7.2.2.) Let $\{\tilde{f}_n\}$ be a Cauchy sequence in $\mathscr{L}^p$, i.e.

$$\|f_m - f_n\| \to 0 \quad \text{as} \quad m, n \to \infty.$$

There is an integer $N(1) > 0$ such that

$$\|f_n - f_{N(1)}\| < 2^{-1} \quad \text{for} \quad n \geqslant N(1);$$

then there is an integer $N(2) > N(1)$ such that

$$\|f_n - f_{N(2)}\| < 2^{-2} \quad \text{for} \quad n \geqslant N(2),$$

and so on. By induction we find a subsequence $\{f_{N(k)}\}$ of $\{f_n\}$ such that

$$\|f_n - f_{N(k)}\| < 2^{-k} \quad \text{for} \quad n \geq N(k)$$

$(k = 1, 2, \ldots)$. Then the series

$$\|f_{N(1)}\| + \|f_{N(2)} - f_{N(1)}\| + \|f_{N(3)} - f_{N(2)}\| + \ldots$$

is convergent by the Comparison Test. Now, by Theorem 1, the series

$$f_{N(1)} + (f_{N(2)} - f_{N(1)}) + (f_{N(3)} - f_{N(2)}) + \ldots$$

converges strongly to a function f in L^p. In other words,

$$\|f_{N(k)} - f\| \to 0 \quad \text{as} \quad k \to \infty.$$

Given any $\epsilon > 0$, we find k so that

$$\|f_{N(k)} - f\| < \epsilon$$

and also

$$2^{-k} < \epsilon.$$

Then

$$\|f_n - f\| \leq \|f_n - f_{N(k)}\| + \|f_{N(k)} - f\| < 2\epsilon$$

for $n \geq N(k)$, and so $\{f_n\}$ converges strongly to f in L^p. In other words $\{\tilde{f}_n\}$ converges strongly to $\tilde{f}$ in $\mathscr{L}^p$, as required.

It is interesting to note the pattern of this proof: nearly all of the working is done in L^p though the result is interpreted in the quotient space $\mathscr{L}^p$.

Among the spaces $\mathscr{L}^p$ $(p \geq 1)$, $\mathscr{L}^2$ is particularly well behaved. From one point of view this is because there is a scalar product on $\mathscr{L}^2$. According to Hölder's Inequality (with $p = q = 2$), if $f, g \in L^2$, then $fg \in L^1$ and

$$\int |fg| \leq \|f\| \, \|g\|.$$

We may therefore define the *scalar product* (or *inner product*)

$$(\tilde{f}, \tilde{g}) = \int fg$$

of any two functions $\tilde{f}, \tilde{g}$ in $\mathscr{L}^2$. Once again the definition is independent of the choice of representative f, g for the equivalence classes $\tilde{f}$, $\tilde{g}$. It is immediately verified that, for any $\tilde{f}, \tilde{g}, \tilde{h}$ in $\mathscr{L}^2$ and any a, b in $\mathbf{R}$,

SP 1. $(\tilde{f}, \tilde{g}) = (\tilde{g}, \tilde{f})$,

SP 2. $(a\tilde{f} + b\tilde{g}, \tilde{h}) = a(\tilde{f}, \tilde{h}) + b(\tilde{g}, \tilde{h})$,

SP 3. $(\tilde{f}, \tilde{f}) \geq 0$,

SP 4. $(\tilde{f}, \tilde{f}) = 0$ if and only if $\tilde{f} = \tilde{0}$.

The norm on $\mathscr{L}^2$ is related to this scalar product by the simple rule

$$\|\tilde{f}\| = \sqrt{(\tilde{f},\tilde{f})}.$$

In this case Hölder's Inequality becomes the familiar Schwarz' Inequality:

$$|(\tilde{f},\tilde{g})| \leqslant \|\tilde{f}\|\,\|\tilde{g}\| \quad (\tilde{f},\tilde{g}\in\mathscr{L}^2)$$

(cf. Proposition 6.3.3). In the terminology of § 7.7, $\mathscr{L}^2$ is a *Hilbert space.* For any two non-zero elements $\tilde{f}$, $\tilde{g}$ in $\mathscr{L}^2$ we could define the angle θ between $\tilde{f}$ and $\tilde{g}$ by the rule

$$\cos\theta = \frac{(\tilde{f},\tilde{g})}{\|\tilde{f}\|\,\|\tilde{g}\|} \quad (0 \leqslant \theta \leqslant \pi)$$

so that the scalar product

$$(\tilde{f},\tilde{g}) = \|\tilde{f}\|\,\|\tilde{g}\|\cos\theta.$$

For any $\tilde{f}, \tilde{g}$ in $\mathscr{L}^2$ we say that $\tilde{f}, \tilde{g}$ are *orthogonal* if $(\tilde{f},\tilde{g}) = 0$. Among the most important geometrical properties of the Hilbert space $\mathscr{L}^2$ we mention three.

***Proposition 3* (*Pythagoras' Theorem*).** *If $\tilde{f},\tilde{g}$ are orthogonal, then*

$$\|\tilde{f}\pm\tilde{g}\|^2 = \|\tilde{f}\|^2+\|\tilde{g}\|^2.$$

***Proposition 4* (*The Parallelogram Law*).**

$$\|\tilde{f}+\tilde{g}\|^2+\|\tilde{f}-\tilde{g}\|^2 = 2\|\tilde{f}\|^2+2\|\tilde{g}\|^2.$$

Theorem 3. *Let S be a closed convex subset of $\mathscr{L}^2$ and $\tilde{f}$ any point of $\mathscr{L}^2$. Then there is a unique point of S which is nearest to $\tilde{f}$* (Fig. 22).

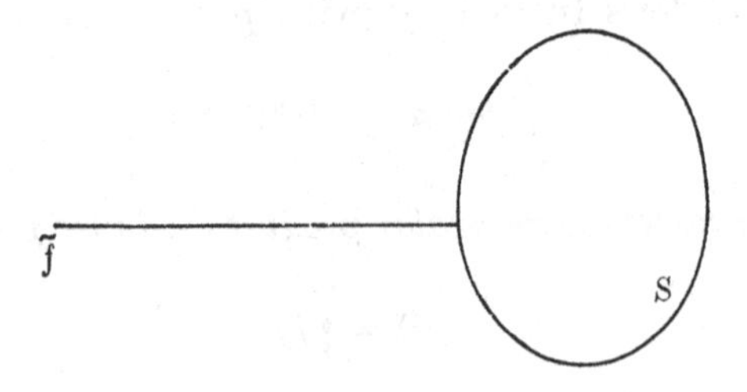

Fig. 22

The first two of these results are trivial consequences of the rules SP 1, SP 2, SP 3, but Theorem 3 makes full use of the completeness of $\mathscr{L}^2$. Recall that a *convex* set S in $\mathscr{L}^2$ is a non-empty subset of $\mathscr{L}^2$ which contains with any two points $\tilde{f}$, $\tilde{g}$, the whole segment

$$\{(1-t)\tilde{f}+t\tilde{g}\colon 0 \leqslant t \leqslant 1\}$$

joining them.

Proof of Theorem 3. As the distance in $\mathscr{L}^2$ is invariant under translations we may as well assume that $\tilde{f}$ is the origin $\tilde{0}$ and write

$$d = \inf\{\|\tilde{s}\| : \tilde{s} \in S\}.$$

Suppose that $\tilde{g}, \tilde{h} \in S$ and $\|\tilde{g}\| = \|\tilde{h}\| = d$. As S is convex, $\frac{1}{2}(\tilde{g}+\tilde{h}) \in S$ and so $\|\frac{1}{2}(\tilde{g}+\tilde{h})\| \geqslant d$. By the parallelogram law

$$\|\tilde{g}-\tilde{h}\|^2 = 2\|\tilde{g}\|^2 + 2\|\tilde{h}\|^2 - \|\tilde{g}+\tilde{h}\|^2 \leqslant 0$$

and so $\|\tilde{g}-\tilde{h}\| = 0$. This establishes the uniqueness.

To prove the existence of $\tilde{g}$ in S with $\|\tilde{g}\| = d$, we need the completeness of $\mathscr{L}^2$ and the fact that S is closed (i.e. contains all its limit points). By the definition of d as an infimum, there is a sequence $\{\tilde{s}_n\}$ of points of S for which $\|\tilde{s}_n\| \to d$. Using the parallelogram law as above

$$\|\tilde{s}_m - \tilde{s}_n\|^2 \leqslant 2\|\tilde{s}_m\|^2 + 2\|\tilde{s}_n\|^2 - 4d^2$$

and so
$$\|\tilde{s}_m - \tilde{s}_n\| \to 0 \quad \text{as} \quad m, n \to \infty.$$

As $\mathscr{L}^2$ is complete, $\{\tilde{s}_n\}$ converges strongly to a point $\tilde{g}$ in $\mathscr{L}^2$. But $\tilde{s}_n \in S$ for all n, and S is closed, thus $\tilde{g} \in S$ (cf. Appendix to Volume 1). As $n \to \infty$ the inequality

$$\|\tilde{g}\| \leqslant \|\tilde{s}_n\| + \|\tilde{g} - \tilde{s}_n\|$$

gives
$$\|\tilde{g}\| \leqslant d,$$

from which the theorem follows.

As we might expect from our experience with Euclidean space, Theorem 3 gives a method of 'dropping a perpendicular' from a point $\tilde{f}$ to any closed *linear* subspace S of $\mathscr{L}^2$. This is followed up in Ex. 4.

Our final result in this section is Riesz's famous representation theorem for bounded linear functionals on $\mathscr{L}^2$.

Theorem 4. *If F is a bounded linear functional on $\mathscr{L}^2$, then there is a unique element $\tilde{h}$ of $\mathscr{L}^2$ such that*

$$F(\tilde{f}) = (\tilde{h}, \tilde{f})$$

for all $\tilde{f}$ in $\mathscr{L}^2$.

Proof. If $F = 0$ then we may obviously take $\tilde{h} = \tilde{0}$, so assume without loss of generality that $\|F\| = 1$. Recall from § 7.4, p. 184 that

$$\|F\| = \sup\{|F(\tilde{f})|/\|\tilde{f}\| : \|\tilde{f}\| \neq 0\}.$$

It is immediately checked that the 'hyperplane'

$$H = \{\tilde{f} \in \mathscr{L}^2 : F(\tilde{f}) = 1\}$$

is a closed convex subset of $\mathscr{L}^2$ whose minimum distance from the origin is 1 (Ex. 5). Thus there is a (unique) point $\tilde{h}$ of H with $\|\tilde{h}\| = 1$.

For all $\tilde{f}$ in $\mathscr{L}^2$ and t in $\mathbf{R}$, let

$$\phi_{\tilde{f}}(t) = \|\tilde{h}+t\tilde{f}\|$$

and note that
$$\phi_{\tilde{f}}(t)^2 = 1+2(\tilde{h},\tilde{f})\,t+\|\tilde{f}\|^2 t^2:$$

it follows that the derivative $\phi_{\tilde{f}}'(0)$ exists and equals $(\tilde{h},\tilde{f})$. By the definition of $\|F\|$,
$$F(\tilde{h}+t\tilde{f}) \leqslant \|\tilde{h}+t\tilde{f}\|$$

and so
$$F(\tilde{f}) = \frac{F(\tilde{h}+t\tilde{f})-F(\tilde{h})}{t} \leqslant \frac{\phi_{\tilde{f}}(t)-\phi_{\tilde{f}}(0)}{t}$$

for $t > 0$; and this inequality is reversed if $t < 0$. Let $t \to 0$ through positive and negative values separately and deduce that

$$(\tilde{h},\tilde{f}) \leqslant F(\tilde{f}) \leqslant (\tilde{h},\tilde{f}),$$

i.e.
$$F(\tilde{f}) = (\tilde{h},\tilde{f}) \quad \text{for all } \tilde{f} \text{ in } \mathscr{L}^2.$$

This establishes the existence of $\tilde{h}$. If also

$$F(\tilde{f}) = (\tilde{g},\tilde{f}) \quad \text{for all } \tilde{f} \text{ in } \mathscr{L}^2,$$

then we should have $(\tilde{g}-\tilde{h},\tilde{f}) = 0$ for all $\tilde{f}$ in $\mathscr{L}^2$, and in particular $(\tilde{g}-\tilde{h}, \tilde{g}-\tilde{h}) = 0$ from which it would follow that $\tilde{g} = \tilde{h}$.

This result may be stated in terms of the *dual space* of $\mathscr{L}^2$ which is the normed linear space consisting of the bounded linear functionals on $\mathscr{L}^2$ with the norm $\|F\|$ as defined above.

Corollary. *For any $\tilde{h}$ in $\mathscr{L}^2$, let $F_{\tilde{h}}$ denote the bounded linear functional defined by*
$$F_{\tilde{h}}(\tilde{f}) = (\tilde{h},\tilde{f}) \quad (\tilde{f}\in\mathscr{L}^2).$$

Then the mapping
$$\theta\colon \tilde{h} \to F_{\tilde{h}}$$

is a norm-preserving linear mapping of $\mathscr{L}^2$ onto its dual space.

This corollary is expressed more briefly by saying that $\mathscr{L}^2$ is *self-dual*.

Proof. First of all, by Schwarz' Inequality,

$$|F_{\tilde{h}}(\tilde{f})| \leqslant \|\tilde{h}\|\,\|\tilde{f}\|.$$

Thus $F_{\tilde{h}}$ is indeed a bounded linear functional and

$$\|F_{\tilde{h}}\| \leqslant \|\tilde{h}\|.$$

It is then clear that θ is a linear mapping, and Riesz's Theorem shows that θ is one–one and onto.

Finally, $|F_{\tilde{h}}(\tilde{h})| = \|\tilde{h}\|^2$, which implies that

$$\|F_{\tilde{h}}\| \geqslant \|\tilde{h}\|.$$

Thus $\|F_{\tilde{h}}\| = \|\tilde{h}\|$, and this completes the proof.

Exercises

1. Let μ be the counting measure on $\{1, 2, 3, \ldots\}$ described in Ex. 8.4.1. What are the spaces $L^p(\mu)$ and $\mathscr{L}^p(\mu)$ in this case?

2. Let μ be Lebesgue measure on the real line $\mathbf{R}$ and define f_n in $L^1(\mu)$ as follows:

$$\begin{aligned} f_n(x) &= n \quad \text{for } x \text{ in } (0, 1/n), \\ &= 0 \quad \text{for all other } x \text{ in } \mathbf{R}. \end{aligned}$$

Show that $\{f_n\}$ converges pointwise (everywhere) to zero but that there is no function f in $L^1(\mu)$ to which $\{f_n\}$ converges strongly.

3. Let μ be Lebesgue measure on the interval $[0, 1]$ and consider the sequence $\{\chi_{I_n}\}$ in $L^p(\mu)$, where

$$I_n = [n2^{-k} - 1, (n+1)\,2^{-k} - 1]$$

and k is the unique integer satisfying $2^k \leqslant n < 2^{k+1}$. Show that this sequence converges strongly to zero but does not converge pointwise at any point of $[0, 1]$.

4. (i) Let S be a closed linear subspace of $\mathscr{L}^2$, $\tilde{f}$ any point of $\mathscr{L}^2$ and $\tilde{g}$ the unique point of S nearest to $\tilde{f}$ (given by Theorem 3). Show that $\tilde{f} - \tilde{g}$ is orthogonal to every element of S.

(ii) Let $S^{\perp} = \{\tilde{h} \in \mathscr{L}^2 : (\tilde{h}, \tilde{s}) = 0 \text{ for all } \tilde{s} \text{ in } S\}$.

Show that any element $\tilde{f}$ of $\mathscr{L}^2$ may be expressed uniquely in the form $\tilde{f} = \tilde{g} + \tilde{h}$, where $\tilde{g} \in S$, $\tilde{h} \in S^{\perp}$.

(iii) Write $P\tilde{f} = \tilde{g}$, $Q\tilde{f} = \tilde{h}$. Show that P, Q are continuous linear mappings of $\mathscr{L}^2$ onto S, $S^{\perp}$, respectively, and that

$$P^2 = P, \quad Q^2 = Q, \quad PQ = QP = 0.$$

(P and Q are called the 'projections' of $\mathscr{L}^2$ on S, $S^{\perp}$, respectively.)

5. Let F be a bounded linear functional on $\mathscr{L}^2$ for which $\|F\| = 1$. Let H be the hyperplane

$$H = \{\tilde{f} \in \mathscr{L}^2 : F(\tilde{f}) = 1\},$$

and S the unit sphere

$$S = \{\tilde{f} \in \mathscr{L}^2 : \|\tilde{f}\| = 1\}.$$

Show that $S \cap H$ consists of precisely one point.

17.2 The Radon–Nikodym Theorem (measure theoretic)

Theorem 1 (Radon–Nikodym). *Let $\mathscr{A}$ be a σ-algebra of subsets of X and μ, ν two finite measures on $\mathscr{A}$. Then there is a positive function h in $L^1(\mu)$ satisfying*

$$\nu(E) = \int_E h\,d\mu \tag{1}$$

for all E in $\mathscr{A}$, if and only if

$$\mu(E) = 0 \Rightarrow \nu(E) = 0 \tag{2}$$

for all E in $\mathscr{A}$. The function h is uniquely determined by condition (1) *to within a μ-null function.*

The integral in (1) may be regarded either as a Daniell integral with corresponding measure μ, or as the integral obtained from μ by the classical approach of Chapter 12.

Proof. The uniqueness of h may be proved right away: if h, h' both satisfy (1), then

$$\int_E (h-h')\,d\mu = 0$$

for all E in $\mathscr{A}$. Let $\quad S = \{x \in X: h(x) > h'(x)\}$

and deduce, from the Corollary to the Monotone Convergence Theorem, that $\mu(S) = 0$; similarly let

$$T = \{x \in X: h(x) < h'(x)\}$$

and deduce that $\mu(T) = 0$. Thus $h = h'$ almost everywhere with respect to μ.

It is clear that (1) implies (2). In proving the converse let us begin by assuming the stronger condition that μ dominates ν, i.e.

$$\nu(E) \leqslant \mu(E)$$

for all E in $\mathscr{A}$. It follows that $L^1(\mu) \subset L^1(\nu)$ (cf. p. 202), and as μ is finite,

$$L^2(\mu) \subset L^1(\mu).$$

Thus $$F(f) = \int f\,d\nu$$

defines a linear functional F on $L^2(\mu)$. Moreover,

$$|F(f)| \leqslant \int |f|\,d\nu \leqslant \int |f|\,d\mu \leqslant \|1\|\,\|f\|,$$

where the 'norms' refer to $L^2(\mu)$, and so F is bounded. By Riesz's

Theorem 17.1.4 applied to the Hilbert Space $\mathscr{L}^2(\mu)$, there is an element h in $L^2(\mu)$ such that

$$\int f d\nu = \int f h d\mu \tag{3}$$

for all f in $L^2(\mu)$. Now $\chi_E \in L^2(\mu)$ for any E in $\mathscr{A}$ and so

$$\nu(E) = \int_E h d\mu$$

as required.

In particular, if $$U = \{x \in X : h(x) < 0\},$$

we deduce that $$0 \leqslant \nu(U) = \int_U h d\mu \leqslant 0$$

so that $\mu(U) = 0$. Thus h is positive almost everywhere with respect to μ. In the same way, if

$$V = \{x \in X : h(x) > 1\},$$

we have $$0 \leqslant \mu(V) - \nu(V) = \int_V (1-h) d\mu \leqslant 0,$$

whence $\mu(V) = 0$. Thus $$0 \leqslant h \leqslant 1$$

almost everywhere with respect to μ, and we may as well truncate h so that these inequalities hold everywhere in X as this does not affect the fundamental relation (1).

Let us now assume only that condition (2) holds and define the finite measure

$$\lambda = \mu + \nu$$

on $\mathscr{A}$ which dominates both μ and ν. By the above argument there is an element g of $L^2(\lambda)$ which satisfies

$$\int f d\nu = \int f g d\lambda \tag{4}$$

for all f in $L^2(\lambda)$, and $$0 \leqslant g \leqslant 1.$$

As $\lambda = \mu + \nu$, this equation may be rewritten

$$\int f(1-g) d\nu = \int f g d\mu \tag{5}$$

for all f in $L^2(\lambda)$. We should like to substitute $f = \chi_E(1-g)^{-1}$ in (5) and so deduce our result: this idea is good, but requires some justification. Let

$$A = \{x \in X : 0 \leqslant g(x) < 1\},$$

$$B = \{x \in X : g(x) = 1\},$$

(so that A and B consist of the 'good' and 'bad' points, respectively, for $(1-g)^{-1}$). We may substitute $f = \chi_B$ in (5) and deduce that

$$\mu(B) = 0. \tag{6}$$

We may substitute $\quad f = \chi_E(1+g+\ldots+g^{n-1})$

in (5) (see Ex. 1) and deduce that

$$\int_E (1-g^n)\,d\nu = \int_E (g+\ldots+g^n)\,d\mu \tag{7}$$

for any E in $\mathscr{A}$. If $x \in A$, $1-g(x)^n \uparrow 1$, and if $x \in B$, $1-g(x) = 0$. Thus the left hand side of (7) is increasing and converges to $\nu(E \cap A)$. The integrand on the right hand side is increasing and converges to a positive limit h, say, outside the μ-null set B, so that

$$\nu(E \cap A) = \int_E h\,d\mu \tag{8}$$

for all E in $\mathscr{A}$. As the integral in (8) exists for all E in $\mathscr{A}$ we see in particular, with $E = X$, that $h \in L^1(\mu)$.

Finally, we use condition (2) to deduce from (6) that $\nu(B) = 0$; thus ν is concentrated on the set A (in the sense defined in § 16.1) and (8) gives

$$\nu(E) = \int_E h\,d\mu$$

for all E in $\mathscr{A}$, as required.

The above proof gives a good deal more than we have stated. Let us not assume that (2) holds. The proof goes through without change until the last paragraph, where we are now unable to conclude that $\nu(B) = 0$. We have therefore expressed X as the union of two disjoint sets A, B such that μ is concentrated on A; moreover the *contraction* of ν by A satisfies the condition

$$\nu_A(E) = \int_E h\,d\mu$$

for all E in $\mathscr{A}$. (Recall from § 16.1 that $\nu_A(E) = \nu(E \cap A)$, by definition.) In general, this decomposition $X = A \cup B$ is not unique, but the corresponding decomposition of measures

$$\nu = \nu_A + \nu_B$$

is unique. (Cf. the Jordan and Hahn Decompositions of Theorems 16.1.1, 2.) To state this we introduce a standard terminology. If condition (2) holds we say that ν is *absolutely continuous* with respect to μ and write

$$\nu \ll \mu$$

(not to be confused with the stronger condition $\nu \ll \mu$). If μ, ν are concentrated on disjoint subsets of X we say that μ, ν are *mutually singular* and write $\mu \perp \nu$.

Theorem 2 (Lebesgue Decomposition Theorem). *Let $\mathscr{A}$ be a σ-algebra of subsets of X and μ, ν two finite measures on $\mathscr{A}$. Then ν may be expressed uniquely as*

$$\nu = \nu_1 + \nu_2,$$

where $\nu_1 \ll \mu$ and $\nu_2 \perp \mu$.

Proof. All that remains to prove is the uniqueness. If we also have

$$\nu = \nu_1' + \nu_2',$$

where
$$\nu_1' \ll \mu \quad \text{and} \quad \nu_2' \perp \mu,$$
then
$$\nu_1 - \nu_1' = \nu_2' - \nu_2 = \lambda,$$
say, satisfies
$$\lambda \ll \mu \quad \text{and} \quad \lambda \perp \mu,$$
whence $\lambda = 0$. (The reader is asked to give a more detailed verification of these last three statements in Ex. 2.)

In both of these theorems we have assumed that $\mu(X)$, $\nu(X)$ are finite. If X is *σ-finite* with respect to μ, ν then we may find $A_n \uparrow X$ with $\mu(A_n)$ finite and $B_n \uparrow X$ with $\nu(B_n)$ finite; and hence

$$A_n \cap B_n = C_n \uparrow X$$

with both $\mu(C_n)$, $\nu(C_n)$ finite. Theorem 1 applies to the contractions μ_{C_n}, ν_{C_n} and yields a positive function h_n which vanishes outside C_n and satisfies

$$\nu(E \cap C_n) = \int_E h_n \, d\mu$$

for all E in $\mathscr{A}$. As $C_n \subset C_{n+1}$ we may assume without loss of generality that $h_n \leqslant h_{n+1}$ for $n \geqslant 1$. Any x in X belongs to C_n for some n depending on x and so $h_n \uparrow h$, say, where $h: X \to \mathbf{R}$. The Monotone Convergence Theorem gives

$$\nu(E) = \int_E h \, d\mu$$

for all E in $\mathscr{A}$, with the usual understanding that both sides are infinite if $h\chi_E$ is not in $L^1(\mu)$. The functions h_n are uniquely determined to within a null function, and so the same is true of h. The function h will not lie in $L^1(\mu)$ unless $\nu(X)$ is finite, but $h\chi_{C_n} \in L^1(\mu)$ for all n and we may therefore say that h is *locally integrable* with respect to μ. With these modifications Theorem 1 holds in the σ-finite case. A similar extension of Theorem 2 is left as an exercise (Ex. 3).

The reader may recall from § 3.5 that the function F was said to be *absolutely continuous* on $[a, b]$ if, to any $\epsilon > 0$, there exists a $\delta > 0$ such that

$$|\Sigma\{F(b_i) - F(a_i)\}| < \epsilon \quad \text{whenever} \quad \Sigma(b_i - a_i) < \delta \tag{9}$$

(where (a_i, b_i) are disjoint subintervals of $[a, b]$ for $i = 1, \ldots, r$). As we saw in Ex. 3.5.5, condition (9) is equivalent to the apparently stronger condition: given any $\epsilon > 0$, there is a $\delta > 0$ such that

$$\Sigma|F(b_i) - F(a_i)| < \epsilon \quad \text{whenever} \quad \Sigma(b_i - a_i) < \delta \tag{10}$$

((a_i, b_i) are disjoint subintervals of $[a, b]$). Either of these conditions applied to a single interval (a_i, b_i) shows that F is continuous. In this classical setting the term 'absolutely continuous' is quite natural – but what has it to do with the above measure theoretic definition $\nu \ll \mu$? To answer this question we prove a long-promised result.

Theorem 3. *The function F: $[a, b] \to \mathbf{R}$ is expressible as an indefinite integral:*

$$F(x) = \int_a^x f + C \tag{11}$$

for some f in $L^1[a, b]$, if and only if F is absolutely continuous.

Proof. The easier half has already been mentioned in § 3.5. Suppose that F is given by (11). First of all, if

$$|f(x)| \leqslant K$$

for all x in $[a, b]$ then F satisfies the Lipschitz' Condition

$$|F(x) - F(y)| \leqslant K|x - y|$$

for all x, y in $[a, b]$, from which (10) follows immediately. Secondly, if $f \in L^{\text{inc}}[a, b]$, there is an increasing sequence $\{\phi_n\}$ of step functions which converges to f almost everywhere in $[a, b]$. To any $\epsilon > 0$, we may find an integer N such that

$$0 \leqslant \int_a^b (f - \phi_N) < \epsilon/2.$$

Once N is chosen, there is a constant K (depending on N and ultimately on ϵ) such that

$$|\phi_N(x)| \leqslant K$$

for all x in $[a, b]$. Now let S be the union of the disjoint intervals $(a_i, b_i) \subset [a, b]$ for $i = 1, \ldots, r$: then

$$\int_S f = \int_S (f - \phi_N) + \int_S \phi_N,$$

whence
$$\left|\int_S f\right| < \epsilon/2 + Kl(S),$$
which gives (9) if we take $\delta = \epsilon/2K$. The extension to an arbitrary function f in $L^1[a,b]$ follows immediately on taking differences.

Conversely, let us assume that F is absolutely continuous on $[a,b]$. As F is continuous, let
$$\nu[a,x) = \nu[a,x] = F(x) - F(a)$$
and extend ν to an additive real valued function ν on the ring $\mathscr{R}$ generated by the subintervals of $[a,b]$ as in Theorem 11.3.2. Let μ denote the restriction of Lebesgue measure to $[a,b]$. Then the condition of absolute continuity for F becomes
$$|\nu(E)| < \epsilon \quad \text{for all } E \text{ in } \mathscr{R} \text{ with } \mu(E) < \delta. \tag{12}$$
Subdivide $[a,b]$ into N disjoint intervals $I_1, \ldots, I_N$ each of length less than δ; then
$$\nu(A) = \sum_{i=1}^{N} \nu(A \cap I_i)$$
and
$$|\nu(A)| < N\epsilon \quad \text{for any } A \text{ in } \mathscr{R}.$$
The absolute continuity of F therefore implies that ν is bounded and so by Proposition 16.1.1, $\nu = \nu^+ - \nu^-$ where ν^+, ν^- are the positive additive functions on $\mathscr{R}$ defined by
$$\nu^+(E) = \sup\{\nu(A) : A \in \mathscr{R}, A \subset E\},$$
$$\nu^-(E) = \sup\{-\nu(A) : A \in \mathscr{R}, A \subset E\}.$$
(Cf. the classical proof in Exx. 3.5.2, 4, 8 that F has bounded variation on $[a,b]$ and so is the difference of two increasing functions.) Condition (12) implies that
$$\nu^+(E) \leqslant \epsilon \quad \text{for all } E \text{ in } \mathscr{R} \text{ with } \mu(E) < \delta,$$
with a similar condition for ν^-. Without loss of generality we may therefore consider the case where ν is positive and the corresponding function F is increasing.

The additive function ν extends to a Lebesgue–Stieltjes measure ν on the σ-algebra $\mathscr{A}$ generated by the subintervals of $[a,b]$. Certainly μ is defined on $\mathscr{A}$. Let E be any element of $\mathscr{A}$ satisfying $\mu(E) = 0$. Then E may be covered by a sequence $\{I_n\}$ of subintervals of $[a,b]$ for which
$$\Sigma\mu(I_n) < \delta.$$
By a simple induction argument we may arrange that these intervals are disjoint (cf. the proof of Theorem 3.2.1). If
$$E_n = I_1 \cup \ldots \cup I_n,$$

this implies that $\mu(E_n) < \delta$, whence $\nu(E_n) < \epsilon$ for all $n \geqslant 1$. Thus

$$\Sigma \nu(I_n) \leqslant \epsilon$$

for arbitrarily small ϵ and so $\nu(E) = 0$.

Now the Radon–Nikodym Theorem guarantees the existence of a positive function f in $L^1[a,b]$ for which

$$\nu(E) = \int_E f\,d\mu$$

for all E in $\mathscr{A}$. In particular, with $E = [a,x]$, we deduce that

$$F(x) - F(a) = \int_a^x f\,d\mu$$

as required.

The ring $\mathscr{R}$ described in the above proof is much smaller than the σ-algebra $\mathscr{A}$ (which consists of the Borel subsets of $[a,b]$). If E is restricted to $\mathscr{R}$, the condition

$$\mu(E) = 0 \Rightarrow \nu(E) = 0 \qquad [(2)]$$

though true, is of no practical value, as E has to be a finite set of points in this case. We should expect a condition of type (12) as the null sets in $\mathscr{A}$ are expressed in terms of sets of arbitrarily small measure in $\mathscr{R}$. On the other hand if E is allowed to vary in $\mathscr{A}$ then conditions (2) and (12) are equivalent.

Proposition 1. *Let μ, ν be measures on a σ-algebra $\mathscr{A}$ and assume that ν is finite. Then the following two conditions are equivalent:*

(*a*) $\nu \ll \mu$;

(*b*) *to any $\epsilon > 0$, there is a $\delta > 0$ such that*

$$\nu(E) < \epsilon \textit{ for all } E \textit{ in } \mathscr{A} \textit{ with } \mu(E) < \delta.$$

Proof. (i) Assume that (*b*) holds. If $\mu(E) = 0$ then $\mu(E) < \delta$ for every $\delta > 0$; hence $\nu(E) < \epsilon$ for every $\epsilon > 0$ and this implies that $\nu(E) = 0$.

(ii) Assume that (*b*) is false. Then there is an $\epsilon > 0$ and sets A_n in $\mathscr{A}$ such that

$$\mu(A_n) < 2^{-n}, \quad \nu(A_n) \geqslant \epsilon$$

$(n = 1, 2, \ldots)$. In accordance with the construction of $\limsup \chi_{A_n}$ (cf. Theorem 16.1.2) let

$$B_n = \bigcup\{A_m : m > n\}, \quad B = \bigcap\{B_n : n \geqslant 1\}.$$

Thus $B_n \downarrow B$. But $\mu(B_n) < 2^{-n}$ and so $\mu(B) = 0$; whereas $\nu(B_n) \geqslant \epsilon$ and so $\nu(B) \geqslant \epsilon$. (At the very last step we use the finiteness of ν.) This shows that (*a*) is false and completes the proof.

Proposition 1 is vitally dependent on the finiteness of ν: for example, let $\mathscr{A}$ be the σ-algebra of all Lebesgue measurable sets on $(0, 1)$ and define

$$\nu(E) = \int_E \frac{1}{t}\, dt \quad \text{if this integral exists,}$$

$$= \infty \text{ otherwise.}$$

Then ν is absolutely continuous with respect to Lebesgue measure μ on $(0, 1)$. On the other hand $\mu(0, 1/n) \to 0$ and $\nu(0, 1/n) = \infty$. (See also Ex. 11 below.)

The Radon–Nikodym Theorem extends without trouble to the case where ν is a real or complex measure (and μ remains a σ-finite measure). This is sufficiently illustrated for us by expressing any real measure as an indefinite integral. More precisely, let ν be a real measure on the σ-algebra $\mathscr{A}$ of subsets of X. According to the Jordan Decomposition Theorem,

$$\nu = \nu^+ - \nu^-, \quad |\nu| = \nu^+ + \nu^-$$

and it is clear that ν^+, ν^- are absolutely continuous with respect to the measure $|\nu|$ which dominates them. Applying the Radon–Nikodym Theorem to ν^+, ν^- and subtracting, we find an element h of $L^1(|\nu|)$ which satisfies

$$\nu(E) = \int_E h\, d|\nu|$$

for all E in $\mathscr{A}$. As we saw in § 16.1 this implies that

$$\nu^+(E) = \int_E h^+ d|\nu|, \quad \nu^-(E) = \int_E h^- d|\nu|.$$

Let

$$S = \{x \in X: h(x) \geqslant 0\},$$

$$T = \{x \in X: h(x) < 0\}.$$

Then $S, T \in \mathscr{A}$ and

$$\nu^+(E) = \nu(E \cap S), \quad \nu^-(E) = -\nu(E \cap T).$$

This gives an independent proof of the Hahn Decomposition Theorem 16.1.2.

Exercises

1. In the proof of Theorem 1, justify the substitution of

$$f = \chi_E(1 + g + \ldots + g^{n-1})$$

in equation (5).

2. In the notation used in the proof of Theorem 2, show that the real measure λ satisfies the conditions:

(i) $\mu(E) = 0 \Rightarrow \lambda(E) = 0$ for all E in $\mathscr{A}$;

(ii) there exist disjoint subsets S, T in $\mathscr{A}$ such that

$$\mu(E) = \mu(E \cap S) \quad \text{and} \quad \lambda(E) = \lambda(E \cap T) \quad \text{for all } E \text{ in } \mathscr{A}.$$

Hence show that $\lambda(E) = 0$ for all E in $\mathscr{A}$.

3. Extend Theorem 2 to the case where μ, ν are measures on $\mathscr{A}$ for which X is σ-finite.

4. Extend Theorem 1 to the case where μ is a σ-finite measure on $\mathscr{A}$ and ν is a real measure on $\mathscr{A}$.

5. Let X be the interval $[0, 1]$ and let $\mathscr{A}$ consist of the subsets of X which are either countable or have countable complements. Let μ be the counting measure on $\mathscr{A}$ (i.e. for any A in $\mathscr{A}$, $\mu(A)$ is the number of elements in A, possibly infinite), and let $\nu(A) = 0$ if A is countable, and $\nu(A) = 1$ otherwise $(A \in \mathscr{A})$. Show that ν is absolutely continuous with respect to μ but there is no function h in $L^1(\mu)$ such that

$$\nu(A) = \int_A h\,d\mu \quad \text{for all } A \text{ in } \mathscr{A}.$$

6. Let X be the interval $[0, 1]$ and $\mathscr{B}$ the σ-algebra of Lebesgue measurable subsets of X. Let μ be the counting measure on X and ν Lebesgue measure on X. Show that ν is absolutely continuous with respect to μ but there is no function h in $L^1(\mu)$ such that

$$\nu(B) = \int_B h\,d\mu \quad \text{for all } B \text{ in } \mathscr{B}.$$

7. Let $\nu \ll \mu$ be the finite measures described in Theorem 1 and h the corresponding function in $L^1(\mu)$: we may write

$$h = \frac{d\nu}{d\mu}$$

but note that h is only determined to within a μ-null function.

Suppose also that λ is a finite measure on $\mathscr{A}$ and $\mu \ll \lambda$. Show that $\nu \ll \lambda$ and

$$\frac{d\nu}{d\lambda} = \frac{d\nu}{d\mu}\frac{d\mu}{d\lambda}$$

to within a λ-null function.

8. In the notation of Ex. 7 suppose that $\nu \ll \mu$ and $\mu \ll \nu$. Show that

$$\frac{d\nu}{d\mu} = \left(\frac{d\mu}{d\nu}\right)^{-1}$$

to within a μ-null function.

9. Let $\mathscr{A}$ be a σ-algebra of subsets of X and let $\lambda = \mu + i\nu$ be a complex measure on $\mathscr{A}$.

(i) Apply the Radon–Nikodym Theorem to the positive measure $|\mu| + |\nu|$

to deduce the existence of a positive measure τ on $\mathscr{A}$ and a measurable function $\zeta: X \to \mathbf{C}$ with $|\zeta| = 1$ almost everywhere with respect to τ, such that

$$\lambda(A) = \int_A \zeta \, d\tau$$

for all A in $\mathscr{A}$.

(ii) Show that τ is unique and that ζ is unique to within a τ-null function.

(iii) Show that τ is the measure $|\lambda|$ of Theorem 16.2.1.

10. Let $\{\mu_n\}$ be a sequence of finite measures on a σ-algebra $\mathscr{A}$. Show that there is a finite measure μ on $\mathscr{A}$ such that each of the measures μ_n is absolutely continuous with respect to μ.

11. Let $X = \{1, 2, 3, \ldots\}$, $\mathscr{A}$ be the collection of all subsets of X,

$$\mu(E) = \sum_{n \in E} 2^{-n} \quad \text{and} \quad \nu(E) = \sum_{n \in E} 2^n$$

for all E in $\mathscr{A}$. Show that $\nu \ll \mu$ but that condition (*b*) of Proposition 1 is not satisfied.

17.3 The Radon–Nikodym Theorem for Daniell integrals

The results of this section are more fundamental than those of § 17.2 in the sense that the Radon–Nikodym Theorem of § 17.2 is a special case of the Radon–Nikodym Theorem for Daniell Integrals as stated below (see Ex. 1). At the cost of some repetition we have written the present section so that it may be read before § 17.2 if the student wishes.

In § 6.4 we considered an open subset U of $\mathbf{R}^k$ and a mapping $T: U \to \mathbf{R}^k$ for which there is a *density function* h, viz. a positive function $h: U \to \mathbf{R}$ satisfying

$$m(TI) = \int_I h$$

for any bounded interval I with $\bar{I} \subset U$. Any step function ϕ on $\mathbf{R}^k$ whose support is contained in U may be expressed as a linear combination of characteristic functions χ_I of such intervals I. If ψ is the corresponding 'distorted' step function defined by

$$\psi(Tx) = \phi(x)$$

for x in U, then we verify immediately that

$$\int_{TU} \psi = \int_U \phi h.$$

The main point of Theorem 6.4.1 is that this simple formula extends to a general transformation formula

$$\int_{TU} g = \int_U fh.$$

Quite generally, let L be a linear lattice of functions $\phi: X \to \mathbf{R}$ and P an elementary integral on L which extends by the standard construction of § 8.1 to a Daniell integral; it is convenient to denote this Daniell integral by the same symbol P. If h is a positive real valued function on X and $\phi h \in L^1(P)$ for all ϕ in L, we say that h is *locally integrable* with respect to P. In these circumstances we may define a positive linear functional Q on L by the rule

$$Q(\phi) = P(\phi h) \tag{1}$$

for all ϕ in L, and verify at once that Q is an elementary integral: for if $\phi_n \downarrow 0$ ($\phi_n \in L$) then $\phi_n h \downarrow 0$ and we may apply the Monotone Convergence Theorem for P to equation (1) with ϕ_n in place of ϕ. We may now extend Q to a Daniell integral which is related to the Daniell integral P by the simple density formula (1) at the level of elementary functions. There is a general theorem analogous to Theorem 6.4.1 which extends (1) to all functions for which one side makes sense.

We make two simplifying assumptions about the linear lattice L. First of all, L satisfies the *stronger Stone condition* of p. 23:

$$\min\{1, \phi\} \in L \quad \text{for any positive } \phi \text{ in } L.$$

This implies that 1 is measurable for the Daniell integral derived from *any* elementary integral on L. Secondly, the underlying space X is *σ-finite* with respect to L as defined on p. 125: this requires the existence of an increasing sequence $\{X_n\}$ of elementary sets

$$X_n = \{x \in X: \phi_n(x) \geqslant 1\}$$

($\phi_n \in L$) whose union is X, and ensures the σ-finiteness of the measure derived from *any* elementary integral on L.

It is essential for the truth of the Radon–Nikodym Theorem that we make some assumption rather like the σ-finiteness of X (see Zaanen [22], Ch. 7, where this question is discussed in some detail); on the other hand the stronger Stone Condition may be replaced quite simply by a suitable assumption about the measurability of 1 (Exx. 3, 5).

Theorem 1. *Let L be a linear lattice of functions $\phi: X \to \mathbf{R}$ satisfying the stronger Stone condition, and suppose that X is σ-finite with respect* *to L. Let P, Q be Daniell integrals obtained by extending elementary integrals on L, and let h be a positive function on X such that*

$$Q(\phi) = P(\phi h) \qquad [(1)]$$

for all ϕ in L. Then $f \in L^1(Q)$ if and only if $fh \in L^1(P)$, and in this case

$$Q(f) = P(fh). \qquad (2)$$

In the course of the proof we shall see that the locally integrable function h is measurable (with respect to P) and so there is no need to make this assumption in the statement of the theorem.

Before proving Theorem 1 we note one or two obvious implications for the measures μ, ν corresponding to P, Q, respectively. First of all, we may rewrite (2) as

$$\int f d\nu = \int f h d\mu. \qquad (3)$$

Set $f = \chi_E$ in (3) and deduce that

$$\nu(E) = \int_E h d\mu. \qquad (4)$$

As usual we interpret (4) as saying that E has finite measure with respect to ν if and only if h is integrable with respect to μ on E and that equality holds in this case. The points at which h vanishes make no contribution to this integral. Let

$$S = \{x \in X : h(x) \neq 0\}.$$

Then

$$\nu(E) = \int_{E \cap S} h d\mu, \qquad (5)$$

with the same kind of interpretation as (4). As we might expect, the set S is strategically important. In fact, it follows quite easily from the proof that E is measurable with respect to ν if and only if $E \cap S$ is measurable with respect to μ (infinite values being allowed): see Ex. 4.

Proof. For ease of reading we split the proof into four smaller parts. Note that the σ-finiteness condition is not used in parts (i) and (ii).

(i) *If A is Q-null, then $A \cap S$ is P-null.*

Let A be Q-null. There is an increasing sequence $\{\psi_n\}$ of positive functions in L diverging at each point of A and for which $Q(\psi_n)$ is bounded. Now $\{\psi_n h\}$ diverges at each point of $A \cap S$. In view of equation (1) and the Monotone Convergence Theorem for P we deduce that $A \cap S$ is P-null.

(ii) *If $f \in L^1(Q)$, then $fh \in L^1(P)$ and (2) holds.*

First of all, let $f \in L^{\text{inc}}(Q)$ and let $\{\phi_n\}$ be an increasing sequence of functions in L which converges to f outside a Q-null set A. Then $\{\phi_n h\}$ converges to fh outside the set $A \cap S$. It follows from (i) and the Monotone Convergence Theorem that $fh \in L^1(P)$ and (2) holds.

The extension to $L^1(Q)$ follows immediately on taking differences.

(iii) *If A is P-null, then A is Q-null.*

To begin with, assume that $\nu(X)$ is finite. Let A_c denote the elementary set

$$A_c = \{x \in X : \phi(x) \geqslant c\}.$$

Recall from § 8.4, p. 27 the function

$$n(\min\{\phi, c\} - \min\{\phi, c - 1/n\}),$$

which belongs to L (by the stronger Stone condition) and decreases to χ_{A_c} as $n \to \infty$. The Monotone Convergence Theorem applies to (1) and shows that

$$\nu(A_c) = \int_{A_c} h \, d\mu. \tag{6}$$

Thus (4) holds for any elementary set A_c.

Suppose that A is P-null, i.e. $\mu(A) = 0$. There is an increasing sequence $\{\psi_n\}$ of positive functions in L which diverges at each point of A and for which

$$\int \psi_n \, d\mu \leqslant K,$$

where $K > 0$ without loss of generality. For any $\epsilon > 0$, define the elementary set

$$E_n^{\epsilon} = \{x \in X : \psi_n(x) \geqslant K/\epsilon\}$$

(just as in the proof of our first main Theorem 3.2.1). Then

$$\chi_{E_n^{\epsilon}} \leqslant (\epsilon/K)\, \psi_n$$

(as ψ_n is positive) and so $\quad \mu(E_n^{\epsilon}) \leqslant \epsilon.$

Now $E_n^{\epsilon} \uparrow E^{\epsilon}$, where $\quad \mu(E^{\epsilon}) \leqslant \epsilon.$

As $\nu(X)$ is finite we may apply the Monotone Convergence Theorem to (6) and deduce that

$$\nu(E^{\epsilon}) = \int_{E^{\epsilon}} h \, d\mu.$$

As $\{\psi_n\}$ diverges at each point of A, $A \subset E^{\epsilon}$ for each $\epsilon > 0$. But $\epsilon > \epsilon' > 0$ implies that $E_n^{\epsilon} \supset E_n^{\epsilon'}$ for all n, and so implies that $E^{\epsilon} \supset E^{\epsilon'}$. Now let $\epsilon \downarrow 0$ through a sequence of values. Then $E^{\epsilon} \downarrow E$ where $A \subset E$ and $\mu(E) = 0$ as $\mu(E) \leqslant \epsilon$ for all $\epsilon > 0$. The Monotone Convergence Theorem gives

$$\nu(E) = \int_E h \, d\mu,$$

which is zero as $\mu(E) = 0$. As the measure ν is complete and $A \subset E$, we conclude that $\nu(A) = 0$, i.e. A is Q-null.

Even if $\nu(X) = \infty$, there is an increasing sequence $\{X_n\}$ of elementary sets whose union is X and for which all $\nu(X_n)$ are finite. The above argument shows that $\nu(A \cap X_n) = 0$ for all n and so $\nu(A) = 0$, as before.

(iv) *If $fh \in L^1(P)$, then $f \in L^1(Q)$ and* (2) *holds.*

In this part we assume to begin with that $1 \in L^1(P) \cap L^1(Q)$. Let $fh \in L^1(P)$ and define

$$\begin{aligned} f_0(x) &= f(x) \quad \text{if} \quad x \in S, \\ &= 0 \quad \text{otherwise.} \end{aligned}$$

Then obviously $f_0 h = fh$. As $1 \in L^1(Q)$ it follows from (ii) that $h \in L^1(P)$ and so h is certainly P-measurable. Thus f_0 is P-measurable as f_0 is the product of the P-measurable functions $f_0 h$, $1/h$. As $1 \in L^1(P)$ the truncated function

$$\bar{f}_0 = \text{mid}\{-K, f_0, K\}$$

is integrable with respect to P and so there is a sequence $\{\phi_n\}$ of elements of L which converges to $\bar{f}_0$ almost everywhere with respect to P. We may as well truncate each ϕ_n and assume that $|\phi_n| \leqslant K$ for all n. By the critically important part (iii), $\{\phi_n\}$ converges to $\bar{f}_0$ almost everywhere with respect to Q and we apply the Dominated Convergence Theorem to the equation

$$Q(\phi_n) = P(\phi_n h)$$

to deduce that

$$Q(\bar{f}_0) = P(\bar{f}_0 h).$$

Now let $K \uparrow \infty$ through a sequence of values and obtain

$$Q(f_0) = P(f_0 h). \tag{7}$$

As a special case with $f = \chi_X, f_0 = \chi_S$, we deduce that

$$\nu(S) = \int_S h \, d\mu.$$

But we also know from (ii) that

$$\nu(X) = \int_X h \, d\mu.$$

Thus $\nu(S) = \nu(X)$ (finite) and so $f = f_0$ almost everywhere with respect to Q. Equation (2) now follows from (7) as $f_0 h = fh$.

To complete the proof we express X as the union of an increasing sequence $\{X_n\}$ of elementary sets with finite $\mu(X_n)$, $\nu(X_n)$. Still supposing that $fh \in L^1(P)$ the above proof shows that $f\chi_{X_n} \in L^1(Q)$ and

$$Q(f\chi_{X_n}) = P(fh\chi_{X_n})$$

for all $n \geqslant 1$. If f is positive the Monotone Convergence Theorem gives (2) immediately; in general we apply the same argument to f^+, f^- separately and deduce (2) by taking differences.

The reader will find it instructive to compare the proof of Theorem 1 with the one given for Theorem 6.4.1. The analogy is clear if we write $\nu(A) = \mu(TA)$. One additional step in the earlier proof was to establish the fact that

$$\int_V g\,d\mu = \int_U f\,d\nu,$$

which is not given by Theorem 1.

In the light of Theorem 1 there is now a more general form of the Radon–Nikodym Theorem.

Theorem 2 (Radon–Nikodym). *Let L be a linear lattice of functions $\phi: X \to \mathbf{R}$ satisfying the stronger Stone condition, and suppose that X is σ-finite with respect to L. Let P, Q be Daniell integrals obtained by extending elementary integrals on L. Then the following conditions are equivalent.*

(a) There is a positive function $h: X \to \mathbf{R}$ such that

$$Q(\phi) = P(\phi h) \qquad [(1)]$$

for all ϕ in L.

(b) Every P-null subset of X is Q-null.

If h, h' both satisfy (a) then $h - h'$ is P-null.

Proof. Let μ, ν denote the measures corresponding to P, Q, respectively. To begin with (for all four parts) we assume that

$$1 \in L^1(P) \cap L^1(Q),$$

i.e. $\mu(X)$ and $\nu(X)$ are both finite.

(i) Suppose that h, h' both satisfy (*a*). As $1 \in L^1(Q)$ it follows by Theorem 1 that $h, h' \in L^1(P)$; moreover, if E is any P-measurable set, then

$$P\{\chi_E(h-h')\} = 0.$$

In particular we may substitute

$$E = \{x \in X: h(x) > h'(x)\} \quad \text{or} \quad \{x \in X: h(x) < h'(x)\}$$

and deduce from the Corollary to the Monotone Convergence Theorem that $\mu(E) = 0$. Thus $h = h'$ almost everywhere with respect to μ.

(ii) It is clear from Theorem 1 that (*a*) implies (*b*): this is the substance of part (iii) of the proof.

(iii) In proving that (*b*) implies (*a*) let us first suppose that P dominates Q on L, i.e.

$$Q(\phi) \leqslant P(\phi)$$

for all positive ϕ in L. It follows that

$$L^2(P) \subset L^1(P) \subset L^1(Q)$$

and so Q acts as a linear functional on $L^2(P)$. Moreover

$$|Q(f)| \leqslant Q(|f|) \leqslant P(|f|) \leqslant \{P(1)\}^{\frac{1}{2}}\{P(|f|^2)\}^{\frac{1}{2}}.$$

Thus Q is a bounded linear functional on $L^2(P)$ and so, by Riesz's Theorem 17.1.4 there is a function h in $L^2(P)$ for which

$$Q(f) = P(fh) \tag{8}$$

for all f in $L^2(P)$. Let $E = \{x \in X: h(x) < 0\}$; then $\chi_E \in L^2(P)$ and so

$$0 \leqslant \nu(E) = P(\chi_E h) \leqslant 0,$$

from which we deduce that $\mu(E) = 0$. In the same way, if

$$G = \{x \in X: h(x) > 1\}$$

we deduce that $0 \leqslant \mu(G) - \nu(G) = P\{\chi_G(1-h)\} \leqslant 0$,

whence $\mu(G) = 0$. Thus $\qquad 0 \leqslant h \leqslant 1$

almost everywhere with respect to μ. Without altering (8) we may truncate h, if necessary, and assume that these inequalities hold everywhere. In fact (8) holds for all f in $L^1(P)$: to see this we apply the Dominated Convergence Theorem to the bounded functions

$$f_n = \operatorname{mid}\{-n, f, n\}$$

which lie in $L^2(P)$ and are dominated by $|f|$. Condition (a) then follows by restricting f to L.

(iv) Suppose that (b) holds. We may define an elementary integral R on L by the rule

$$R(\phi) = P(\phi) + Q(\phi) \tag{9}$$

for all ϕ in L. The elementary integral R extends to a Daniell integral R on the space $L^1(R) = L^1(P) \cap L^1(Q)$, and

$$R(f) = P(f) + Q(f) \tag{10}$$

for all f in $L^1(R)$ (see Ex. 13.3.4). The argument of (iii) now yields a function g in $L^2(R)$ which satisfies

$$0 \leqslant g \leqslant 1$$

and

$$Q(f) = R(fg) \tag{11}$$

for all f in $L^1(R)$. Combining (10) and (11) we obtain

$$Q\{f(1-g)\} = P(fg) \tag{12}$$

for all f in $L^1(R)$.

We should like to substitute $f(1-g)^{-1}$ for f in (12): to justify this, consider the set

$$B = \{x \in X : g(x) = 1\}$$

(consisting of the 'bad' points for the quotient function $f(1-g)^{-1}$). As $g \in L^1(R)$, the function χ_B is R-measurable and so we may legitimately substitute $f = \chi_B$ in (12) to give $\mu(B) = 0$. By condition (*b*) we also have $\nu(B) = 0$ and we may alter g (on an R-null set) so that (12) holds and

$$0 \leqslant g(x) < 1 \tag{13}$$

for all x in X. In view of (13), $g^n \downarrow 0$ and $g + g^2 + \ldots + g^n \uparrow h$, say (where, of course, $h = g(1-g)^{-1}$). As g is bounded,

$$f(1+g+\ldots+g^{n-1}) \in L^1(R)$$

and
$$Q\{f(1-g^n)\} = P\{f(g+g^2+\ldots+g^n)\}$$

for all f in $L^1(R)$. The Monotone Convergence Theorem gives

$$Q(f) = P(fh)$$

for all positive f in $L^1(R)$, then for all f in $L^1(R)$ by taking differences. Finally, (1) follows if we restrict f to L.

In the general case we may express X as the union of an increasing sequence $\{X_n\}$ of elementary sets where $\mu(X_n)$, $\nu(X_n)$ are finite. The above argument gives an increasing sequence $\{h_n\}$ of positive functions in $L^1(P)$ where h_n vanishes outside X_n and satisfies

$$Q(f\chi_{X_n}) = P(fh_n)$$

for all f in $L^1(R)$, and so certainly for all f in L. The sequence $\{h_n\}$ converges to a positive function h which satisfies (1), in the first place for positive ϕ, and then by taking differences for all ϕ in L. Each h_n is uniquely determined to within a P-null function, and so the same is true of h. This completes the proof.

Exercises

1. Deduce Theorem 17.2.1 from Theorem 2. (Some care is required if μ is not a complete measure on $\mathscr{A}$.)

2. Prove directly the following analogue of Theorem 1.

Let μ, ν be σ-finite measures on a σ-algebra $\mathscr{A}$, let L consist of the $\mathscr{A}$-simple functions ϕ on X and let h be a positive function on X such that

$$\int \phi \, d\nu = \int \phi h \, d\mu$$

for all ϕ in L. Then $f \in L^1(\nu)$ if and only if $fh \in L^1(\mu)$, and in this case

$$\int f d\nu = \int fh d\mu.$$

3. At what points in the proof of Theorem 1 have we used the stronger Stone Condition? Could the proof be adapted to give the same conclusion if we merely assume that 1 is measurable for both P and Q?

4. Use the notation introduced after the statement of Theorem 1: show that E is ν-measurable if and only if $E \cap S$ is μ-measurable (cf. Ex. 6.4.5).

5. At what points in the proof of Theorem 2 have we used the stronger Stone Condition? Could the proof be adapted to give the same conclusion if we merely assume that 1 is measurable for both P and Q?

17.4 The dual spaces $\mathscr{L}^p$, $\mathscr{L}^q$

As a final illustration of the Radon–Nikodym Theorem we prove the promised generalisation of Riesz's Theorem 17.1.4 to the spaces $\mathscr{L}^p$. For definiteness let us assume that the spaces L^p and $\mathscr{L}^p$ are defined in terms of a Daniell integral $\int$ with corresponding measure μ. We shall see in Ex. 1. that the proofs may be adapted very simply to take care of the classical case where we begin with a (not necessarily complete) measure μ on a σ-algebra $\mathscr{A}$. In order to apply the Radon–Nikodym Theorem we assume that X is σ-finite. In fact the main result – the duality of $\mathscr{L}^p$ and $\mathscr{L}^q$ – may be established when $p > 1$ by a geometric argument on the lines of the proof already given for Theorem 17.1.4 (Exx. 10–15): this makes no mention of σ-finiteness. On the other hand, the geometrical proof does not apply to the space $\mathscr{L}^1$, in which case we need the Radon–Nikodym Theorem and its consequent assumption of σ-finiteness.

First of all we consider the spaces L^p, L^q.

Theorem 1 (X *σ-finite*). *Let F be a bounded linear functional on $L^p (p \geqslant 1)$. Then there is a measurable function h which satisfies*

$$F(f) = \int fh$$

for all f in L^p. When $p > 1$, $h \in L^q$, where $1/p + 1/q = 1$; and when $p = 1$, h is equal almost everywhere to a bounded function.

Proof. There is a constant M such that

$$|F(f)| \leqslant M \|f\|$$

for all f in L^p. If $|f| \leqslant g$ this gives

$$|F(f)| \leqslant M\|f\| \leqslant M\|g\|,$$

so that F is relatively bounded in the sense of § 10.2. By Proposition 10.2.1 we may express F as the difference of two positive linear functionals which are both bounded by the same constant M. This reduces the proof to the case where F is positive.

If $\phi_n \in L^p$ and $\phi_n \downarrow 0$ then $\int \phi_n{}^p \downarrow 0$ and so $\|\phi_n\| \downarrow 0$. As

$$0 \leqslant F(\phi_n) \leqslant M\|\phi_n\|$$

this implies that $F(\phi_n) \downarrow 0$. In other words F is an elementary integral on L^p. Furthermore, if E is null, then $\chi_E \in L^p$, and $\|\chi_E\| = 0$ gives $F(\chi_E) = 0$. The Radon–Nikodym Theorem of the previous section may now be applied. When $p = 1$, we take $L = L^1$ and the result follows immediately. More generally, let L be the linear lattice of simple functions in L^1: then there is a positive measurable function h such that

$$F(f) = \int fh$$

for all f in L. This equation extends to all positive f in L^p by the Monotone Convergence Theorem and then to all f in L^p by taking differences.

As X is σ-finite, there is an increasing sequence $\{X_n\}$ of sets of finite measure whose union is X.

The case $p > 1$. Define

$$h_n = \min\{h, n\chi_{X_n}\}$$

so that $h_n \uparrow h$ everywhere. As X_n has finite measure, $h_n{}^r \in L^1$ for all $r > 0$; thus

$$\int h_n{}^q \leqslant \int h_n{}^{q-1}h = F(h_n{}^{q-1}) \leqslant M\,\|h_n{}^{q-1}\|_p = M\left\{\int h_n{}^q\right\}^{1/p}.$$

From this it follows that $\left\{\int h_n{}^q\right\}^{1/q} \leqslant M,$

and by the Monotone Convergence Theorem that $h \in L^q$ and $\|h\| \leqslant M$.

The case $p = 1$. We are given that

$$\left|\int fh\right| \leqslant M\int |f|$$

for all f in L^1, and we shall show that $0 \leqslant h \leqslant M$ almost everywhere. Let

$$E = \{x \in X: h(x) > M\}.$$

If $1 \in L^1$ then $\chi_E \in L^1$ and the substitution $f = \chi_E$ gives

$$0 \leqslant \int_E (h - M) \leqslant 0,$$

whence $\mu(E) = 0$. More generally, as X is σ-finite, we deduce in the same way that $\mu(E \cap X_n) = 0$ for all n, from which $\mu(E) = 0$. Thus

$$0 \leqslant h \leqslant M$$

outside the null set E.

In order to interpret the last part of Theorem 1 as a statement about normed linear spaces, we introduce some further notation: a function $g: X \to \mathbf{R}$ is *essentially bounded* if g is equal almost everywhere to a bounded function, and in this case the *essential supremum* of g is

$$\text{ess sup}\, g = \inf\{\sup f \colon f \in \tilde{g}\}.$$

We denote by L^∞ the space of all essentially bounded measurable functions g on X and write

$$\|g\|_\infty = \text{ess sup}\, |g|.$$

In this notation it follows from the most elementary properties of the space L^1 that

$$\left|\int fg\right| \leqslant \|f\|_1 \|g\|_\infty$$

for any f in L^1 and any g in L^∞ (Ex. 3). This may be regarded as an extension of Hölder's Inequality when $p = 1$, $q = \infty$. The most plausible reason for the notation L^∞ is that the relation $1/p + 1/q = 1$ is formally satisfied when we substitute $p = 1$, $q = \infty$.

In line with our previous definition of $\mathscr{L}^p$ we denote by $\mathscr{L}^\infty$ the normed linear space of equivalence classes $\tilde{g}$ with g in L^∞, where

$$\tilde{g} + \tilde{h} = (g + h)^\sim, \quad a\tilde{g} = (ag)^\sim,$$

and

$$\|\tilde{g}\|_\infty = \|g\|_\infty$$

for any representatives g, h of the equivalence classes $\tilde{g}$, $\tilde{h}$ and any a in $\mathbf{R}$.

We may now interpret Theorem 1 as a statement about duality.

Theorem 2 (X *σ-finite*). *Let $p \geqslant 1$, $q > 1$ satisfy the relation*

$$1/p + 1/q = 1.$$

(*Note that we allow $p = 1$, $q = \infty$, but not $p = \infty$, $q = 1$.*) *For any $\tilde{h}$ in $\mathscr{L}^q$ let $F_{\tilde{h}}$ denote the bounded linear functional on $\mathscr{L}^p$ defined by*

$$F_{\tilde{h}}(\tilde{f}) = \int fh$$

for all $\tilde{f}$ in $\mathscr{L}^p$. Then the mapping

$$\theta: \tilde{h} \to F_{\tilde{h}}$$

is a norm-preserving linear mapping of $\mathscr{L}^q$ onto the dual space of $\mathscr{L}^p$.

Proof. It is simpler to do most of the working in L^p and L^q! Given h in L^q, let F_h denote the linear functional on L^p defined by

$$F_h(f) = \int fh$$

for all f in L^p. By Hölder's Inequality,

$$|F_h(f)| \leqslant \|h\|_q \|f\|_p;$$

thus F_h is bounded and $\quad \|F_h\| \leqslant \|h\|_q.$

When $p > 1$, $|h|^q \in L^1$ and $|h|^{q-1} \operatorname{sgn} h \in L^p$. Thus

$$\int |h|^q = F_h(|h|^{q-1} \operatorname{sgn} h) \leqslant \|F_h\| \left\{ \int |h|^q \right\}^{1/p},$$

from which $\quad \|h\|_q \leqslant \|F_h\|.$

When $p = 1$, $q = \infty$, write M as shorthand for $\|F_h\|$; we are given

$$\left| \int fh \right| \leqslant M \|f\|$$

for all f in L^1. Let $\quad E = \{x \in X: |h(x)| > M\}$

and argue as in the proof of Theorem 1, replacing χ_E by $\chi_E \operatorname{sgn} h$, that E is null. Thus $|h| \leqslant M$ almost everywhere in X; in other words

$$\|h\|_\infty \leqslant M.$$

This establishes in all cases that $\|F_h\| = \|h\|_q$. In terms of the quotient spaces $\mathscr{L}^p$ and $\mathscr{L}^q$ we have

$$\|F_{\tilde{h}}\| = \|\tilde{h}\|_q.$$

Thus the mapping θ is norm-preserving.

It is clear that θ is linear and so it only remains to prove that θ is *onto* the dual space of $\mathscr{L}^p$. Suppose that $\tilde{F}$ is a bounded linear functional on $\mathscr{L}^p$ and let F be the bounded linear functional on L^p defined by

$$F(f) = \tilde{F}(\tilde{f})$$

for all f in L^p. According to Theorem 1, $F = F_h$ for some h in L^q and hence $\tilde{F} = F_{\tilde{h}}$. This completes the proof.

It is customary to use the canonical mapping θ to 'identify' $\mathscr{L}^q$ with the dual space of $\mathscr{L}^p$. When $p, q > 1$ this allows us to think of $\mathscr{L}^p$ and

$\mathscr{L}^q$ as dual spaces of each other (Ex. D). It is important to note that Theorem 2 is not true in general when $p = \infty$, $q = 1$: for example, when μ is Lebesgue measure on $\mathbf{R}^k$ or the counting measure on $\{1, 2, \ldots\}$ (Exx. E, F).

Exercises

1. Let μ be a measure on a σ-algebra $\mathscr{A}$ of subsets of X. Establish Theorem 2 for the spaces $\mathscr{L}^p(\mu)$, $\mathscr{L}^q(\mu)$.

2. Suppose that $g(x) \leqslant M$ for almost all x in X; then we call M an *essential upper bound* of g. Show that ess sup g is the smallest essential upper bound of g.

3. Establish the inequality

$$\left|\int fg\right| \leqslant \|f\|_1 \|g\|_\infty$$

for any f in L^1 and any g in L^∞.

4. Let X consist of two points a, b and define $\mu(\varnothing) = 0$, $\mu(\{a\}) = 1$, $\mu(\{b\}) = \infty$, $\mu(X) = \infty$. In this case find the spaces $L^1(\mu)$, $L^\infty(\mu)$ and their dual spaces.

5. Use Ex. 4 to show that the σ-finiteness of X cannot be dropped from Theorem 2 when $p = 1, q = \infty$. At what point(s) does the proof break down?

6. We say that the measure μ has the *finite subset property* (Zaanen [22]) if any set E for which $\mu(E) = \infty$ contains a subset A for which $0 < \mu(A) < \infty$.

Let $h \in L^\infty(\mu)$ and define the linear functional F_h on $L^1(\mu)$ as before, viz.

$$F_h(f) = \int fh\, d\mu$$

for all f in $L^1(\mu)$. Show that $\|F_h\| \leqslant \|h\|_\infty$.

Show also that $\|F_h\| = \|h\|_\infty$ for all h in $L^\infty(\mu)$ if and only if μ has the finite subset property. (Cf. Exx. 4, 5.)

7. For any h in L^1 let F_h denote the bounded linear functional on L^∞ defined by

$$F_h(f) = \int fh$$

for all f in L^∞. Show that $\|F_h\| = \|h\|_1$.

Let $F_{\tilde{h}}$ be the induced linear functional on $\mathscr{L}^\infty$ $(F_{\tilde{h}}(\tilde{f}) = F_h(f))$ and show that the mapping $\theta \colon \tilde{h} \to F_{\tilde{h}}$ is a norm-preserving linear mapping of $\mathscr{L}^1$ *into* the dual space of $\mathscr{L}^\infty$.

8. Let X be σ-finite. Show that a bounded linear functional F on L^∞ is of the form F_h $(h \in L^1)$, as in Ex. 7, if and only if F satisfies the Daniell condition

$$\phi_n \downarrow 0 \Rightarrow F(\phi_n) \to 0 \quad (\phi_n \in L^\infty).$$

9. When $1 < p, q < \infty$, Theorems 1 and 2 may be extended to the non-σ-finite case. Establish the steps as follows.

(i) Let F be a bounded linear functional on L^p. For any σ-finite measurable subset E of X there is a function h_E in L^q vanishing outside E which satisfies

$$F(f) = \int f h_E$$

for all f in L^p vanishing outside E.

(ii) If A is a σ-finite measurable subset of E, then $h_A = h_E$ almost everywhere on A.

(iii) $\|h_E\|_q \leqslant \|F\|$.

(iv) Let $M = \sup \|h_E\|$ where the supremum is taken over all σ-finite measurable subsets E of X. There is an increasing sequence $\{E_n\}$ of such sets whose union S is σ-finite and $M = \|h_S\|$.

(v) If E is a σ-finite measurable set containing S, then $h_E = h_S$ almost everywhere on X.

(vi) The function $h = h_S$ satisfies the conclusion of Theorem 1.

(vii) $\|F\| = \|h\|_q$ and the conclusion of Theorem 2 holds.

Clarkson's Inequalities

In the following exercises $p, q > 1$ *and* $1/p + 1/q = 1$.

10. Suppose that $0 < s \leqslant 1$ and let L^s consist of all measurable functions f for which $|f|^s \in L^1$. Show that L^s is a linear space. If $\|f\| = (\int |f|^s)^{1/s}$ show that

$$\|f+g\| \geqslant \|f\| + \|g\|$$

for any *positive* f, g in L^s.

(Note that the sense of Minkowski's Inequality has been reversed.)

11. Let
$$f(x) = (1+x^{1/p})^p + (1-x^{1/p})^p$$
for $0 < x < 1$.

(i) If $p \geqslant 2$ show that $f''(x) \leqslant 0$ for $0 < x < 1$ and deduce that

$$f(x) \leqslant f(c) + (x-c)f'(c)$$

for $0 < x, c < 1$. Now let $c \uparrow 1$ and deduce that

$$f(x) \leqslant 2^{p-1}(x+1)$$

for $0 < x < 1$.

Substitute $c = x^q$ and deduce that

$$f(x) \leqslant 2(1+x^{q/p})^{p/q}$$

for $0 < x < 1$.

(ii) If $1 < p \leqslant 2$ show that these inequalities hold in the reverse sense.

12. (i) Use the inequalities of Ex. 11 to prove that, if $p \geqslant 2$, then

$$2\{|a|^p + |b|^p\} \leqslant |a+b|^p + |a-b|^p \leqslant 2^{p-1}\{|a|^p + |b|^p\},$$
$$|a+b|^p + |a-b|^p \leqslant 2\{|a|^q + |b|^q\}^{p/q},$$
$$2\{|a|^p + |b|^p\}^{q/p} \leqslant |a+b|^q + |a-b|^q$$

for all real numbers a, b. Deduce *Clarkson's Inequalities* [6] for L^p ($p \geqslant 2$):

$$2\{\|f\|^p + \|g\|^p\} \leqslant \|f+g\|^p + \|f-g\|^p \leqslant 2^{p-1}\{\|f\|^p + \|g\|^p\},$$

$$\|f+g\|^p + \|f-g\|^p \leqslant 2\{\|f\|^q + \|g\|^q\}^{p/q},$$

$$2\{\|f\|^p + \|g\|^p\}^{q/p} \leqslant \|f+g\|^q + \|f-g\|^q \quad (f, g \in L^p).$$

(ii) Show that these inequalities hold in the reverse sense if $1 < p \leqslant 2$.

13. Let S be a closed convex set in $\mathscr{L}^p$ ($p > 1$). Use Clarkson's Inequalities to prove that S contains a unique element of smallest norm. Does this result hold for $\mathscr{L}^1$?

14. Let F be a bounded linear functional on $\mathscr{L}^p$ for which $\|F\| = 1$. Show that there is a unique point $\tilde{h}$ of $\mathscr{L}^p$ on the unit sphere $\{\tilde{f} \in \mathscr{L}^p: \|\tilde{f}\| = 1\}$ such that $F(\tilde{h}) = 1$. Show that $\tilde{h}$ is the unique point of the hyperplane $\{\tilde{f} \in \mathscr{L}^p: F(\tilde{f}) = 1\}$ nearest to the origin.

15. In the light of Exx. 13, 14 adapt the proof of Theorem 17.1.4 and hence establish Theorems 1 and 2 of the present section without the restriction that X should be σ-finite (but $1 < p, q < \infty$).

The Hahn–Banach Theorem

Any thoroughgoing discussion of duality for normed linear spaces requires the Hahn–Banach Theorem. The general proof requires a transfinite argument but the proof given in Ex. A may be adapted easily to give a proof by 'finite induction' when the space B in question contains a countable dense subset.

A. ***The Hahn–Banach Theorem.*** *Let M be a linear subspace of the normed linear space B. Then any bounded linear functional F on M may be extended to a bounded linear functional T on B with the same norm.*

Prove this theorem using the following steps.

(i) Suppose that $\|F\| = 1$ and that $y \in B \setminus M$. Show that

$$-\|x_1+y\| - F(x_1) \leqslant \|x_2+y\| - F(x_2)$$

for all x_1, x_2 in M. Deduce that there is a real number r satisfying

$$-\|x+y\| - F(x) \leqslant r \leqslant \|x+y\| - F(x)$$

for all x in M.

Let N be the linear space spanned by M and $\{y\}$, and define the linear functional G on N by setting

$$G(x+ay) = F(x) + ar$$

for all x in M and all a in $\mathbf{R}$. Show that G extends F and $\|G\| = 1$.

(ii) Consider the class $\mathscr{C}$ of all bounded linear functionals G which extend F to some linear subspace M_G containing M and for which $\|G\| = 1$. We partially order $\mathscr{C}$ by writing

$$G_1 \prec G_2 \quad \text{if} \quad G_2 \text{ extends } G_1$$

(more precisely $G_1 \prec G_2$ if and only if $M_{G_1} \subset M_{G_2}$ and $G_1(x) = G_2(x)$ for all x in M_{G_1}). Use Zorn's Lemma to find a maximal element T of $\mathscr{C}$ with respect to this partial ordering.

(iii) Use (i) to prove that $M_T = B$.

(iv) Deduce the Hahn–Banach Theorem.

B. The Hahn–Banach Theorem ensures the existence of many bounded linear functionals on B: for example, to each point b of B, other than the origin, there is at least one linear functional T on B with $\|T\| = 1$ such that $T(b) = \|b\|$. Verify this.

C. Let B be a normed linear space (other than $\{0\}$) and denote by B^* the *dual space* of B, viz. the linear space of all bounded linear functionals F on B with the norm given by

$$\|F\| = \sup\{F(x): x \in B, \|x\| \leqslant 1\}.$$

Likewise B^{**} denotes the dual space of B^* and is called the *second dual* of B.

To any x in B and any F in B^* there is a real number $F(x)$. Show that the mapping ϕ_x defined by

$$\phi_x(F) = F(x)$$

for all F in B^* is an element of B^{**}. Show that the mapping ϕ defined by $\phi(x) = \phi_x$ for all x in B, is a linear mapping of B into B^{**}, and that $\|\phi_x\| = \|x\|$ for all x in B.

D. The mapping ϕ of Ex. C is of fundamental importance: when $\phi(B) = B^{**}$, i.e. when ϕ is *onto* B^{**}, we say that the space B is *reflexive*. Show that the spaces $\mathscr{L}^p$ for $1 < p < \infty$ are reflexive. [It is strictly in terms of the mapping ϕ that we may regard $\mathscr{L}^p$ and $\mathscr{L}^q$ as dual spaces of each other when $1 < p, q < \infty$.]

E. Let μ be the counting measure on $X = \{1, 2, 3, \ldots\}$ and recall that $l^\infty = L^\infty(\mu) = \mathscr{L}^\infty(\mu)$ consists of all bounded sequences $\{s_n\}$ of real numbers. Denote by M the linear subspace of l^∞ consisting of all convergent sequences $\{s_n\}$. Show that

$$F(\{s_n\}) = \lim s_n$$

defines a bounded linear functional on M. Extend F by the Hahn–Banach Theorem to a bounded linear functional T on l^∞ with the same norm and show that T does not satisfy the Daniell condition. What can we deduce from Ex. 8?

F. Extend Ex. E to the case where X is infinite but σ-finite with respect to a measure μ. Apply in particular to Lebesgue measure on $\mathbf{R}^k$.

G. Let C denote the linear space of continuous real valued functions on the compact interval $[-1, 1]$ with norm given by $\|f\| = \sup |f(x)|$, and let μ be Lebesgue measure on $[-1, 1]$. Show that C may be regarded as a linear subspace of $\mathscr{L}^\infty(\mu)$.

The Dirac δ-function F on C is defined by setting

$$F(f) = f(0)$$

(as in Example 3 of § 8.1). Extend F to a bounded linear functional T on $\mathscr{L}^\infty$ with norm 1 and show that T does not satisfy the Daniell condition. Deduce that the mapping θ of Ex. 7 carries $\mathscr{L}^1$ onto a proper subspace of $(\mathscr{L}^\infty)^*$. (In the context of this example why could we not extend F to $\mathscr{L}^\infty$ by means of the Daniell construction?)

APPENDIX

In the text there are several closely related ideas involving collections of sets. For convenience we list their definitions and some of the corresponding extension theorems.

Let $\mathscr{C}$ be a non-empty collection of subsets of X and let $\overline{\mathscr{C}}$ denote the collection of sets

$$E_1 \cup \ldots \cup E_s,$$

where $E_1, \ldots, E_s$ are disjoint elements of $\mathscr{C}$; by convention we agree to include in $\overline{\mathscr{C}}$ the empty union $\varnothing$ (corresponding to $s = 0$). The *Disjointness Condition* for $\mathscr{C}$ now says (p. 86):

$$A, B \in \mathscr{C} \Rightarrow A \cap B, A \setminus B \in \overline{\mathscr{C}}.$$

If $\mathscr{C}$ satisfies the stronger condition:

$$A, B \in \mathscr{C} \Rightarrow A \cap B \in \mathscr{C} \quad \text{and} \quad A \setminus B \in \overline{\mathscr{C}},$$

then $\mathscr{C}$ is a *semi-ring* (p. 86). [For example, the intervals in $\mathbf{R}^k$ form a semi-ring.]

A *ring* $\mathscr{R}$ of subsets of X satisfies the condition (p. 79):

$$A, B \in \mathscr{R} \Rightarrow A \cup B, A \setminus B \in \mathscr{R}.$$

[For example, the elementary figures in $\mathbf{R}^k$ form a ring: Ex. 3.1.6.]

According to Proposition 11.2.2, $\mathscr{C}$ satisfies the Disjointness Condition if and only if $\overline{\mathscr{C}}$ is a ring. A ring of sets is a ring in the customary algebraic sense if we interpret $\triangle$ and $\cap$ as addition and multiplication respectively (Ex. 11.1.C).

A *σ-ring* of subsets of X is a ring which is closed with respect to countable unions (p. 92).

An *algebra* or a *σ-algebra* of subsets of X is a ring or a σ-ring, respectively, which includes the 'universal' set X as one of its members (pp. 79, 92). [For example, the Lebesgue measurable sets form a σ-algebra of subsets of $\mathbf{R}^k$: Theorem 6.2.1.]

A collection $\mathscr{M}$ of subsets of X which satisfies the following condition (p. 114) is said to be *monotone*:

$$A_n \in \mathscr{M} \ (n \geq 1) \quad \text{and} \quad A_n \uparrow A \Rightarrow A \in \mathscr{M},$$

$$A_n \in \mathscr{M} \ (n \geq 1) \quad \text{and} \quad A_n \downarrow A \Rightarrow A \in \mathscr{M}.$$

A monotone collection need not be a ring (Ex. 13.1.1), but a monotone ring is a σ-ring (Proposition 13.1.1).

Given any collection $\mathscr{C}$ of subsets of X, the intersection of all rings containing $\mathscr{C}$ is itself a ring, viz. the smallest ring containing $\mathscr{C}$: we call this ring the *ring generated by* $\mathscr{C}$ (Theorem 11.1.1). [For example, the intervals in $\mathbf{R}^k$ generate the ring of elementary figures: Proposition 11.2.2 (ii).] In exactly the same way we may establish the existence of a smallest algebra, σ-ring, σ-algebra and monotone collection containing $\mathscr{C}$ and call these respectively the *algebra*, *σ-ring*, *σ-algebra* and *monotone collection generated by* $\mathscr{C}$ (see also Theorem 11.4.1). This argument does not apply to semi-rings: Ex. 11.4.8. [The intervals in $\mathbf{R}^k$ generate the σ-algebra of Borel sets (p. 158) which is smaller than the σ-algebra of Lebesgue measurable sets (p. 159).] Given a ring $\mathscr{R}$ of subsets of X; the σ-ring generated by $\mathscr{R}$ coincides with the monotone collection generated by $\mathscr{R}$ (Theorem 13.1.1). This result is used in proving the uniqueness theorems at the end of the Appendix.

A function $\mu: \mathscr{C} \to \mathbf{R}$ is *additive* if

$$\mu(A_1 \cup \ldots \cup A_m) = \mu(A_1) + \ldots + \mu(A_m)$$

for any disjoint sets $A_1, \ldots, A_m$ in $\mathscr{C}$ whose union is also in $\mathscr{C}$ (p. 87). If $\mathscr{C}$ satisfies the Disjointness Condition then $\overline{\mathscr{C}}$ is the ring generated by $\mathscr{C}$, and any additive function μ on $\mathscr{C}$ may be extended uniquely to an additive function on $\overline{\mathscr{C}}$. This may be done directly by defining

$$\mu(E_1 \cup \ldots \cup E_s) = \mu(E_1) + \ldots + \mu(E_s)$$

for disjoint sets $E_1, \ldots, E_s$ in $\mathscr{C}$ (and checking the consistency), or by integrating *$\mathscr{C}$-simple functions*

$$\phi = a_1\chi_{A_1} + \ldots + a_m\chi_{A_m}$$

($a_1, \ldots, a_m \in \mathbf{R}$; $A_1, \ldots, A_m \in \mathscr{C}$) as in § 11.2, 3.

Let $[0, \infty]$ denote the 'interval' $[0, \infty) \cup \{\infty\}$. A function

$$\mu: \mathscr{C} \to [0, \infty]$$

is *σ-additive* (or *countably additive*) if

$$\mu(\bigcup A_i) = \Sigma\mu(A_i)$$

for any countable collection $\{A_i\}$ of disjoint sets in $\mathscr{C}$ whose union is also in $\mathscr{C}$ (with a suitable interpretation of infinite values on both sides). A σ-additive function $\mu: \mathscr{C} \to [0, \infty]$ which takes at least one finite value is a *measure* on $\mathscr{C}$ (p. 92).

If $\mathscr{C}$ satisfies the Disjointness Condition, then any measure on $\mathscr{C}$

may be extended uniquely to a measure on the ring $\overline{\mathscr{C}}$ generated by $\mathscr{C}$ (Proposition 11.4.1). A measure μ on $\mathscr{C}$ is *complete* if any subset of a null set is null; more precisely, if

$$B \subset A,\ A \in \mathscr{C} \text{ and } \mu(A) = 0 \Rightarrow B \in \mathscr{C} \text{ (and hence } \mu(B) = 0).$$

[For example, Lebesgue measure m on $\mathbf{R}^k$ is complete, but the restriction of m to the ring of elementary figures is incomplete for $k \geqslant 2$ (p. 96).] The main extension theorem states that any measure μ on a ring $\mathscr{R}$ may be extended to a complete measure $\overline{\mu}$ on a σ-algebra $\mathscr{M}$ containing $\mathscr{R}$ (Theorem 11.4.2). The proof in § 11.4 uses the Daniell construction to extend an elementary integral on the $\mathscr{R}$-simple functions. The purely measure theoretic construction of Carathéodory is given in § 13.5, and this produces exactly the same extension $\overline{\mu}$.

Let μ be a measure on the ring $\mathscr{R}$. A subset A of X is *σ-finite* with respect to μ if A is the union of a sequence $\{A_n\}$ of sets A_n in $\mathscr{R}$ for each of which $\mu(A_n)$ is finite. The measure μ is *σ-finite* if every A in $\mathscr{R}$ is σ-finite with respect to μ. If X is σ-finite with respect to μ, then it follows that μ is σ-finite (p. 116). [For example, $\mathbf{R}^k$ is σ-finite with respect to Lebesgue measure.] There are two fundamental uniqueness theorems.

Let $\mathscr{A}$ be the σ-algebra of subsets of X generated by $\mathscr{R}$. If X is σ-finite with respect to μ, then there is one and only one measure extending μ to $\mathscr{A}$ (Theorem 13.2.1).

Let $\mathscr{S}$ be the σ-ring generated by $\mathscr{R}$. If μ is σ-finite, then there is one and only one measure extending μ to $\mathscr{S}$ (Ex. 13.2.5).

SOLUTIONS

§8.1

1. (i) Suppose that $\phi_n \downarrow 0$ and that $\phi_1(x) = 0$ for all $x > M$; then $\phi_n(x) = 0$ for all $x > M$. Given any $\epsilon > 0$, there is an integer N such that $\phi_n(x) < \epsilon/M$ for $x = 1, 2, \ldots, M$ and all $n \geqslant N$; thus $\int \phi_n < \epsilon$ for all $n \geqslant N$.

(ii) Suppose that f is positive and $\Sigma f(x)$ is convergent. For $n \geqslant 1$ we may define $\phi_n(x) = f(x)$ if $x \leqslant n$, $= 0$ if $x > n$. Then $\phi_n \uparrow f$ and so $f \in L^{\text{inc}}$. We may also allow f to have a finite number of strictly negative terms. We shall now show that L^{inc} consists of sequences f with finitely many strictly negative terms and for which $\Sigma f(x)$ is convergent.

Suppose that $\phi_n \uparrow f$ and $\int \phi_n \leqslant K$. As $\phi_1(x) = 0$ for $x > M$, we have $\phi_n(x) \geqslant 0$ and $f(x) \geqslant 0$ for $x > M$. Now

$$\phi_n(1) + \phi_n(2) + \ldots + \phi_n(x) \leqslant K$$

for any $n \geqslant 1$ and any $x > M$. Keep x fixed and let $n \to \infty$: then

$$f(1) + f(2) + \ldots + f(x) \leqslant K.$$

From this it follows that $\Sigma f(x)$ is (absolutely) convergent to a sum s, say. From the definition of $\int f$ as the limit of $\int \phi_n$ it is clear that $\int f \leqslant s$. To any $\epsilon > 0$, find $m > M$ such that

$$\sum_{x=1}^{m} f(x) > s - \epsilon.$$

For each integer $x \leqslant m$ we find an integer n_x such that

$$\phi_n(x) > f(x) - \epsilon/m$$

for $n \geqslant n_x$ and let $N = \max\{n_1, \ldots, n_m\}$. Then

$$\sum_{x=1}^{m} \phi_N(x) > s - 2\epsilon$$

and so

$$s - 2\epsilon < \int \phi_n \leqslant s$$

for all $n \geqslant N$. This shows finally that

$$s = \int f = \lim \int \phi_n.$$

(iii) Immediate from (ii).

2. Let $\phi_n \uparrow f$ outside the null set S_0. The sets $S_n = \{x: \phi_n(x) \neq 0\}$ are null for $n \geqslant 1$. Thus f vanishes outside the null set $\bigcup_{n=0}^{\infty} S_n$.

3. Clearly $\{0\}$ is not null. The increasing sequence $\{\phi_n\}$ defined by $\phi_n(x) = n\,\|x\|$ diverges on the set $A = \mathbf{R}^k \setminus \{0\}$ and $\int \phi_n = 0$: thus A is null. The rest is easy.

4. Cf. §8.3.

5. (i) Consider $(\phi - \phi_n)^+$ as in the proof of Lemma 2.
(ii) let $\phi_n = \psi_1 + \ldots + \psi_n$ in (i).

6. (i) The function ψ defined by $\psi(x^3) = \phi(x)$ for all x in $\mathbf{R}$ is again a step function and $3P(\phi) = \int \psi$ by Proposition 5.1.3 (or first principles in this simple case). If $\phi_n \downarrow 0$ then $\psi_n \downarrow 0$ and the Daniell condition for P follows from the same condition for the Lebesgue integral (Lemma 3.2.1).

(ii) For any subset B of $\mathbf{R}$ let $A = \{x\colon x^3 \in B\}$. Then A is P-null if and only if B is Lebesgue null: for

$$\phi_n(x) \uparrow \infty \quad (x \in A) \quad \text{if and only if} \quad \psi_n(x^3) \uparrow \infty \quad (x^3 \in B).$$

The rest follows the construction of $L^1(P)$ via $L^{\text{inc}}(P)$. (It may help to consult §8.3.)

7. Let $f = g - h$, where $g, h \in L^{\text{inc}}$. There exist $\theta_n \uparrow g$ a.e. and $\psi_n \uparrow h$ a.e. Thus $\phi_n = \theta_n - \psi_n \to f$ a.e. and

$$\int |\phi_n - f| \leqslant \int |\theta_n - g| + \int |\psi_n - h| \to 0.$$

If f is positive then we may as well replace ϕ_n by ϕ_n^+.

A. In the proof of Lemma 2 simply drop the assumption that the integrals are bounded and allow infinite values.

B. Suppose that $P^*(f)$ and $P^*(g)$ are both finite. Then there exist $\phi_n \uparrow u \geqslant f$, $\psi_n \uparrow v \geqslant g$, where $P(u)$, $P(v)$ are finite. As u, v take their values in $\mathbf{R} \cup \{\infty\}$, we naturally let $(u+v)(x) = \infty$ if $u(x) = \infty$ or $v(x) = \infty$ (or both): with this understanding $u + v \in L^+$ and $u + v \geqslant f + g$ no matter what value in   is given to $(f+g)(x)$ when $f(x)$ or $g(x)$ is infinite. By Ex. A,

$$P(u+v) = P(u) + P(v)$$

and so
$$P^*(f+g) \leqslant P^*(f) + P^*(g). \tag{1}$$

As a particular case, if $P^*(f)$ and $P_*(f)$ are both finite, we may substitute $g = -f$ and deduce that

$$P_*(f) \leqslant P^*(f).$$

[These inequalities may be extended quite simply to allow for infinite values: the only proviso is that the right hand side of (1) must not be of the form $-\infty + \infty$ or $\infty + (-\infty)$.]

When $f, g \in L_1$ we combine these two results to obtain

$$P_*(f) + P_*(g) \leqslant P_*(f+g) \leqslant P^*(f+g) \leqslant P^*(f) + P^*(g).$$

As the extremes are equal we deduce that $f + g \in L_1$ and

$$Q(f+g) = Q(f) + Q(g).$$

For any positive real number a,

$$P^*(af) = aP^*(f)$$

and for any negative real number a,

$$P^*(af) = aP_*(f).$$

Thus, for any f in L_1 and any a in $\mathbf{R}$, $af \in L_1$ and

$$Q(af) = aQ(f).$$

C. Let h vanish outside the null set A. To any $\epsilon > 0$, there exists an increasing sequence $\{\phi_n\}$ of positive elementary functions such that $\phi_n(x) \to \infty$ for all x in A and $P(\phi_n) < \epsilon$. Let $\phi_n \to u$; then $u \in L^+$, $u \geqslant h$ and $P(u) \leqslant \epsilon$, from which $P^*(h) \leqslant 0$. In the same way $P_*(h) \geqslant 0$. So $h \in L_1$ and $Q(h) = 0$.

D. Suppose that $f \in L^{\text{inc}}$ in the extended sense that $f: X \to \bar{\mathbf{R}}$, $\phi_n \uparrow f$ almost everywhere and $P(\phi_n) \leqslant K$. Let $\phi_n \uparrow u$ everywhere; then $f = u$ almost everywhere and $u \in L^+$. Now $P(u) = \lim P(\phi_n)$ is finite and $P^*(u) = P(u)$; moreover $\phi_n \uparrow u$ and $-\phi_n \in L^+$, from which

$$P_*(u) \geqslant \lim P(\phi_n) = P(u).$$

Hence $u \in L_1$.

Now $f = u+h$, where h is null and so belongs to L_1 by Ex. C. Thus $L^{\text{inc}} \subset L_1$ and on taking differences, $L^1 \subset L_1$.

E. To each $n \geqslant 1$ there is an element u_n in L^+ such that $u_n \geqslant f$ and

$$P(u_n) - P^*(f) < 1/2n.$$

As $P(u_n)$ is finite it follows that $u_n \in L^{\text{inc}}$ (allowing values in $\mathbf{R} \cup \{\infty\}$). In the same way we find $-l_n$ in L^{inc} (or, if you like, l_n in L^{dec}) such that $l_n \leqslant f$ and

$$P_*(f) - P(l_n) < 1/2n.$$

The result follows as $P_*(f) = P^*(f)$.

Replace u_n by $\min\{u_1, \ldots, u_n\}$ and l_n by $\max\{l_1, \ldots, l_n\}$. Let $u_n \downarrow h$ and $l_n \uparrow g$, say, so that $g \leqslant f \leqslant h$. The Monotone Convergence Theorem 8.2.2 now shows that $g, h \in L^1$ and $P(h-g) = 0$ so that $g = h$ almost everywhere and $f \in L^1$ by Theorem 2.

F. Let $f \in L_1$. According to the last part of Ex. E,

$$\lim P(l_n) = P(f) = \lim P(u_n).$$

But
$$\lim P(l_n) \leqslant P_*(f) \leqslant P^*(f) \leqslant \lim P(u_n)$$

and so
$$Q(f) = P(f).$$

G. First of all assume that $A^*(f), B^*(f) \in \mathbf{R}$.

$$C(u) = A(u) + B(u) \quad \text{for any } u \text{ in } L^+.$$

Take the infimum for all $u \geqslant f$ to obtain

$$C^*(f) \geqslant A^*(f)+B^*(f).$$

On the other hand, to any $\epsilon > 0$, there exist functions u, v in L^+ such that $u \geqslant f$, $v \geqslant f$ and

$$A(u) < A^*(f)+\epsilon, \quad B(v) < B^*(f)+\epsilon.$$

It is also clear that $w = \min\{u, v\} \in L^+$ (taking the limit of an increasing sequence $\min\{\phi_n, \psi_n\}$). But then $w \geqslant f$ and so

$$C^*(f) \leqslant C(w) = A(w)+B(w) < A^*(f)+B^*(f)+2\epsilon.$$

As ϵ is arbitrarily small,

$$C^*(f) \leqslant A^*(f)+B^*(f).$$

The various interpretations with infinite values may be verified from first principles. For example, if $A^*(f) \in \mathbf{R}$ and $B^*(f) = \infty$, then as u runs through all elements of L^+ satisfying $u \geqslant f$ (there *are* some as $A^*(f) \in \mathbf{R}$), we have $\{A(u)\}$ bounded below, $\{B(u)\} = \{\infty\}$ and so $\{C(u)\} = \{\infty\}$; our equation reads

$$\infty = a+\infty \quad (a \in \mathbf{R})$$

in this case.

For the last part, suppose that $f \in L_1(C)$; then

$$C_*(f) = A_*(f)+B_*(f) \leqslant A^*(f)+B^*(f) = C^*(f),$$

from which it follows by the squeezing argument that $f \in L_1(A) \cap L_1(B)$. Thus

$$L_1(C) \subset L_1(A) \cap L_1(B).$$

The reverse inclusion is proved in almost the same way.

§8.2

1. Let $\phi_n \downarrow 0$. There is a countable subset of $[0, 1]$ outside which each ϕ_n takes a constant value c_n; thus $P(\phi_n) = c_n \downarrow 0$.

The null subsets are the countable subsets of $[0, 1]$ and $L^{\text{inc}}(P) = L^1(P)$ consist of the functions which are constant outside a countable subset of $[0, 1]$.

2. Use the Absolute Convergence Theorem.

3. Apply the Monotone Convergence Theorem to $\{f^+\chi_{S_n}\}$, $\{f^-\chi_{S_n}\}$ and subtract.

4. Apply the Dominated Convergence Theorem with $g = K$.

5. In the last paragraph of the proof of Theorem 5, $\{l_n\}$ converges to $\liminf f_n$.

6. As $\liminf f_n = 0$, the left hand side is zero, but the right hand side is $\min\{m(E), 1-m(E)\}$ which is generally different from zero. Thus the inequality in Fatou's Lemma may be strict.

§8.3

2. The Daniell condition for P is a special case of the Monotone Convergence Theorem for Q. Let $L(Q)$ be the linear lattice of step functions on $\mathbf{R}^k$ and apply Theorem 1.

3. As in Ex. 2 let Q be the Lebesgue integral on R and $L(Q)$ the lattice of step functions. The functions described in Ex. 1 belong to $L = L(P)$ and so we may apply Theorem 1.

§8.4

2. Let $g \in L^1$: according to Ex. 8.1.2, g vanishes outside a countable subset of X and so the same is true of $\text{mid}\{-g, f, g\}$ for any $f: X \to \mathbf{R}$.

5. Null = empty; $L^{\text{inc}} = L^1 = L$ as before, but the measurable functions may take any (real) value at 0. The only measurable sets are $\varnothing$ and $\{0\}$ whose measures are $0, \infty$, respectively.

6. Any subset of $(0, 1]$ is null but $\{0\}$ is not null. $L^{\text{inc}} = L^1$ consist of all real valued functions f on $[0, 1]$ with $f(0) = 0$. Any real valued function on $[0, 1]$ is measurable, so all subsets of $[0, 1]$ are measurable, $\mu(0, 1] = 0$ and $\mu(\{0\}) = \infty$.

8. fg is measurable and $|fg| \leqslant \frac{1}{2}(f^2 + g^2)$. The given inequality is obvious if $f = 0$ a.e. so we may assume without loss of generality that $\int f^2 > 0$. Then

$$t^2 \int f^2 - 2t \int fg + \int g^2 = \int (tf - g)^2 \geqslant 0$$

for all t in $\mathbf{R}$. The substitution $t = \int fg / \int f^2$ gives the required inequality. We can only have equality if $g = tf$ a.e. (with a similar conclusion if we assume that $\int g^2 > 0$).

9. Suppose that $|g| \leqslant K$. Then fg is measurable and

$$|fg| \leqslant K|f| \in L^1.$$

10. Let $\{\phi_n\}$ be a sequence of simple functions converging everywhere to f. Then

$$\frac{\phi_n}{1 + |\phi_n|} \to \frac{f}{1 + |f|},$$

so that this limit is measurable and dominated by 1.

11. $f|g|$ is measurable and $p|g| \leqslant f|g| \leqslant q|g|$ a.e. so

$$p \int |g| \leqslant \int f|g| \leqslant q \int |g|.$$

No: in the case of the Lebesgue integral on $\mathbf{R}$ take $f = g = -\chi_{(-1,0)} + \chi_{(0,1)}$, $p = -1, q = 1$.

13. (i) Argue as on p. 29.

(ii) Each f_i vanishes outside a set S_i of finite measure. Let $S = S_1 \cup \ldots \cup S_k$; then S has finite measure and the simple function $F - h(0, \ldots, 0)\chi_X$ vanishes outside S.

15. Let A be a measurable subset of $[0, 1]$ and $TA = B$ a non-measurable subset of $[0, 2]$. Extend T^{-1} to a strictly increasing continuous function $f: \mathbf{R} \to \mathbf{R}$ and take $h = \chi_A$. Then $h \circ f = \chi_B$ is not measurable.

16. Let $\{S_n\}$ be a sequence of disjoint sets belonging to $\mathscr{M}$; their union S also belongs to $\mathscr{M}$. We have to show that

$$\nu(S) = \Sigma \nu(S_n).$$

For any $A \subset S$ with χ_A in L^1 let $A_n = A \cap S_n$. Then $\chi_{A_n} \in L^1$ and so by Theorem 2

$$\nu(A) = \Sigma \nu(A_n) \leqslant \Sigma \nu(S_n),$$

from which

$$\nu(S) \leqslant \Sigma \nu(S_n).$$

The reverse inequality is obvious when $\nu(S) = \infty$, so assume that $\nu(S) < \infty$ and therefore all $\nu(S_n) < \infty$. To any $\epsilon > 0$ and any $n \geqslant 1$ we may find a set $A_n \subset S_n$ with $\chi_{A_n} \in L^1$ such that

$$\nu(A_n) > \nu(S_n) - \epsilon/2^n.$$

Let $A = \bigcup A_n$; then

$$\nu(S) \geqslant \nu(A) = \Sigma \nu(A_n) > \Sigma \nu(S_n) - \epsilon.$$

As ϵ is arbitrarily small,

$$\nu(S) \geqslant \Sigma \nu(S_n).$$

17. (i) Let m be fixed. To any x in X, there is an integer n such that

$$|f_r(x) - f(x)| < 1/m \quad \text{for all } r \geqslant n,$$

i.e. $x \notin A_{mn}$. Thus $\bigcap_n A_{mn} = \varnothing$.

(ii) $\mu(A_{mn}) \downarrow 0$ as $n \to \infty$. $\mu(A) \leqslant \sum_m \mu(A_{m\,n(m)}) < \epsilon$. If $x \notin A$ then $x \notin A_{m\,n(m)}$ and so

$$|f_r(x) - f(x)| < 1/m \quad \text{for} \quad r \geqslant n(m).$$

(iii) Let $f_n \to f$ outside a null set S. Replace the set A of (ii) by $A \cup S$.

18. Clearly $f_n(x) \to 1$ for all x. The only set of measure less than 1 is $\varnothing$ and f_n does not converge uniformly to 1 because $\|f_n - 1\| = 1$ for all n. This means that the condition $\mu(X) < \infty$ cannot be dropped.

§8.5

1. The Daniell condition for P follows from the Monotone Convergence Theorem for the Lebesgue integral. Let m denote Lebesgue measure on $\mathbf{R}$. The null sets in X are the subsets A of $\mathbf{R}$ with $m(A) = 0$. The integrable functions f on X vanish at ω and are Lebesgue integrable when restricted to $\mathbf{R}$. The measurable functions need not vanish at ω but are Lebesgue

measurable when restricted to $\mathbf{R}$. $\mathcal{M}$ consists of all sets A and $A \cup \{\omega\}$, where A is a Lebesgue measurable subset of $\mathbf{R}$: $\mu(A) = m(A)$ and $\mu(A \cup \{\omega\}) = \infty$ (in particular $\mu(\{\omega\}) = \infty$).

We can also define $\nu(A) = m(A)$ and $\nu(A \cup \{\omega\}) = m(A)$ (in particular $\nu(\{\omega\}) = 0$).

2. In establishing Stone's Theorem we only used sets E for which $\chi_E \in L^1$. For any such E, $\lambda(E) = \mu(E) = \nu(E) = P(\chi_E)$. For any other E in $\mathcal{M}$, $\mu(E) = \infty$ and $\nu(E)$ is as small as possible compatible with the conclusion of Stone's Theorem.

§9.1

1. The proof is virtually the same as for Lemma 3.2.1, but the length of an interval I is replaced by the variation of G on I.

2. The discontinuities p of F form a countable set. Let the values of F be bounded below by A. The open intervals $(F(p-0), F(p))$ for $p \leqslant x$ are disjoint and are contained in the interval $[A, F(x)]$. Thus the series for $H(x)$ is convergent and $H(x) \leqslant F(x) - A$. By the same argument, for any $y < x$,

$$H(x) - H(y) = \sum_{p \in (y, x]} \{F(p) - F(p-0)\} \leqslant F(x) - F(y).$$

From this inequality we deduce all that we need. First of all, $G(y) \leqslant G(x)$ so that G is increasing. Then as $x \searrow y$ we have $H(y+0) - H(y) \leqslant 0$. But H is clearly increasing so that H is continuous on the right, and the same is true of G as $G = F - H$. Finally, $H(x) - H(y) \geqslant F(x) - F(x-0)$ for $y < x$ and so as $y \nearrow x$ we have $H(x) - H(x-0) = F(x) - F(x-0)$ from which $G(x) = G(x-0)$ i.e. G is continuous on the left.

When F is unbounded below we may take our lead from p. 36 and define H_0 by the formula

$$\begin{aligned} H_0(x) &= \sum_{p \in (0, x]} \{F(p) - F(p-0)\} \quad \text{for} \quad x \geqslant 0 \\ &= -\sum_{p \in (x, 0]} \{F(p) - F(p-0)\} \quad \text{for} \quad x < 0. \end{aligned}$$

Then $H_0(b) - H_0(a)$ is the sum of the jumps of F at the discontinuities which lie in the interval $(a, b]$. We may then verify as above that $G_0 = F - H_0$ is increasing and continuous.

3. f is continuous if and only if $\{x \in \mathbf{R}: a < f(x) < b\}$ is an open subset of $\mathbf{R}$ for all $a < b$ in $\mathbf{R}$. The rest is direct verification.

4. This is immediate from the fact that the union of any collection of open subsets of $\mathbf{R}$ is open.

5. (i) F is constant on $(-\infty, 0)$ and $(0, \infty)$ and has the jump

$$F(0+0) - F(0-0) = 1$$

at the origin.

(ii) $F(x) = x^3/3 + C$ for all x in $\mathbf{R}$. To verify this take $\phi = \chi_{(a, b]}$.

6. (i) Let $A_c = \{x \in \mathbf{R}: F(x) \geqslant c\}$. Then A_c has the property:

$$x \in A_c \quad \text{and} \quad y > x \quad \Rightarrow \quad y \in A_c.$$

We may have $A_c = \varnothing$ or $A_c = \mathbf{R}$, but the only other possibilities are $A_c = [d, \infty)$ or $A_c = (d, \infty)$. The continuity of F on the right, rules out the last one; for if $F(x) \geqslant c$ for all $x > d$, we let $x \searrow d$ and deduce that $F(d) \geqslant c$. (Alternatively use the fact that F is upper semicontinuous (Ex. 3) and so A_c is closed.) Now

$$c \leqslant F(b) \Leftrightarrow b \in A_c \Leftrightarrow G(c) \leqslant b.$$

(ii) Immediate from (i).

(iii) Use (ii) and the Daniell condition for the Lebesgue integral on J.

7. It is helpful (though not essential) to note that an increasing function is continuous on the right (left) if and only if it is upper (lower) semicontinuous: see Ex. 3.

It is clear from the definition in Ex. 6 (i) that G is increasing: moreover $G(c) > a \Leftrightarrow c > F(a)$ and so G is lower semicontinuous.

The relationship between the functions F and G may be described pictorially in terms of their graphs. First of all, on the graph of F, draw in vertical segments at the jumps (dotted line in Fig. S. 1) and reflect this augmented graph in the line $x = y$; if we now ignore the vertical segments on the reflected graph (dotted line again in Fig. S. 1) we have the graph of G. Bearing in mind that F is continuous on the right and G is continuous on the left, this picture is now unambiguous.

From the picture we see that $G(F(x))$ is the left hand end point of the interval on which F takes the value $F(x)$: this interval consists of a single point unless $(x, F(x))$ lies on a horizontal segment on the graph of F and the left hand end point $G(F(x))$ will exist for all x in $\mathbf{R}$ provided F is not constant on an interval $(-\infty, a)$. In general we must restrict x so that $F(x) \in J$.

Similarly $F(G(y))$ is the right hand end point of the interval on which G takes the value $G(y)$ and this is defined for any y in J.

When F is strictly increasing and continuous there are no horizontal segments on either graph and so F and G are inverses of each other.

8. (i) Use 6 (ii).

(ii) This is just the proof of Proposition 1 parts (iii), (iv).

(iii) Adapting ψ_n does not alter $\int \psi_n$. Also

$$\int_{I_r} \psi_n \to \int_{I_r} g,$$

where ψ_n, g are both constant on I_r. The identity $\psi_n \circ F \circ G = \psi_n$ follows by Ex. 7.

(iv) Suppose that B is a subset of J not containing any points of the intervals I_r. If $\psi_n \uparrow \infty$ on B and $\int \psi_n \leqslant K$ then we may adapt ψ_n as in (iii) without altering this condition. Let $\phi_n = \psi_n \circ F$ (on the set $F^{-1}(J)$) so that $\psi_n = \phi_n \circ G$. It follows that $\phi_n \uparrow \infty$ on $F^{-1}(B)$ and $\int \phi_n dF \leqslant K$. The rest of the argument now follows the pattern of Proposition 1.

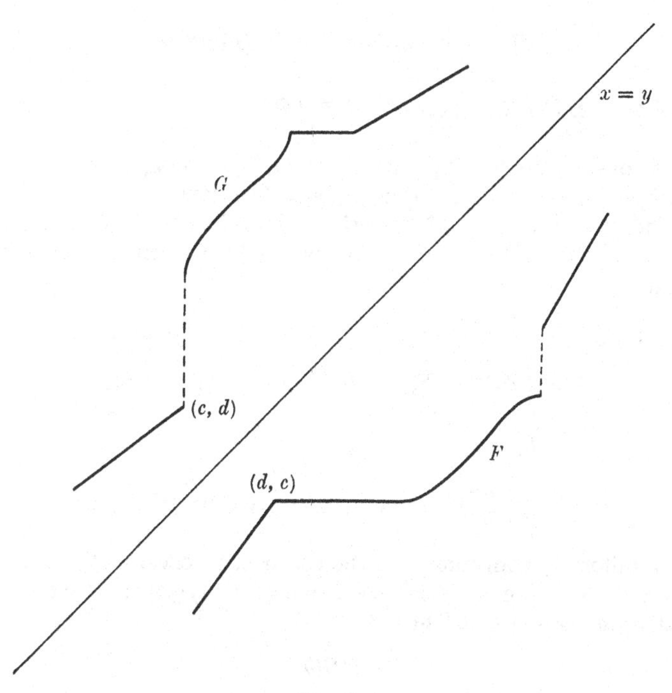

Fig. S. 1

9. $G(y) = y^{\frac{1}{3}}$ as in Ex. 7 and the interval J is the whole of $\mathbf{R}$. In the notation of Ex. 8 let $f(x) = x^2$ for $1 \leqslant x \leqslant 2$; then $g(y) = y^{\frac{2}{3}}$ for $1 \leqslant y \leqslant 8$ and the given integral equals

$$\int_1^8 y^{2/3}dy = \tfrac{3}{5}[y^{5/3}]_1^8 = \tfrac{3}{5}(2^5 - 1) = \tfrac{93}{5}.$$

According to Proposition 1

$$\int_{[1,\,2]} x^2 dx^3 = 3\int_1^2 x^4 dx = \tfrac{3}{5}(2^5 - 1) = \tfrac{93}{5}.$$

10. $G(F(x)) = F(x)$ and $F(G(y)) = G(y)$.
In the notation of Ex. 8 let $f(x) = x^{-2}$ for $x > 0$ and $g(y) = 1/G(y)^2$ for $y > 1$. The given integral equals

$$\int_1^\infty \frac{1}{G(y)^2}dy = \sum_{n=1}^\infty \frac{1}{n^2}.$$

It is simpler to note that the Lebesgue–Stieltjes measure μ_F concentrates unit mass at each integer point: cf. Ex. 8.4.3.

11. Adapt the proof of Proposition 1.

The Riemann–Stieltjes Integral

A. Let $\phi = \sum_{i=1}^{r} f(t_i)\,\chi_{(x_{i-1},x_i]}$; then $S = \int \phi\, dG$.

Since f is continuous, $\phi \to f$ pointwise on $(a,b]$ as $\max(x_i - x_{i-1}) \to 0$, and we may apply the Dominated Convergence Theorem.

(Consider first a *sequence* of dissections of $[a,b]$ as in Ex. 5. 2.12.)

The contribution of the single point a is zero by our convention about the end points.

B. In this case

$$S' = \int \phi\, dG = \sum_{i=1}^{r} f(t_i)\{G(x_i+0) - G(x_{i-1}+0)\},$$

where $G(b+0) = G(b)$.

$$S - S' = \sum_{i=1}^{r} \{f(t_i) - f(t_{i-1})\}\{G(x_{i-1}+0) - G(x_{i-1})\}.$$

Now f is uniformly continuous on the compact interval $[a,b]$ (Ex. 7.6.A), i.e. to any $\epsilon > 0$, there is a $\delta > 0$ such that $|f(x) - f(y)| < \epsilon$ for any x, y in $[a,b]$ satisfying $|x-y| < \delta$. Thus

$$|S - S'| < \epsilon\{G(b) - G(a)\}$$

provided $\max(x_i - x_{i-1}) < \delta/2$ and this shows that S and S' tend to the same limit as $\max(x_i - x_{i-1}) \to 0$.

§9.2

1. μ_F concentrates unit mass at the origin.

2. $F(1,1) - F(-1,1) - F(1,-1) + F(-1,-1) < 0$.

3. As at the beginning of the section let I_x denote the interval of type $(\,,]^k$ with one vertex at 0 and the diagonally opposite vertex at

$$x = (x_1, \ldots, x_k).$$

In the one-dimensional case,

$$\chi_{(a,b]} = (\operatorname{sgn} b)\,\chi_{I_b} - (\operatorname{sgn} a)\,\chi_{I_a}$$

for $a < b$. If $I_1, \ldots, I_k$ are intervals in $\mathbf{R}$, then $I_1 \times \ldots \times I_k$ is an interval in $\mathbf{R}^k$ and

$$\chi_I(x) = \chi_{I_1}(x_1) \ldots \chi_{I_k}(x_k).$$

These two equations give an expression for χ_I which yields equation (3) when we integrate both sides.

4. In the one-dimensional case, any discontinuity p of the increasing function F would give non-zero measure $\mu_F(\{p\})$: the translation invariance

rules this out. Consider the intervals $(0, a]$, $(b, a+b]$: as they have the same measure,

$$F(a)-F(0) = F(a+b)-F(b).$$

Suppose that $F(0) = 0$; then

$$F(a+b) = F(a)+F(b)$$

for any a, b in $\mathbf{R}$. It follows from this that

$$F(ra) = rF(a)$$

for any integer r, then for any rational number r, and then by continuity for any real number r. Thus F is a linear function, and this is equivalent to saying that μ_F is a constant multiple of Lebesgue measure.

The result extends to $\mathbf{R}^k$ by the method of Ex. 3.

5. The extension to $\mathbf{R}^k$ is facilitated by the last two equations displayed in the solution of Ex. 3.

§9.3

1. Yes, because $d(x, I_n{}^c) \geqslant d(x, I_{n+1}{}^c)$.

$$f_1(x) = 1 \quad \text{if} \quad x \in I \quad \text{and} \quad f_1(x) < 1 \quad \text{otherwise.}$$

2. Approximate $\chi_{(a,b]}$ using continuous functions and then verify as in Ex. 9.1.5 that $F(x) = x^3/3 + C$.

3. Approximate χ_I as in the proof of Theorem 1.

§10.1

1. Let C be a closed subset of X and suppose that the open sets G_i $(i \in I)$ cover C. Then the same open sets together with the open set $G = X \setminus C$ cover X. As X is compact, a finite selection of these, $G_{i_1}, \ldots, G_{i_n}, G$, say, cover X and so $G_{i_1}, \ldots, G_{i_n}$ cover C.

Let K be a compact subset of the Hausdorff space X and suppose that $p \in X \setminus K$ (if there are no such points p then $K = X$ is closed). To any x in K there exist disjoint open sets A_x, B_x such that $x \in A_x$, $p \in B_x$. From the open covering A_x $(x \in K)$ of K, we select a finite covering $A_{x_1}, \ldots, A_{x_n}$, say. Then the open set $B_{x_1} \cap \ldots \cap B_{x_n}$ contains p but no point of K. This shows that the set $X \setminus K$ is open and so K is closed.

2. The open sets

$$G_n = \{x \in X : f(x) < n\} \quad (n = 1, 2, \ldots)$$

cover X. By the compactness of X, a finite selection $G_{n_1}, \ldots, G_{n_r}$ cover X. Let $N = \max\{n_1, \ldots, n_r\}$; then $X = G_N$, i.e. $f(x) < N$ for all x.

In the proof of Dini's Theorem we only used the upper semicontinuity of ϕ_n.

3. As the proof of Theorem 4, $\psi^n \downarrow \chi_K$.

4. Each point of X is open (even an open ball as $d(a,x) < 1 \Rightarrow x = a$). Thus every subset of X is open and therefore closed. The compact subsets are the finite subsets of X. Now X is locally compact as $a \in \{a\}$ where $\{a\}$ is open and closed and compact.

Each function of $C_c(X)$ vanishes outside a finite set; thus each function of L^1 vanishes outside a countable set and all integrals are zero. The measure μ is zero on all countable sets and infinite on uncountable sets.

5. Any continuous function ϕ on $\mathbf{R}$ which vanishes outside $[a,b]$ must vanish at a and b. Thus ϕ vanishes outside (a,b) and

$$\left|\int \phi \, dF_0\right| \leqslant (F_0(b-0)-F_0(a+0)) \, \|\phi\|.$$

On the other hand we may find a continuous function ϕ_ϵ (of the shape shown in Fig. 8 on p. 51) which vanishes at a, b and which equals 1 on an interval $[a+\epsilon, b-\epsilon]$. Now $\|\phi_\epsilon\| = 1$ and $\int \phi_\epsilon \, dF_0 \to F_0(b-0)-F_0(a+0)$ as $\epsilon \searrow 0$.

§ 10.2

1. If F is relatively bounded, modify the argument on p. 60 using the inequality

$$|\phi| \leqslant \|\phi\| \, \psi_K.$$

On the other hand suppose that ψ is a positive continuous function on X with compact support K and $|F(\phi)| \leqslant M_K \|\phi\|$ for all ϕ in $C_K(X)$. Then any ϕ in $C_c(X)$ which satisfies $|\phi| \leqslant \psi$ belongs to $C_K(X)$ and so

$$|F(\phi)| \leqslant M_K \|\phi\| \leqslant M_K \|\psi\|.$$

Thus F is relatively bounded.

2. Proposition 1 gives one half. Let $F = G-H$ where G and H are positive. If $|\phi| \leqslant \psi$ then

$$|F(\phi)| \leqslant |G(\phi)| + |H(\phi)| \leqslant G(\psi) + H(\psi).$$

According to (3),

$$\begin{aligned} P(\psi) &= \sup\{G(\phi)-H(\phi) : 0 \leqslant \phi \leqslant \psi\} \\ &\leqslant \sup\{G(\phi) : 0 \leqslant \phi \leqslant \psi\} = G(\psi) \end{aligned}$$

for any positive ψ in L. Thus $P \leqslant G$. Also $G-P = H-N$.

3. A may also be written in either of the (less symmetric) forms

$$A = F + (G-F)^+ = G + (F-G)^+.$$

These show clearly that $A \geqslant F$ and $A \geqslant G$. But if $C \geqslant F$ and $C \geqslant G$ it follows that $C-F \geqslant G-F$ and as $C-F$ is positive $C-F \geqslant (G-F)^+$; in other words $C \geqslant A$.

Likewise B is the largest linear functional on L that is less than or equal to F and G.

4. By equation (3) and Ex. 3,

$$\begin{aligned}A(\psi) &= F(\psi)+\sup\{G(\phi)-F(\phi): 0 \leqslant \phi \leqslant \psi\}\\ &= \sup\{F(\theta)+G(\phi): 0 \leqslant \phi \leqslant \psi, \theta+\phi = \psi\}.\end{aligned}$$

Similarly, using the relation $B = F-(F-G)^+$ we deduce that

$$B(\psi) = \inf\{F(\theta)+G(\phi): 0 \leqslant \phi \leqslant \psi, \theta+\phi = \psi\}.$$

$\min\{P, N\} = \frac{1}{2}T-\frac{1}{2}T = 0$. Given $\epsilon > 0$ and a positive ψ in L, there exist positive θ, ϕ in L such that $\theta+\phi = \psi$ and $P(\theta)+N(\phi) < \epsilon$.

5. If $0 \leqslant \phi_1 \leqslant \psi_1$ and $0 \leqslant \phi_2 \leqslant \psi_2$, then $0 \leqslant \phi_1+\phi_2 \leqslant \psi_1+\psi_2$ and so $P(\psi_1+\psi_2) \geqslant F(\phi_1)+F(\phi_2)$. Taking suprema for all such ϕ_1, ϕ_2 we have

$$P(\psi_1+\psi_2) \geqslant P(\psi_1)+P(\psi_2).$$

On the other hand, if $0 \leqslant \phi \leqslant \psi_1+\psi_2$ we may take $\phi_1 = \min\{\phi, \psi_1\}$ and $\phi_2 = \phi-\phi_1$. It is easy to check that $0 \leqslant \phi_1 \leqslant \psi_1$ and $0 \leqslant \phi_2 \leqslant \psi_2$: thus $F(\phi) = F(\phi_1)+F(\phi_2) \leqslant P(\psi_1)+P(\psi_2)$. Take the supremum for all such ϕ and get the reverse inequality

$$P(\psi_1+\psi_2) \leqslant P(\psi_1)+P(\psi_2).$$

It is clear that $P(\psi) \geqslant 0$ and $P(c\psi) = cP(\psi)$ for all positive c in $\mathbf{R}$ and all positive ψ in L. The extension to a linear functional on L follows exactly as in the proof of Proposition 1.

$$\begin{aligned}N(\psi) &= P(\psi)-F(\psi)\\ &= \sup\{F(\theta): 0 \leqslant \theta \leqslant \psi\}-F(\psi)\\ &= \sup\{-F(\phi): 0 \leqslant \phi \leqslant \psi\} \quad (\text{writing } \phi = \psi-\theta).\\ T(\psi) &= P(\psi)+N(\psi)\\ &= \sup\{F(\phi)-F(\theta): 0 \leqslant \phi \leqslant \psi, 0 \leqslant \theta \leqslant \psi\}\\ &= \sup\{F(\tau): |\tau| \leqslant \psi\} \quad (\text{writing } \tau = \phi-\theta).\end{aligned}$$

6. As $F = P-N$, $T = P+N$ on L', it is immediately verified that $F_0 = P_0-N_0$, $T_0 = P_0+N_0$. As P_0, N_0 are increasing functions on $\mathbf{R}$, it follows that F_0 has bounded variation on $[a, b]$.

$$T(\chi_{(a,b]}) = T_0(b)-T_0(a) = \sup\{F(\phi): |\phi| \leqslant \chi_{(a,b]}\}.$$

In evaluating this supremum note that $\phi = \Sigma a_i \chi_{(x_{i-1}, x_i]}$ for some partition $a = x_0 < \ldots < x_r = b$, and $F(\phi) = \Sigma a_i\{F_0(x_i)-F_0(x_{i-1})\}$ will take its largest value for the given partition when $a_i = \pm 1$. It follows that $T_0(b)-T_0(a)$ is the total variation of F_0 on $[a, b]$. The positive and negative variations of F_0 on $[a, b]$ are $P_0(b)-P_0(a)$ and $N_0(b)-N_0(a)$, respectively.

7. The first part is a special case of Ex. 8. The given integral equals

$$\int_{-1}^{2} 2x^4 dx = \tfrac{2}{5}(2^5+1) = \tfrac{66}{5}.$$

8. The function G is defined as in Ex. 9.1.11 by the formula

$$G(x) = \int_c^x g + C$$

(c, C arbitrary), though g is no longer assumed to be positive. We may apply the previous result to the positive functions g^+, g^- separately and then use the relation $g = g^+ - g^-$. It follows that if $f \in L^1(\mu_G)$ (as defined on p. 65) then

$$\int f dG = \int fg.$$

The Riemann–Stieltjes Integral

A. Express G as the difference of two increasing functions (Ex. 3.5.4) and use Ex. 9.1.B.

§10.3

1. Argue as on p. 70.

2. ζ_1, ζ_2 are continuous at any point p where $\psi_1(p) + \psi_2(p) > 0$. Suppose that $\psi_1(p) + \psi_2(p) = 0$ and hence $\zeta(p) = 0$. To any $\epsilon > 0$, there is a $\delta > 0$ such that $\psi_1(x) < \epsilon$ and $\psi_2(x) < \epsilon$ for all x in X satisfying $d(x, p) < \delta$. As $|\zeta_1| \leqslant \psi_1$, $|\zeta_2| \leqslant \psi_2$ it follows that both ζ_1 and ζ_2 are continuous at p.

3. Let $\tau(x) = \mathrm{tr}_{\psi(x)} \zeta(x) = \zeta(x)$ or $\psi(x)\,\zeta(x)/|\zeta(x)|$, whichever is nearer the origin, and let $\epsilon > 0$ be given. If $|\zeta(p)| < \psi(p)$ then $\tau(x) = \zeta(x)$ in some ball $B(p, \delta)$. If $|\zeta(p)| > \psi(p)$ then $\tau(x) = \psi(x)\,\zeta(x)/|\zeta(x)|$ in some ball $B(p, \delta)$. If $\zeta(p) = 0$ then $|\zeta(x)| < \epsilon$ and a fortiori $|\tau(x)| < \epsilon$ in some ball $B(p, \delta)$. If $|\zeta(p)| = \psi(p) > 0$ then both $|\zeta(x) - \zeta(p)| < \epsilon$ and $|\psi(x)\,\zeta(x)/|\zeta(x)| - \zeta(p)| < \epsilon$ in some ball $B(p, \delta)$.

4. $C(t)$ traces out the unit circle centre 0 once in the positive (anticlockwise) sense and then once in the negative sense.

$$\int_C f(z)\,dz = \int_0^{2\pi} f(e^{it})\,ie^{it}dt - \int_{2\pi}^{4\pi} f(e^{-it})\,ie^{-it}dt$$

$$= i\int_0^{2\pi} f(e^{it})\,e^{it}dt - i\int_0^{2\pi} f(e^{iu})\,e^{iu}du = 0.$$

5. Use the inequalities

$$|A(t_i) - A(t_{i-1})| \leqslant |C(t_i) - C(t_{i-1})| \leqslant |A(t_i) - A(t_{i-1})| + |B(t_i) - B(t_{i-1})|$$

and a similar one with A and B interchanged.

6. As in Ex. 9.1.A (without necessarily mentioning Riemann!) the integral

$$H(f) = \int_a^b f(C(t))\,dC(t)$$

is the limit of the sum $\sum_{i=1}^{r} f(C(t_i))\{C(t_i)-C(t_{i-1})\}$,

as $\max(t_i-t_{i-1}) \to 0$, and this sum is dominated by

$$\|f\| \Sigma |C(t_i)-C(t_{i-1})| \leqslant \|f\| l(C).$$

7. Suppose that C is an arc. To any complex valued continuous function g on $[a,b]$ we may define a unique continuous function f on X by setting

$$f(C(t)) = g(t) \quad (a \leqslant t \leqslant b).$$

(In other words $f = g \circ C^{-1}$.) There exists a complex valued step function ϕ on $[a,b]$ whose non-zero values have modulus 1 such that

$$\int_a^b \phi(t)\, dC(t)$$

is the polygonal length displayed in Ex. 5. As in §9.3 approximate ϕ by means of a continuous function g with $\|g\| = 1$ and then define the corresponding function f of X. In this way we can find continuous functions f on X with $\|f\| = 1$ for which $H(f)$ is arbitrarily close to $l(C)$.

8. This is almost immediate from the definitions if we use the approximating sums of Ex. 6.

§11.1

1. $\mathscr{R} = \{\varnothing, [1,2), [1,3), [2,3)\}$.

2. $\mathscr{R} = \{\varnothing, [1,2), [2,3), [3,4), [1,3), [2,4), [1,2) \cup [3,4), [1,4)\}$.

3. Suppose that $\mathscr{C}$ is closed with respect to unions and complements. Then

$$A \in \mathscr{C}, \quad A^c \in \mathscr{C} \Rightarrow A \cup A^c = X \in \mathscr{C};$$

also $$A, B \in \mathscr{C} \Rightarrow A^c, B \in \mathscr{C} \Rightarrow (A^c \cup B)^c \in \mathscr{C} \Rightarrow A \setminus B \in \mathscr{C}$$

and so $\mathscr{C}$ is an algebra. The converse is obvious.

6. Verify that $$f^{-1}(A) \cup f^{-1}(B) = f^{-1}(A \cup B)$$

and $$f^{-1}(A) \setminus f^{-1}(B) = f^{-1}(A \setminus B)$$

from which it follows that $f^{-1}(\mathscr{R})$ is a ring. It also follows that if $\mathscr{S}$ is a ring of subsets of X, then

$$\mathscr{T} = \{A \subset Y : f^{-1}(A) \in \mathscr{S}\}$$

is a ring of subsets of Y. In particular, if the ring $\mathscr{S}$ contains $f^{-1}(\mathscr{C})$ then $\mathscr{T}$ contains $\mathscr{C}$, whence $\mathscr{T}$ contains $\mathscr{R}$ and $\mathscr{S}$ contains $f^{-1}(\mathscr{R})$. These two results show that $f^{-1}(\mathscr{R})$ is the smallest ring containing $f^{-1}(\mathscr{C})$.

7. No: for example, let $\mathscr{R}$ be the ring of Ex. 1 and let

$$\begin{aligned} f(x) &= x \quad \text{if} \quad x \in [1,2) \\ &= 2x-3 \quad \text{if} \quad x \in [2,3). \end{aligned}$$

Boolean Rings and Algebras

A. $$(a+b)^2 = a+b \Rightarrow a^2+ab+ba+b^2 = a+b$$
$$\Rightarrow a+ab+ba+b = a+b$$
$$\Rightarrow ab+ba = 0.$$

Set $b = a$ to obtain $a+a = 0$ for all a, and then $ab = ba$ for all a, b.

D. (iii) $ab = a,\ bc = b \Rightarrow ac = abc = ab = a$.

F. $a \vee b = a+b+ab$ is the analogue of $A \cup B = A \triangle B \triangle (A \cap B)$ and $a' = 1+a$ is the analogue of $A^c = X \triangle A$.

H. The class of Boolean rings with unit is isomorphic to the class of distributive complemented lattices with 0 and 1.

§ 11.2

2. (i) ⇒ (ii): $\chi_{A \cup B} = \max\{\chi_A, \chi_B\}$, $\chi_{A \setminus B} = (\chi_A - \chi_B)^+$.

(ii) ⇒ (i): if $\mathscr{R} = \{A : \chi_A \in L(\mathscr{C})\}$ is a ring then $L(\mathscr{C}) = L(\mathscr{R})$ is a linear lattice by Proposition 1, Corollary.
(ii) ⇒ (iii): $\chi_A \chi_B = \chi_{A \cap B}$.
(iii) ⇒ (ii): $\chi_A, \chi_B \in L(\mathscr{C}) \Rightarrow \chi_{A \cup B} = \chi_A + \chi_B - \chi_{A \cap B} \in L(\mathscr{C})$

and $$\chi_{A \setminus B} = \chi_A - \chi_{A \cap B} \in L(\mathscr{C}).$$

(iii) ⇔ (iv): $\chi_A \chi_B = \chi_{A \cap B}$.

3. By Ex. 2 $\mathscr{R} = \{A : \chi_A \in L(\mathscr{C})\}$ is a ring containing $\mathscr{C}$. Let $\mathscr{R}'$ be any ring containing $\mathscr{C}$. Then $A \in \mathscr{R} \Rightarrow \chi_A \in L(\mathscr{R}') \Rightarrow A \in \mathscr{R}'$ and so $\mathscr{R} \subset \mathscr{R}'$.

4. Use Ex. 2 (iv). If $A, B \in \mathscr{C}$ and $A \neq B$ then
$$\chi_{A \cap B} = \tfrac{1}{2}\chi_A + \tfrac{1}{2}\chi_B - \tfrac{1}{2}\chi_{A \triangle B} \in L(\mathscr{C}).$$

Neither condition is necessary: for example, let A, B be distinct sets with non-empty intersection and take $\mathscr{C} = \{A, B, A \triangle B\}$, $\mathscr{C} = \{A, B, A \cap B\}$, respectively.

5. The intersection of two open triangles is either empty or is an open convex polygon and so may be triangulated, i.e. is a polyhedral set in the sense of p. 86 (Fig. S. 2). All other intersections of simplexes in $\mathbf{R}^2$ are either empty or are line segments or points.
The complement of an open triangle in $\mathbf{R}^2$ is the disjoint union of six open convex sets (each of which is the intersection of either two or three open half-planes), six open half-lines, three open line segments and three points (Fig. S. 3). Using the previous paragraph it follows that the difference of two open triangles is a polyhedral set. The complement of an open line segment in $\mathbf{R}^2$ is the disjoint union of two open half-planes, two open half-lines and two points (Fig. S. 4). Thus the difference of an open triangle and an open

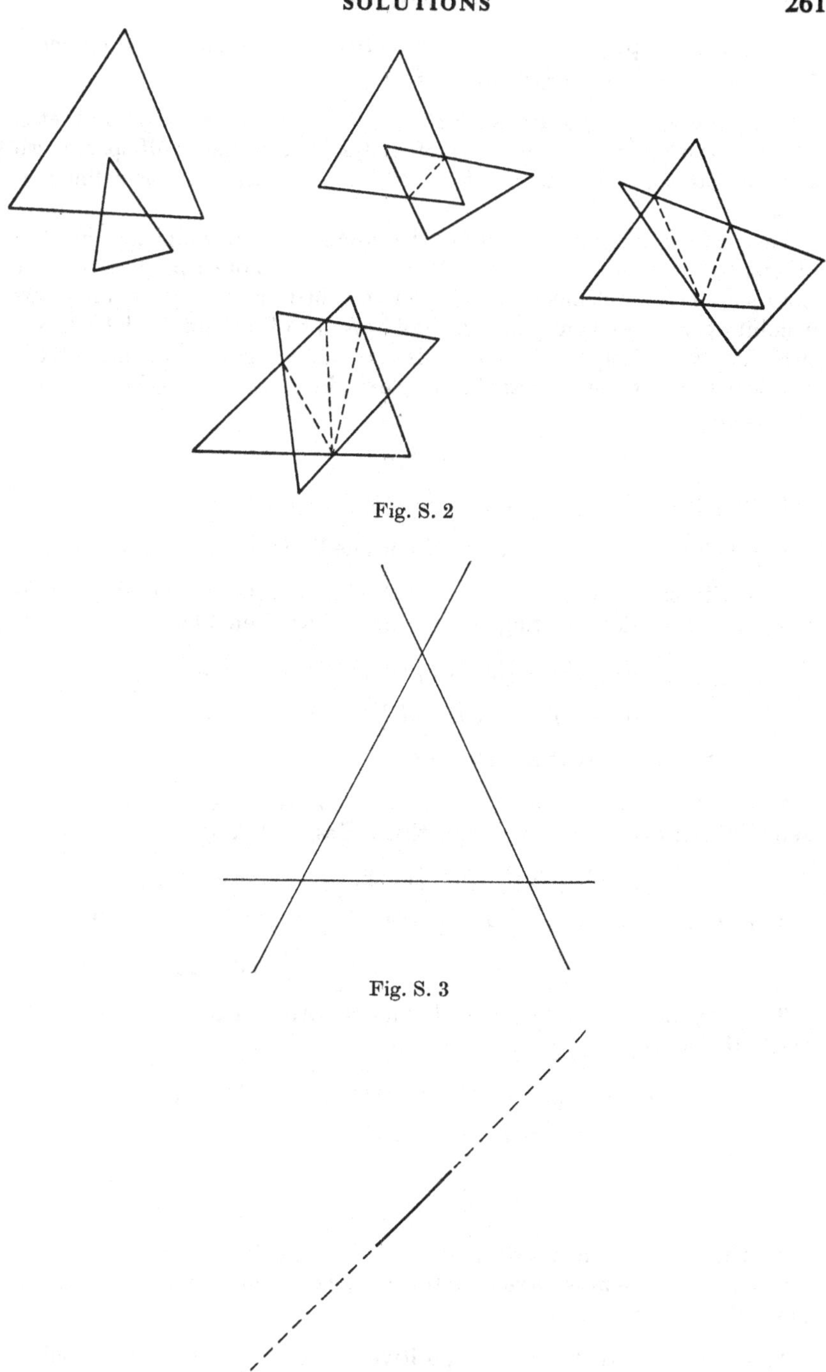

Fig. S. 2

Fig. S. 3

Fig. S. 4

line segment is a polyhedral set. It is easy to check that all other differences of simplexes in $\mathbf{R}^2$ are polyhedral sets.

6. In any linear subspace M of $\mathbf{R}^k$ consider the convex polygonal set S which is the intersection with M of finitely many open half-spaces: we assume that S may be triangulated, i.e. S is a polyhedral set as defined on p. 86.

The argument of Ex. 5 applies to the intersections of simplexes. For the differences of simplexes the numbers are much more complicated. For example, the complement of an open tetrahedron in $\mathbf{R}^3$ has numerous disjoint open components, but we may appeal to Proposition 11.1.1 to see that there are finitely many! With this in mind the whole argument of Ex. 5 now generalises to show that the simplexes in $\mathbf{R}^k$ satisfy the Disjointness Condition.

§ 11.3

1. $\varnothing \cup \varnothing = \varnothing \Rightarrow \mu(\varnothing)+\mu(\varnothing) = \mu(\varnothing) \Rightarrow \mu(\varnothing) = 0.$

$\mu(A \cup B) = \mu(A \cap B)$ and A and B both lie between $A \cap B$ and $A \cup B$.

2. Verify condition (iv) of Ex. 11.2.2. No two sets in $\mathscr{C}$ are disjoint. If μ could be extended to a ring $\mathscr{R}$ containing $\mathscr{C}$ we should have

$$\mu(A \cap B) = \tfrac{1}{2}(\mu(A)+\mu(B)-\mu(A \triangle B)) = 2,$$

$$\mu(A \cap C) = \tfrac{1}{2}(\mu(A)+\mu(C)-\mu(A \triangle C)) = 2\tfrac{1}{2},$$

contrary to the fact that $A \cap B = A \cap C$.

3. Apart from the empty set, the only two disjoint sets are $[0, \frac{1}{4})$, $[\frac{1}{4}, \frac{3}{4})$ and additivity is easily checked for them. Yes, by taking

$$\mu[\tfrac{1}{4}, \tfrac{1}{2}) = 0, \quad \mu[\tfrac{1}{2}, \tfrac{3}{4}) = 2, \quad \mu[\tfrac{3}{4}, 1) = 0.$$

4. $$\mu(A_1 \cup \ldots \cup A_n) = \mu(A_1)+\ldots-\mu(A_1 \cap A_2)-\ldots+\mu(A_1 \cap A_2 \cap A_3)+\ldots +(-1)^{n-1}\mu(A_1 \cap \ldots \cap A_n).$$

This can be proved easily by induction. Alternatively, integrate the algebraic identity:

$$\chi_{A_1 \cup \ldots \cup A_n} = 1-(1-\chi_{A_1})(1-\chi_{A_2})\ldots(1-\chi_{A_n})$$
$$= \chi_{A_1}+\ldots-\chi_{A_1}\chi_{A_2}-\ldots+\ldots.$$

§ 11.4

1. There is a set A in $\mathscr{C}$ with $\mu(A)$ finite. Now use the fact that $A \cup \varnothing = A$. No, because we may have infinite values here and we must rule out the possibility $\mu(\varnothing) = \infty$.

2. Use the fact that a series of positive terms is convergent if and only if the partial sums are bounded.

5. We may 'disjointify' the union

$A_1 \cup A_2 \cup \ldots$ as $A_1 \cup (A_2 \setminus A_1) \cup (A_3 \setminus (A_1 \cup A_2)) \cup \ldots$. See Ex. 3.

7. The collection of all subsets of finite unions of elements of $\mathscr{C}$ is a ring containing $\mathscr{C}$. The collection of all subsets of countable unions of elements of $\mathscr{C}$ is a σ-ring containing $\mathscr{C}$.

8. Let

$X = \{1, 2, 3\}, \mathscr{C}_1 = \{\varnothing, \{1\}, \{2, 3\}, \{1, 2, 3\}\}$ and $\mathscr{C}_2 = \{\varnothing, \{1\}, \{2\}, \{3\}, \{1, 2, 3\}\}$.

9. Any point of X is the intersection of a decreasing sequence of intervals of $\mathscr{C}$ and X is countable.

10. $(0, 1]$ is the union of the intervals $(1/n, 1]$ all of which have measure less than 1.

11. Let L be the linear lattice of functions $\phi: X \to \mathbf{R}$ such that $\phi(x_n) = 0$ for all $n > N(\phi)$ and define

$$\int \phi = \Sigma p_n \phi(x_n).$$

Then $\int$ is an elementary integral and μ is the corresponding measure: cf. Ex. 8.4.3.

12. Use the Monotone Convergence Theorem.

13. Let $B_n = A_n \setminus A_{n+1}$ for $n \geqslant 1$. Then $\mathscr{R}$ consists of all finite unions of the sets A_n, B_n (including $\varnothing$). As the sets A_n are 'nested', any set in $\mathscr{R}$ is either of the form

$$S = B_{n_1} \cup \ldots \cup B_{n_r} \quad \text{with} \quad n_1 < n_2 < \ldots < n_r$$

or

$$A_n \cup S \quad \text{with} \quad n > n_r.$$

Now let $\mu(A_n) = n$, $\mu(B_n) = -1$, $\mu(S) = -r$, $\mu(A_n \cup S) = n - r$. No two of the sets A_n are disjoint and so we only have to check additivity for disjoint S, S' or disjoint $A_n \cup S, S'$.

Condition (iii) of Proposition 2 is automatically satisfied in this case because the sets A_n have non-empty intersection and hence $S_n \downarrow \varnothing$ can only occur when $S_n = \varnothing$ for all $n \geqslant N$. (Incidentally, this implies that μ is *σ-additive* according to the remarks after the proof of Proposition 2.) On the other hand, $\phi_n \downarrow 0$ and $\int \phi_n d\mu = 1$ for all n, so the implication (iii) $\Rightarrow$ (iv) in Proposition 2 depends on the assumption that μ is positive.

14. Let $A = \bigcap A_n$. The σ-ring $\mathscr{S}$ generated by $\{A_1, A_2, \ldots\}$ consists of countable unions of the sets A and B_n for $n \geqslant 1$ (including $\varnothing$). As $X = A_1$, $\mathscr{S}$ is also a σ-algebra.

15. The disjointness condition is trivial in this case. $\mathscr{R}$ consists of the finite subsets of X.

16. The σ-ring generated by $\mathscr{C}$ consists of the countable subsets of X and the σ-algebra generated by $\mathscr{C}$ consists of the countable subsets of X and their complements in X.

In each case the σ-algebra $\mathscr{M}$ of measurable sets is the collection of *all* subsets of X. For $c = 0$, $\mu(A) = 0$ for countable A and $\mu(A) = \infty$ otherwise; for $0 < c < \infty$, $\mu(A) = c$ times the number of elements in A; for $c = \infty$, $\mu(\varnothing) = 0$ and $\mu(A) = \infty$ for all other $A \subset X$.

17. $\mathscr{A}$ consists of the finite subsets of X and their complements in X. In each case $\mu(A) = 0$ for finite A and $\mu(A) = \mu(X)$ for finite $X \setminus A$. (i) $\mathscr{M}$ consists of all subsets of X and $\bar{\mu} = 0$. (ii) $\mathscr{M}$ consists of the countable sets and their complements, with $\bar{\mu}(A) = 0$ if A is countable and $\bar{\mu}(A) = 1$ otherwise. (iii) $\mathscr{M}$ consists of all subsets of X, with $\bar{\mu}(A) = 0$ if A is countable and $\bar{\mu}(A) = \infty$ otherwise.

18. To any atom we may adjoin an arbitrary null set and obtain another atom: in each of the examples we give the atoms to within null sets.

8.4.1: each natural number n.
8.4.2: no atoms.
8.4.3: the origin (or, by the above observation, any set containing the origin).
8.4.4: no atoms.
11.4.11: each point x_n with $p_n > 0$.
Theorem 9.1.1: each point of discontinuity of F.

19. (i) As E is not an atom, there is a subset A of E satisfying

$$0 < \mu(A) < \mu(E): \quad \text{either} \quad \mu(A) \leqslant \tfrac{1}{2}\mu(E) \quad \text{or} \quad \mu(E \setminus A) \leqslant \tfrac{1}{2}\mu(E).$$

This process may be repeated to find B.

(ii) The inductive step is to let

$$s_n = \sup\{\mu(A) : A \in \mathscr{A}, A \supset A_{n-1}, \mu(A) \leqslant t\}$$

and then there exists $A_n \supset A_{n-1}$ such that

$$s_n - s_1/2^n < \mu(A_n) \leqslant s_n. \tag{*}$$

From this definition it is clear that $\{s_n\}$ is decreasing and $\{A_n\}$ is increasing. Also, by (*),

$$\mu(A) = \lim \mu(A_n) = \lim s_n = s, \quad \text{say.}$$

(iii) Clearly $s \leqslant t$. Suppose that $s < t$. Then by (i) we may find $B \in \mathscr{A}$ contained in $X \setminus A$ such that $s < \mu(A \cup B) < t$. But then $A \cup B \supset A_{n-1}$ for all $n \Rightarrow \mu(A \cup B) \leqslant s_n$ for all $n \Rightarrow \mu(A \cup B) \leqslant s$, a contradiction. Thus $\mu(A) = t$ as required.

(iv) All $s_n = t$.

§ 12.1

1. $-f(n) \geqslant 0$ if and only if n is odd.

2. If f is $\mathscr{A}$-measurable then each set $\{x : f(x) = c\}$ belongs to $\mathscr{A}$.

5. If $f(x)+g(x) > c$, there is a rational number r in the open interval $(c-g(x), f(x))$. Thus

$$\{x: f(x)+g(x) > c\} = \bigcup_r (\{x: f(x) > r\} \cap \{x: g(x) > c-r\})$$

and the rationals are countable.

6. $\{x: f(x)^2 \geqslant c\} = \{x: f(x) \geqslant c^{\frac{1}{2}}\} \cup \{x: f(x) \leqslant -c^{\frac{1}{2}}\}$ if $c > 0$

$= X$ if $c \leqslant 0$.

§ 12.2

1. Suppose that ϕ is integrable. By considering ϕ^+, ϕ^- we may assume without loss of generality that all $a_i > 0$. Then

$$\chi_{A_i} \leqslant \frac{1}{a_i}\phi \quad \text{and so} \quad \mu(A_i) \leqslant \frac{1}{a_i}\int \phi\, d\mu.$$

2. Let $A_c = \{x: |f(x)| \geqslant c\}$, where $c > 0$. Then

$$\chi_{A_c} \leqslant \frac{1}{c}|f| \quad \text{and so} \quad \mu(A_c) \leqslant \frac{1}{c}\int |f|\, d\mu.$$

3. Prove first for positive f and then use $f = f^+ - f^-$.

§ 12.3

1. If E is locally measurable, then $E \cap X_n \in \mathscr{A}$ for all n, and so

$$E = \bigcup (E \cap X_n) \in \mathscr{A}.$$

2. $\bigcup (E_n \cap A) = (\bigcup E_n) \cap A$ and $(E \setminus F) \cap A = (E \cap A) \setminus (F \cap A)$.

σ-additivity for infinite values is obvious. If E is locally measurable with respect to $\bar{\mu}$ then E is locally measurable with respect to μ: hence $\bar{\mu}$ is saturated.

3. Every subset E of X is locally measurable; so μ is saturated if and only if $\mathscr{A}$ consists of *all* subsets of X.

4. Every subset E of X is locally measurable. The extended measure $\bar{\mu}$ is defined on the σ-algebra $\mathscr{M}$ of *all* subsets of X and

$$\bar{\mu}(A) = 0 \quad \text{if } A \text{ is countable,}$$

$$\bar{\mu}(A) = \infty \quad \text{if } A \text{ is uncountable.}$$

This measure $\bar{\mu}$ *is* saturated.

§ 13.1

2. (i), (ii) $\mathscr{C}$ is already a σ-algebra. (iii) $\mathscr{C}$ is a σ-ring. The σ-algebra generated by $\mathscr{C}$ consists of all A and all $X \setminus A$ with A in $\mathscr{C}$.

§ 13.2

1. $(A \cup B) \cap E = (A \cap E) \cup (B \cap E)$ and $(A \setminus B) \cap E = (A \cap E) \setminus (B \cap E)$.

2. Let $A_1, A_2, \ldots$ be disjoint. Then

$$\mu\{(\bigcup A_n) \cap E)\} = \mu\{\bigcup(A_n \cap E)\} \Rightarrow \mu_E(\bigcup A_n) = \Sigma \mu_E(A_n).$$

3. The subsets of finite unions of sets in $\mathscr{C}$ form a ring containing $\mathscr{C}$ and hence containing $\mathscr{R}$ (cf. Ex. 11.4.7). In other words any element of $\mathscr{R}$ is contained in a finite union of sets in $\mathscr{C}$ and as such has finite measure.

4. Let $\mathscr{R}$ consist of the sets in $\mathscr{S}$ of finite measure. If $A, B \in \mathscr{R}$ then

$$\mu(A \cup B) \leqslant \mu(A) + \mu(B) \quad \text{and} \quad \mu(A \setminus B) \leqslant \mu(A)$$

are both finite. Thus $\mathscr{R}$ is a ring. A similar argument holds for countable unions of σ-finite sets.

Assume that μ is σ-finite: any element A of $\mathscr{S}$ is the union of a sequence $\{A_n\}$ of elements of $\mathscr{R}$. If $\mathscr{R}$ is a σ-ring then $A \in \mathscr{R}$ and $\mu(A)$ is finite. On the other hand, if μ is finite then $\mathscr{R} = \mathscr{S}$ is a σ-ring.

No: for example, let $\mathscr{S}$ consist of all subsets of the interval $[0, 1]$ and let $\mu(A) = 0$ if A is countable and $\mu(A) = \infty$ otherwise. Then the sets of finite measure form a σ-ring and μ is not a finite measure.

5. The extension procedure of § 11.4 gives at least one measure λ extending μ to $\mathscr{S}$. Let ν be another measure on $\mathscr{S}$ which agrees with λ on $\mathscr{R}$ and let $\mathscr{M} = \{A \in \mathscr{S} : \lambda(A) = \nu(A)\}$.

(i) Assume that λ and ν are finite. We argue as in Theorem 1 that $\mathscr{M}$ is monotone and hence coincides with $\mathscr{S}$.

(ii) In general, let E be an arbitrary set in $\mathscr{S}$: by Ex. 11.4.7 there is a sequence of elements of $\mathscr{R}$ whose union contains E. As μ is σ-finite there is a sequence $\{A_n\}$ of elements of $\mathscr{R}$ of finite μ measure whose union contains E. By replacing A_n by $A_1 \cup \ldots \cup A_n$, if necessary, we may arrange that $\{A_n\}$ is increasing. Apply part (i) to the contractions λ_{A_n}, ν_{A_n} and deduce that

$$\lambda(E \cap A_n) = \nu(E \cap A_n)$$

for all n. As $n \to \infty$ this gives $\lambda(E) = \nu(E)$ and completes the proof. (Incidentally, this proof shows that λ is σ-finite.)

Let X be an uncountable set, $\mathscr{R}$ the ring of finite subsets of X and μ the zero measure on $\mathscr{R}$. Then $\mathscr{S}$ consists of the countable subsets of X, whereas $\mathscr{A}$ consists of the countable subsets of X together with their complements in X. We may take $\lambda(A) = 0$ for all A in $\mathscr{A}$ and $\nu(A) = 0$ if $A \in \mathscr{S}$, $= \infty$ if $X \setminus A \in \mathscr{S}$.

6. Argue as for Ex. 5 only with $\mathscr{M} = \{A \in \mathscr{S} : \mu_1(A) \leqslant \mu_2(A)\}$.

7. Suppose that $A \subset E \subset B$, where $A, B \in \mathscr{A}$ and $\lambda(B \setminus A) = 0$. Then $B \setminus E \in \mathscr{A}'$, which implies that $E \in \mathscr{A}'$ and

$$\lambda'(E) = \lambda'(B) = \lambda(B) = \bar{\lambda}(E).$$

8. Suppose that $A \in \mathscr{A}$ and $B \subset C$ where $C \in \mathscr{A}$ and $\lambda(C) = 0$. Then $A \cup B$ is squeezed between A and $A \cup (C \setminus A)$, where $\lambda(C \setminus A) = 0$. On the other hand suppose that $A \subset E \subset D$, where $A, D \in \mathscr{A}$ and $\mu(D \setminus A) = 0$; then we may write $E = A \cup B$, where $B \subset D \setminus A$ and $\mu(D \setminus A) = 0$.

The argument with the symmetric difference is very similar, only $A \triangle B$ is squeezed between $A \setminus C$ and $A \cup C$.

§ 13.3

1. Write $A \sim B$ if and only if $\mu(A \triangle B) = 0$ and verify that $\sim$ is an equivalence relation. Define $d(\tilde{A}, \tilde{B}) = d(A, B)$, where A, B are any representatives of the equivalence classes $\tilde{A}, \tilde{B}$. It is then easy to verify that this definition is unambiguous and that $d(\tilde{A}, \tilde{B}) = 0$ if and only if $\tilde{A} = \tilde{B}$.

2. Let $\mathscr{R}_0$ be the subring of $\mathscr{R}$ on which μ and ν are both finite. By the uniqueness theorem of Ex. 13.2.5 we may apply the Approximation Theorem to μ and ν with $L = L(\mathscr{R}_0)$. Thus we may find increasing sequences $\{S_n^\epsilon\}, \{T_n^\epsilon\}$ of sets in $\mathscr{R}_0$ whose unions S^ϵ, T^ϵ contain E and such that

$$\mu(S^\epsilon) \leqslant \mu(E) + \epsilon/3, \quad \nu(T^\epsilon) \leqslant \nu(E) + \epsilon/3.$$

Let $R_n^\epsilon = S_n^\epsilon \cap T_n^\epsilon$, $R^\epsilon = S^\epsilon \cap T^\epsilon$; then $R_n^\epsilon \uparrow R^\epsilon$, where R^ϵ contains E. As in the proof of Theorem 2, take N large enough to ensure that

$$\mu(R^\epsilon \setminus R_N^\epsilon) \leqslant \epsilon/3, \quad \nu(R^\epsilon \setminus R_N^\epsilon) \leqslant \epsilon/3$$

and then take $A = R_N^\epsilon$.

3. By Lemma 8.2.1, $f = g - h$, where $g, h \in L^{\text{inc}}$, h is positive and $\int h < \epsilon/3$. Find ϕ_n, ψ_n and θ_n as in the proof of Theorem 1.

4. As R dominates P and Q it is easy to see that an R-null set is both P-null and Q-null, that

$$L^{\text{inc}}(R) \subset L^{\text{inc}}(P) \cap L^{\text{inc}}(Q) \quad \text{and} \quad L^1(R) \subset L^1(P) \cap L^1(Q).$$

On the other hand, suppose that A is both P-null and Q-null. Then there exist increasing sequences $\{\phi_n\}, \{\psi_n\}$ which diverge on A and such that $P(\phi_n)$ and $Q(\psi_n)$ are bounded. Let $\tau_n = \min\{\phi_n, \psi_n\}$; then $\{\tau_n\}$ is increasing, diverges on A and $R(\tau_n)$ is bounded: thus A is R-null. Now suppose that $f \in L^1(P) \cap L^1(Q)$. By Ex. 3, to any $\epsilon > 0$, there exist increasing sequences $\{\phi_n\}, \{\psi_n\}$ of elementary functions such that

$$P(\phi_n) < P(f) + \epsilon, \quad Q(\psi_n) < Q(f) + \epsilon$$

for all n, and $\qquad f \leqslant \lim \phi_n, \quad f \leqslant \lim \psi_n.$

Let $\tau_n = \min\{\phi_n, \psi_n\}$. Then $\{\tau_n\}$ is an increasing sequence of elementary functions which satisfies

$$R(\tau_n) < P(f) + Q(f) + 2\epsilon \quad \text{and} \quad f \leqslant \lim \tau_n.$$

Write f^ϵ for $\lim \tau_n$. (In the present context, as in Exx. 8.1.A–G, it is natural to let $f^{\epsilon}(x) = \infty$ if $\tau_n(x) \to \infty$: the set of all such x is, of course, R-null.) It follows that

$$f^{\epsilon} \in L^1(R), f \leqslant f^{\epsilon}, P(f) \leqslant P(f^{\epsilon}) \leqslant P(f)+\epsilon, Q(f) \leqslant Q(f^{\epsilon}) \leqslant Q(f)+\epsilon,$$

and

$$P(f)+Q(f) \leqslant R(f^{\epsilon}) \leqslant P(f)+Q(f)+2\epsilon.$$

Let $g_n = \min\{f^1, f^{1/2}, \ldots, f^{1/n}\}$; then $\{g_n\}$ is a decreasing sequence of elements of $L^1(R)$ whose limit g belongs to $L^1(R)$. Moreover $f \leqslant g$, $P(f) = P(g)$, $Q(f) = Q(g)$, from which it follows that $f = g$ almost everywhere with respect to R, $f \in L^1(R)$ and $R(f) = P(f)+Q(f)$.

5. Consider $\operatorname{mid}\{-\phi, f, \phi\}$, where ϕ is a positive elementary function.

§ 13.4

1. If $\phi_n \in L$ then $X_n = \{x \in X: \phi_n(x) \geqslant 1\}$ is a closed subset of the support of ϕ_n and as such is compact. On the other hand, by Proposition 10.1.1, any compact subset K of X is expressible as $K = \{x \in X: \psi(x) \geqslant 1\}$, where ψ is a positive element of L.

2. (i) Let $X = \bigcup X_n$, where $X_n = \{x: \phi_n(x) \geqslant 1\}$ (positive $\phi_n \in L$): then

$$f = \Sigma \phi_n / 2^n (P(\phi_n)+1)$$

is a strictly positive element of $L^1(P)$.

(ii) Let $S_{mn} = \{x: \theta_n(x) \geqslant 1/m\} = \{x: m\theta_n(x) \geqslant 1\}$. Then $X = \bigcup_{m,n} S_{mn}$ and the elementary sets S_{mn} are countable by the familiar diagonal process of §2.1.

(iii) For any x in X, there exists an integer n such that $\theta_n(x) > 0$ and so (ii) applies.

(iv) X is σ-finite with respect to L if and only if there is a strictly positive f in $L^1(P)$. Note that P is just *one* Daniell integral obtained by extending an elementary integral on L.

3. The σ-finiteness of X cannot be omitted. In the special case cited, $\mu(S) = 0$ if S is countable and $\mu(S) = \infty$ otherwise; $\nu(S) = 0$ for all S.

4. Similar to Ex. 3.

§ 13.5

1. Take $R = S = E$ in the definitions of $c^*(E)$, $c_*(E)$.

2. Let E_1, E_2 be bounded sets and let $S_1, S_2 \in \mathscr{R}$ satisfy $S_1 \supset E_1$, $S_2 \supset E_2$. Then $S_1 \cup S_2 \supset E_1 \cup E_2$ and $S_1 \cap S_2 \supset E_1 \cap E_2$. Thus

$$\begin{aligned} c^*(E_1 \cup E_2)+c^*(E_1 \cap E_2) &\leqslant m(S_1 \cup S_2)+m(S_1 \cap S_2) \\ &= m(S_1)+m(S_2). \end{aligned}$$

Taking infima as S_1, S_2 vary independently we get

$$c^*(E_1 \cup E_2)+c^*(E_1 \cap E_2) \leqslant c^*(E_1)+c^*(E_2).$$

In exactly the same way,

$$c_*(E_1)+c_*(E_2) \leqslant c_*(E_1 \cup E_2)+c_*(E_1 \cap E_2).$$

When E_1, E_2 are contented it follows that both $E_1 \cup E_2$ and $E_1 \cap E_2$ are contented and

$$c(E_1 \cup E_2)+c(E_1 \cap E_2) = c(E_1)+c(E_2).$$

Suppose that E is contained in the bounded interval I. Then

$$c^*(I \setminus E) = m(I)-c_*(E) \quad \text{and} \quad c_*(I \setminus E) = m(I)-c^*(E).$$

Thus, if E is contented it follows that $I \setminus E$ is contented and

$$c(I \setminus E) = m(I)-c(E).$$

Finally, suppose that E_1, E_2 are bounded and contented. Choose a bounded interval I containing $E_1 \cup E_2$: then $I \setminus E_2$ is contented and hence $E_1 \setminus E_2 = E_1 \cap (I \setminus E_2)$ is contented.

Combining these results we see that the bounded contented sets form a ring of subsets of $\mathbf{R}^k$ on which c is an additive function.

3. If A, B are countable we may prove as in § 2.1 that $A \times B$ is countable. It then follows by induction on the dimension k that S is countable.

The only intervals which are contained in S either consist of a single (rational) point or are empty. Thus $c_*(S) = 0$. On the other hand, if $S \subset I_1 \cup \ldots \cup I_n (I_1, \ldots, I_n$ disjoint) then $\bar{S} = I \subset \bar{I}_1 \cup \ldots \cup \bar{I}_n$. The closed intervals $\bar{I}_1, \ldots, \bar{I}_n$ may overlap at their faces but they have the same measures as $I_1, \ldots, I_n$, respectively, and we deduce that $c^*(S) \geqslant 1$.

In the same way $c^*(G) \geqslant 1$. We contend that $c_*(G) \leqslant \frac{1}{2}$. Suppose on the contrary that $c_*(G) > \frac{1}{2}$, then we may find an elementary figure $R \subset G$ such that $m(R) > \frac{1}{2}$ and by taking slightly smaller closed intervals, if necessary, we may arrange that R be closed. By the Heine–Borel Theorem, R is covered by a finite selection of the intervals defining G and so

$$m(R) \leqslant 1/2^2+1/2^3+\ldots=1/2.$$

This contradiction establishes the fact that $c_*(G) \leqslant \frac{1}{2}$ and hence that G is discontented. [Alternatively, use the inequality $c_*(G) \leqslant m_*(G)$ established in Ex. 6 and the fact that $m(G) \leqslant \frac{1}{2}$.]

Let $H = G \cap I$. Then the above argument also shows that $c^*(H) \geqslant 1$ and $c_*(H) \leqslant \frac{1}{2}$: thus H is discontented and the *compact* set $I \setminus G = I \setminus H$ is discontented.

4. This is just another way of saying that $I \cap E$ is contented for all bounded intervals I.

5. Express $\mathbf{R}^k$ as the union of an increasing sequence of bounded intervals and use Ex. 2.

6. For any bounded set E,

$$c_*(E) \leqslant m_*(E) \leqslant m^*(E) \leqslant c^*(E).$$

Then argue as in Ex. 5.

7. (Cf. Ex. 3.3.A.) Suppose that χ_S is Riemann integrable on I. To any $\epsilon > 0$, there are positive step functions ϕ, ψ which vanish outside I such that

$$\phi \leqslant \chi_S \leqslant \psi \quad \text{and} \quad \int(\psi - \phi) < \epsilon.$$

Let $A = \{x\colon \phi(x) > 0\}$ and $B = \{x\colon \psi(x) \geqslant 1\}$: then

$$\phi \leqslant \chi_A \leqslant \chi_S \leqslant \chi_B \leqslant \psi,$$

so that A, B are elementary figures, $A \subset S \subset B$ and $m(B \setminus A) < \epsilon$. Thus S is contented and $c(S) = \int \chi_S$.

Suppose that S is contented. Given $\epsilon > 0$, we find elementary figures A, B satisfying $A \subset S \subset B$ and $m(B \setminus A) < \epsilon$, whence

$$\chi_A \leqslant \chi_S \leqslant \chi_B, \quad \int(\chi_B - \chi_A) < \epsilon$$

and χ_S has Riemann integral equal to $c(S)$.

8. Suppose that f is Riemann integrable on $[a, b]$. To any $\epsilon > 0$, we may find positive step functions ϕ, ψ which vanish outside $[a, b]$ and satisfy the conditions $\phi \leqslant f \leqslant \psi$, $\int(\psi - \phi) < \epsilon$. The corresponding ordinate sets Q_ϕ, Q_ψ are elementary figures in $\mathbf{R}^2$ which satisfy $Q_\phi \subset Q_f \subset Q_\psi$, $m(Q_\psi \setminus Q_\phi) < \epsilon$ (Fig. S. 5). Thus Q_f is contented and

$$c(Q_f) = \int_a^b f(x)\,dx.$$

On the other hand, suppose that A, B are elementary figures satisfying $A \subset Q_f \subset B$, $m(B \setminus A) < \epsilon$. First of all, remove from B any points (x, y) for which $x < a$ or $x > b$ or $y < 0$ (as $A \subset Q_f$, there are no such points in A). Express A and B as disjoint unions of rectangles using all common vertical lines of subdivision (cf. § 3.1, where this kind of argument is given in full). We may then discard some rectangles from B and adjoin some to A, if necessary, so that they are ordinate sets of step functions ϕ, ψ, respectively which satisfy $\phi \leqslant f \leqslant \psi$ and $\int(\psi - \phi) < \epsilon$ (Fig. S. 6). This gives the converse.

9. Each element of $\mathscr{R}$ is a finite union of elements of $\mathscr{C}$ and we adjoin $\varnothing$ to allow finite unions as well as countable unions in the definition of $\mu^*(E)$.

10. P satisfies the Daniell condition because μ is σ-additive (Proposition 11.4.2).

Suppose that $\mu^*(E)$ is finite. To any sequence $\{A_n\}$ of disjoint sets in $\mathscr{R}$, whose union A contains E and for which $\Sigma\mu(A_n)$ is convergent, we may define $\phi_n = \chi_{A_1} + \ldots + \chi_{A_n}$: this gives an increasing sequence $\{\phi_n\}$ whose limit $\chi_A \geqslant \chi_E$. On taking infima we deduce that $P^*(\chi_E) \leqslant \mu^*(E)$. This inequality is also true in a trivial sense when $\mu^*(E) = \infty$.

In establishing the reverse inequality we may as well assume that $P^*(\chi_E)$ is finite. To any $\epsilon > 0$, there is an increasing sequence $\{\phi_n\}$ of positive elementary functions such that $\lim \phi_n \geqslant \chi_E$ and $P(\phi_n) \leqslant P^*(\chi_E) + \epsilon$. Let c be defined by

$$c(P^*(\chi_E) + 2\epsilon) = P^*(\chi_E) + \epsilon,$$

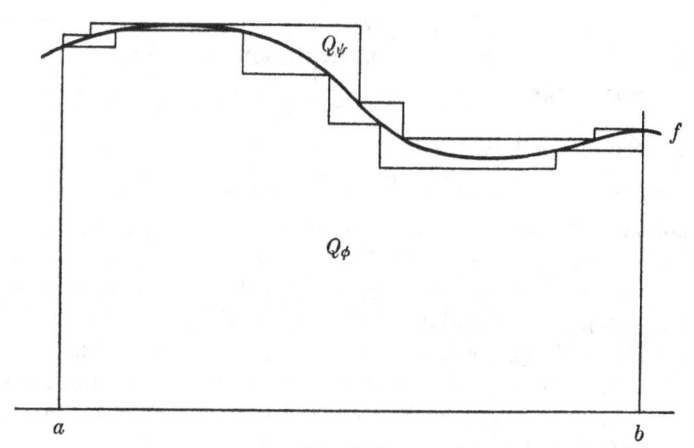

Fig. S. 5

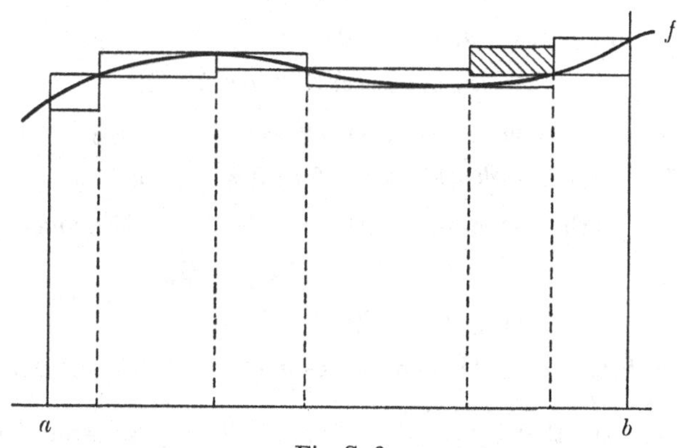

Fig. S. 6

so that $0 < c < 1$, and let

$$S_n^\epsilon = \{x \in X: \phi_n(x) \geqslant c\}.$$

Then $S_n^\epsilon \in \mathscr{R}$ and $S_n^\epsilon \uparrow S^\epsilon$, say, where $S^\epsilon \supset E$ (because $\lim \phi_n \geqslant \chi_E$). Moreover

$$\chi_{S_n^\epsilon} \leqslant \frac{1}{c}\phi_n \Rightarrow \mu(S_n^\epsilon) \leqslant \frac{1}{c} P(\phi_n) \leqslant P^*(\chi_E) + 2\epsilon$$

for all n. Taking $A_n^\epsilon = S_n^\epsilon \setminus S_{n-1}^\epsilon$ for $n \geqslant 1 (S_0^\epsilon = \varnothing)$ we deduce that $\mu^*(E) \leqslant P^*(\chi_E) + 2\epsilon$ for arbitrarily small ϵ and so $\mu^*(E) \leqslant P^*(\chi_E)$.

For any ϕ in L, it follows from the definitions in Exx. 8.1.A, B that

$$P^*(\phi - f) = P(\phi) - P_*(f).$$

In particular with $\phi = \chi_A$ $(A \in \mathscr{R})$ we see that $\chi_{A \cap E} \in L_1(P)$ if and only if

$$P(\chi_A) = P^*(\chi_{A \cap E}) + P^*(\chi_{A \cap E^c}).$$

In view of Ex. 8.1.F, this is equivalent to Theorem 1.

11. In the definition of $\mu^*(E)$ given on p. 129, we may assume that the terms A_n are disjoint. An equivalent definition would be:

$$\mu^*(E) = \inf\{\lim \mu(S_n): S_n \in \mathscr{R}\ (n = 1, 2, \ldots),\ S_n \uparrow S \supset E\}.$$

This suggest the 'dual definition'

$$\mu_*(E) = \sup\{\lim \mu(T_n): T_n \in \mathscr{R}\ (n = 1, 2, \ldots),\ T_n \downarrow T \subset E\}.$$

The proof that $P_*(\chi_E) = \mu_*(E)$ is similar to the proof given in Ex. 10.

To any descending sequence $\{T_n\}$ whose intersection T is contained in E, there corresponds a decreasing sequence $\{\chi_{T_n}\}$ whose limit $\chi_T \leqslant \chi_E$. Thus $\mu_*(E) \leqslant P_*(\chi_E)$. (In particular $\mu_*(E) = \infty \Rightarrow P_*(\chi_E) = \infty$.)

Suppose that $P^*(\chi_E)$ is finite and let $\epsilon > 0$ be given. Then there is a decreasing sequence $\{\psi_n\}$ of elementary functions whose limit v satisfies $v \leqslant \chi_E$ and $P(v) > P_*(\chi_E) - \epsilon$. Without loss of generality we may assume that $0 \leqslant \psi_n(x) \leqslant 1$ and hence $0 \leqslant v(x) \leqslant 1$ for all x. Let

$$K_n = \{x \in X: \psi_n(x) > 0\},$$

then $K_n \in \mathscr{R}$ and $\mu(K_n) \leqslant \mu(K_1)$ for all n. Let

$$T_n = \{x \in X: \psi_n(x) \geqslant \epsilon/(\mu(K_1)+1)\};$$

then $\{T_n\}$ is a decreasing sequence with intersection T, say. Now

$$x \in T \Rightarrow \psi_n(x) \geqslant \epsilon/(\mu(K_1)+1) \quad \text{for all } n \Rightarrow v(x) > 0 \Rightarrow x \in E$$

(using $v \leqslant \chi_E$ at the last step): in other words $T \subset E$. Moreover

$$\chi_{T_n} + (\epsilon/(\mu(K_1)+1))\, \chi_{K_n} \geqslant \psi_n,$$

so that

$$\mu(T_n) + \epsilon \geqslant P(\psi_n) > P_*(\chi_E) - \epsilon$$

and $\mu(T_n) > P_*(\chi_E) - 2\epsilon$. This now gives $\mu_*(E) \geqslant P_*(\chi_E) - 2\epsilon$ for arbitrary $\epsilon > 0$ and so $\mu_*(E) \geqslant P_*(\chi_E)$.

Finally, suppose that $P_*(\chi_E) = \infty$. To an arbitrarily large number N there is a decreasing sequence $\{\psi_n\}$ of elementary functions whose limit v satisfies $v \leqslant \chi_E$ and $P(v) > N$. The above argument (with $\epsilon = 1$, say) gives $\mu(T_n) > N - 1$ and so $\mu_*(E) = \infty$.

12. It is immediately verified that ν satisfies conditions (8), (9) and (10). The measurable sets in the sense of (11) are just X and $\varnothing$: let $X = \{p, q, r\}$; the singleton set $E = \{p\}$ is not measurable as we see by taking $A = \{p, q\}$ so that $\nu(A) = 1$, $\nu(A \cap E) = 1$, $\nu(A \cap E^c) = 1$; by the same token $E^c = \{q, r\}$ is not measurable.

In fact $\lambda(E) = \nu(E)$ for all E! Let $A = \{p\}$, $B = \{q\}$; then

$$\lambda(A) = \lambda(B) = \lambda(A \cup B) = 1.$$

(For a detailed treatment of inner measure see Halmos [11] or Royden [18]. Many of the properties discussed in these books may be derived quite simply from the Daniell integral using Exx. 10, 11. For example, we should expect a 'satisfactory' inner measure μ_* to satisfy

$$\mu_*(A) + \mu_*(B) \leqslant \mu_*(A \cup B)$$

for disjoint sets A, B, just as

$$\mu^*(A \cup B) \leqslant \mu^*(A) + \mu^*(B)$$

for all A, B (cf. the solution of Ex. 8.1.B).)

§ 14.1

1. Let $A_1 \subset A_2$ be distinct non-empty sets in $\mathscr{R}$ and $B_1 \subset B_2$ distinct non-empty sets in $\mathscr{S}$. It is easy to verify that $(A_2 \times B_2) \setminus (A_1 \times B_1)$ does not belong to $\mathscr{T}$ (Fig. S. 7).

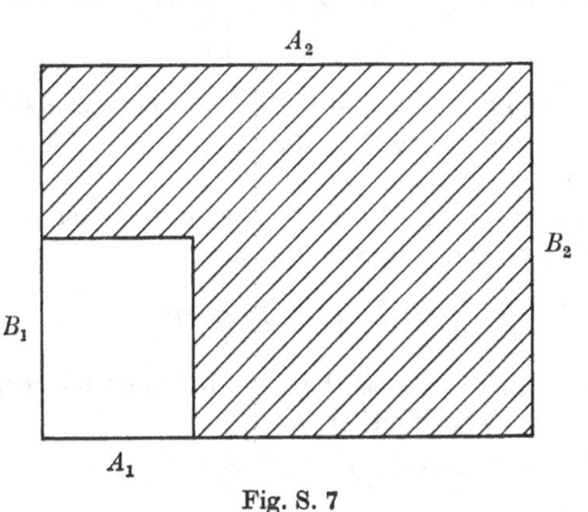

Fig. S. 7

2. Apply the proof of Lemma 1: for semi-rings the first part is simpler.

3. Let $\mathscr{T}'$ denote the collection of all subsets C of $X \times Y$ for which the cross-sections $C_x = \{y \in Y : (x, y) \in C\}$ belong to $\mathscr{S}$ for all x in X. In this notation $(C \setminus D)_x = C_x \setminus D_x$, with a similar identity for countable unions. Thus $\mathscr{T}'$ verifies the conditions for a σ-ring. Moreover $\mathscr{T}'$ contains all $A \times B$ ($A \in \mathscr{R}, B \in \mathscr{S}$), whence $\mathscr{T}' \supset \mathscr{T}$. This proves the first part and the second follows in almost the same way.

Finally, A and B occur as cross-sections of $A \times B$ provided $A \times B$ is not empty.

4. All cross-sections of D consist of a single point and so are measurable, but D does not belong to $\mathscr{T}$ because D is uncountable (cf. Ex. 11.4.7). Thus the natural converse of Ex. 3 does not hold.

5. As we saw in the paragraphs following the proof of Theorem 1, there exists at least one measure λ on $\mathscr{T}$ which satisfies condition (3). It only remains to establish the uniqueness of λ when μ, ν are σ-finite. This is virtually Ex. 13.2.5 as λ is uniquely defined as far as the *ring* generated by $\{A \times B : A \in \mathscr{R}, B \in \mathscr{S}\}$.

6. Let μ, ν be Lebesgue measure on $\mathbf{R}$ and let E be the non-measurable subset of $[0, 1]$ constructed in § 6.2. Then $\{0\} \times E$ is a subset of the null set $\{0\} \times [0, 1] \in \mathscr{T}$, although $\{0\} \times E \notin \mathscr{T}$ by Ex. 3.

7. Verify for sets $(A_1 \times A_2) \times A_3 = A_1 \times (A_2 \times A_3)$ $(A_i \in \mathscr{A}_i)$ and use the uniqueness of extension.

§ 14.2

1. For any subset C of $X \times Y$, $\lambda(C)$ is the number of elements in C (possibly infinite). A function $a: X \times Y \to \mathbf{R}$ is a 'double sequence': to comply with standard notation write $a(i,j) = a_{ij}$. A positive function $a \in L^1(X \times Y)$ if and only if $\sum_{(i,j) \in C} a_{ij}$ is bounded for every finite subset C of $X \times Y$ and $\int a\, d\lambda$ is the supremum of all such sums. An arbitrary

$$a \in L^1(X \times Y)$$

if and only if $a^+, a^- \in L^1(X \times Y)$ and $\int a\, d\lambda = \int a^+ d\lambda - \int a^- d\lambda$: in these circumstances it is customary to say that the 'double series' $\Sigma_{i,j}\, a_{ij}$ is absolutely convergent and has sum $\int a\, d\lambda$.

In this context Fubini's Theorem says: if the double series $\Sigma_{i,j}\, a_{ij}$ is absolutely convergent, then its sum equals

$$\sum_i \left(\sum_j a_{ij}\right) = \sum_j \left(\sum_i a_{ij}\right);$$

these are the 'sum by rows' and the 'sum by columns' respectively.

Tonelli's Theorem says: if either

$$\sum_i \left(\sum_j |a_{ij}|\right) \quad \text{or} \quad \sum_j \left(\sum_i |a_{ij}|\right)$$

is convergent then the double series $\Sigma_{i,j}\, a_{ij}$ is absolutely convergent and hence, by Fubini, its sum by rows and sum by columns are equal.

In this case the sum by columns is $0+0+0+\ldots = 0$ and the sum by rows is $\frac{3}{2}+(-\frac{3}{2}+\frac{7}{4})+(-\frac{7}{4}+\frac{15}{8})+\ldots = 2$. This example shows that we cannot omit the condition $f \in L^1(X \times Y)$ in Fubini's Theorem, and we cannot omit the condition that one of the repeated integrals for $|f|$ should exist in Tonelli's Theorem.

2. Write $f(x,n) = f_n(x)$ for $n \geqslant 1$. If $\Sigma \int |f_n(x)|\, dx$ is convergent, then $\Sigma f_n(x)$ is convergent for almost all x and

$$\int (\Sigma f_n(x))\, dx = \Sigma \int f_n(x)\, dx$$

(cf. the Absolute Convergence Theorem 8.2.3).

3. When f and g are positive this follows from Tonelli's Theorem (bearing in mind that the product of two measurable functions is measurable, and the product of two σ-finite subsets is σ-finite). The general case follows by linearity using the positive and negative parts of f and g.

4. If $y \neq 0$, $\displaystyle\int_{-1}^{1} f(x,y)\, dx = [-\tfrac{1}{2}y(x^2+y^2)^{-1}]_{-1}^{1} = 0.$

Thus each repeated integral is zero.

On the other hand, if $y \neq 0$,

$$\int_0^1 f(x,y)\,dx = -\tfrac{1}{2}y\left(\frac{1}{1+y^2}-\frac{1}{y^2}\right) = \frac{1}{2}\left(\frac{1}{y}-\frac{y}{1+y^2}\right)$$

and the repeated integral

$$\int_0^1 \left(\int_0^1 f(x,y)\,dx\right) dy$$

does not exist. Thus f is not integrable on the unit square

$$\{(x,y): 0 \leqslant x \leqslant 1, 0 \leqslant y \leqslant 1\}$$

and *a fortiori* on the larger given square.

5. (i) If $\phi = \chi_A$ $(A \in \mathscr{A})$ then $P_\phi = A \times [0,1) \in \mathscr{C}$. The extension to a positive $\mathscr{A}$-simple function ϕ is obvious.

(ii) If f is $\mathscr{A}$-measurable, express f as the limit (everywhere) of an increasing sequence $\{\phi_n\}$ of positive $\mathscr{A}$-simple functions and use the fact that $P_{\phi_n} \uparrow P_f$. If P_f is $\mathscr{C}$-measurable, then the cross-section by $y = c$ $(c \geqslant 0)$ is $\mathscr{A}$-measurable: in other words $\{x \in X: f(x) > c\} \in \mathscr{A}$ for all $c \geqslant 0$.

(iii) Verify when f is an $\mathscr{A}$-simple function and extend by means of the Monotone Convergence Theorem.

6. The first part is a simple application of the Monotone Convergence Theorem and Theorem 13.1.1, Corollary. The function λ so defined is σ-additive on $\mathscr{C}$ and satisfies $\lambda(A \times B) = \mu(A)\,\nu(B)$ for all A in $\mathscr{A}$, B in $\mathscr{B}$.

The extension to the σ-finite case is just as for Theorem 14.1.1 (and, of course, we must interpret multiplication by ∞ as for that theorem using equations (4), (5), (6) on p. 142).

7. Since f^+ and f^- are both integrable with respect to λ we may as well assume that f is positive. If f is a $\mathscr{C}$-simple function the result follows at once by linearity from Ex. 6.

Express f as the limit everywhere of an increasing sequence $\{\phi_n\}$ of positive $\mathscr{C}$-simple functions and let

$$\Phi_n(x) = \int \phi_n(x,y)\,dy.$$

This function Φ_n may take the value ∞ but as

$$\int \Phi_n(x)\,dx = \int \phi_n\,d\lambda \leqslant \int f\,d\lambda$$

is finite, $\Phi_n(x)$ is finite for almost all x (with respect to μ). It follows by the Monotone Convergence Theorem that

$$g(x) = \lim \Phi_n(x) = \int f(x,y)\,dy$$

is finite for almost all x (with respect to μ) and

$$\int g(x)\,dx = \int f\,d\lambda.$$

The other repeated integral is treated in exactly the same way.

Even if the measures μ, ν are complete it does not follow that the product measure λ is complete (Ex. 14.1.6) and so the assumption that f should be λ-integrable is more restrictive than the corresponding assumption in Theorem 1. In the present setting, however, the functions f_x: $Y \to \mathbf{R}$ defined by $f_x(y) = f(x, y)$ are measurable with respect to ν for *all* x in X and not merely for *almost all* x in X with respect to μ (cf. Ex. 14.1.3).

8. For each $n \geqslant 1$, D may be covered by 2^n equal squares of side 2^{-n} as in Fig. S. 8. By taking intersections we see that D belongs to the σ-algebra generated by the rectangles in $[0, 1]^2$.

For each x in $[0, 1]$, $\int \chi_D(x, y)\,dy = 1$. For each y in $[0, 1]$, $\int \chi_D(x, y)\,dx = 0$.

In Tonelli's Theorem and in Ex. 6 we cannot drop the σ-finiteness condition.

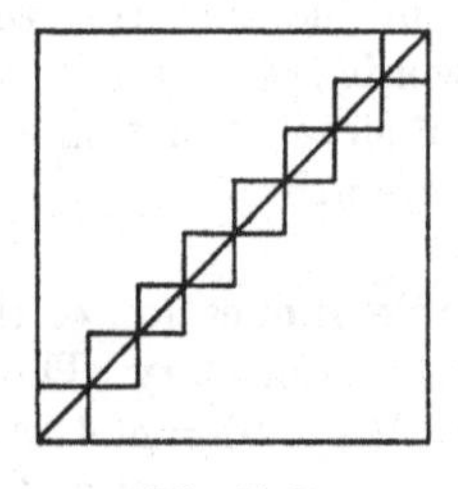

Fig. S. 8

9. (i) $E'' = T(E')$, where $T(x, y) = \frac{1}{2}(x+y,\ -x+y)$. In other words, E'' is obtained by rotating E' about the origin through the angle $\pi/4$ in the clockwise sense and multiplying by $1/\sqrt{2}$.

If E is measurable, then so is E' and hence also E''.

(ii) By part (i), $\{(x, y)\colon F(x, y) \geqslant c\}$ is a measurable subset of $\mathbf{R}^2$.

(iii) For all y in $\mathbf{R}$,

$$\int f(x-y)\,dx = \int f(x)\,dx.$$

In view of (ii), we may apply Tonelli's Theorem (as in Ex. 8) to see that $h(x)$ is defined for almost all x in $\mathbf{R}$ and

$$\int h(x)\,dx = \int f(x)\,dx \int g(y)\,dy.$$

The given inequality follows in almost the same way from

$$|h(x)| \leqslant \int |f(x-y)\,g(y)|\,dy.$$

(iv) For almost all x,

$$\int_{-\infty}^{\infty} f(x-y)\,g(y)\,dy = \int_{-\infty}^{\infty} f(u)\,g(x-u)\,du$$

by the substitution $u = x-y$. The associativity follows similarly by evaluating

$$\int e(t-x)f(x-y)\,g(y)\,d(x,y)$$

in two ways as a repeated integral.

(v) By Fubini's Theorem (applied to the real and imaginary parts)

$$\hat{h}(t) = \int \left(\int f(x-y)\,g(y)\,e^{-ixt}\,dx \right) dy.$$

Using the substitution $u = x-y$ in the first integral:

$$\hat{h}(t) = \int f(u)\,e^{-iut}\,du \int g(y)\,e^{-iyt}\,dy = \hat{f}(t)\,\hat{g}(t).$$

§ 14.3

1. λ_0 is the zero measure on the ring of finite subsets of $[0,1]^2$. The complete measures $\bar{\lambda}$, $\bar{\mu}$, $\bar{\nu}$ are zero on countable subsets and infinite on uncountable subsets. Let $A = \{0\}$, $B = [0,1]$; then $\bar{\mu}(A) = 0$, $\bar{\nu}(B) = \infty$ and $\bar{\lambda}(A \times B) = \infty$.

2. As in the last part of the proof of Theorem 1, if $\bar{\mu}(A) = 0$ and $\bar{\nu}(B) = \infty$, then $\bar{\lambda}(A \times B) = \infty$ or 0.

3. As for Ex. 8.3.2.

4. Given two elements f, g of M; we may express $[0,1]$ as the union of finitely many disjoint subintervals on each of which f and g have a linear graph. It follows that $af+bg \in M$ for any a, b in $\mathbf{R}$ and $|f| \in M$. Likewise, we may express $[0,1]$ as the union of disjoint subintervals on which $f_1, \ldots, f_n$ have linear graphs, and similarly for $g_1, \ldots, g_n$. From this we deduce that L is a linear space and $h(x,y)$ is of the stated form.

Using elementary linear algebra, the coefficients a, b, c, d are uniquely determined by the values of $a+bx+cy+dxy$ at the vertices of a proper rectangle (i.e. a rectangle which has strictly positive area). In the given special case, $h(x,y) = xy - \frac{1}{4}$ for (x,y) in the unit square $I = [0,1]^2$. The points for which $xy = \frac{1}{4}$ lie on a curve (rectangular hyperbola) as in Fig. S. 9. If $|h|$ were in L, then there would be a proper closed rectangle J as in Fig. S. 9 in which

$$|h|\,(x,y) = a+bx+cy+dxy.$$

J contains proper rectangles, as shown, in which

$$|h|\,(x,y) = xy - \tfrac{1}{4}, \quad |h|\,(x,y) = \tfrac{1}{4} - xy,$$

respectively. This contradicts the uniqueness of the coefficients and hence proves that $|h| \notin L$.

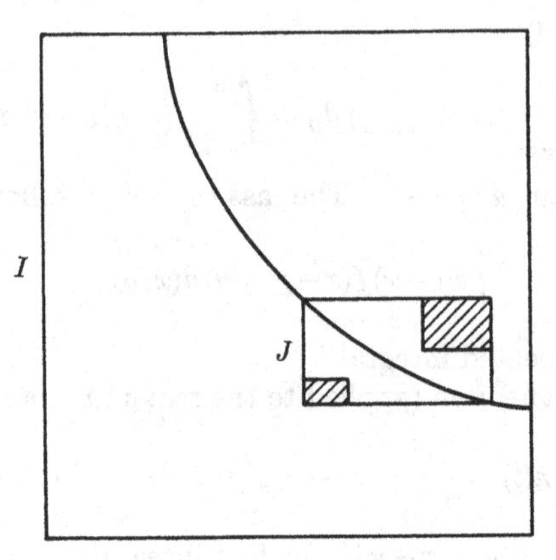

Fig. S. 9

§15.1

1. Argue as in the first paragraph of the present section.

2. In this case $\mathscr{B}(X)$ consists of *all* subsets of X and the compact subsets are just the *finite* subsets. The Borel measures on X are the σ-additive functions μ on $\mathscr{B}(X)$ which are finite on all singleton sets $\{x\}$ ($x \in X$). For any $E \subset X$,

$$\mu(E) = \sum_{x \in E} \mu(x).$$

(This sum may be finite or infinite.)

3. In $\mathbf{R}^1$, $[a, b) = \bigcap (a - 1/n, b) = \bigcup [a, b - 1/n]$, and this generalises easily to $\mathbf{R}^k$.

4. (i) Let $p \in \mathbf{R}$. For each $r > 0$,

$$\inf\{f(x): |x-p| < r\} \leqslant f(p) \leqslant \sup\{f(x): |x-p| < r\}.$$

Thus
$$g(p) \leqslant f(p) \leqslant h(p).$$

(ii) Immediate from (i) and the definition of continuity in terms of distance.

(iii) Suppose that $g(p) > c$. Then there exists $r > 0$ such that

$$\inf\{f(x): |x-p| < r\} > c.$$

Now
$$|y-p| < r/2 \Rightarrow \inf\{f(x): |x-y| < r/2\} > c \Rightarrow g(y) > c.$$

Thus g is lower semicontinuous and similarly h is upper semicontinuous.

We show that $G_n = \{x \in \mathbf{R}: h(x) - g(x) < 1/n\}$ is *open*. This follows (as for similar manipulations with measurable functions in Ex. 12.1.5) by noting that $h(x) - g(x) < 1/n$ if and only if there is a rational number r such that $h(x) < r < g(x) + 1/n$, whence

$$G_n = \bigcup_r (\{x: h(x) < r\} \cap \{x: g(x) > r - 1/n\}).$$

By (ii) we deduce that the points of continuity form the set $\bigcap G_n$.

If f is unbounded below in every neighbourhood $(p-r, p+r)$ of p, then we define $g(p) = -\infty$. There is a similar interpretation of $h(p) = \infty$. The rest is as before – of course, $g(p)$ can only equal $h(p)$ if both are finite!

5. Replace the intervals $(p-r, p+r)$ $(r > 0)$ by arbitrary open sets U containing p. The points of continuity of f still form a G_δ.

§15.2

1. The σ-additivity of μ and m shows that $\mathscr{C}$ is monotone. As in the proof of Theorem 1, $\mathscr{C}$ contains the ring of elementary figures. Now apply Theorem 13.1.1, Corollary to see that $\mathscr{C} = \mathscr{B}$.

2. Without loss of generality T is a linear mapping. By Lemma 6.3.1 T may be expressed as a product of linear mappings PDQ, where P, Q are invertible and D has diagonal matrix with at least one zero on the diagonal. (Matrices are with respect to the standard basis of $\mathbf{R}^k$.) The result then follows from Theorem 2, as $D(\mathbf{R}^k)$ is a closed subset of $\mathbf{R}^k$ with zero measure and $B = PD(\mathbf{R}^k)$.

3. Let T: $\mathbf{R}^k \to \mathbf{R}^k$ be the homeomorphism defined by

$$T(x_1, \ldots, x_k) = (rx_1, \ldots, rx_k),$$

where $r = (x_1^2 + \ldots + x_k^2)^{\frac{1}{2}}$, and let

$$\mu(B) = m(TB)$$

for any Borel set B. Then μ is a Borel measure which is invariant under rotations about the origin, but μ is not of the form cm (witness the regions $r < 1$ and $2 < r < 3$).

For any two points A, B the translation which takes A to B is expressed as a half-turn about A followed by a half-turn about the mid-point of AB. Thus we have translation invariance in this case and Theorem 1 applies.

§15.3

1. Simply adjoint X to $\mathscr{C}$ and apply Lemma 1. As remarked on p. 168, $\mathscr{A} \cap E$ is a σ-algebra of subsets of E.

§15.4

1. Let $\mathscr{B}$ denote the σ-algebra of Borel sets in $\mathbf{R}$. We are given that f is $\mathscr{B}$-measurable. For any B in $\mathscr{B}$, $f^{-1}(B) \in \mathscr{B}$ and so $g^{-1}(f^{-1}(B)) \in \mathscr{A}$; in other words $f \circ g$ is $\mathscr{A}$-measurable.

No: cf. Ex. 8.4.15 where h is Lebesgue measurable, f is continuous and $h \circ f$ is not Lebesgue measurable. Either $\mathscr{A} = \mathscr{B}$ or $\mathscr{A} = \mathscr{L}$ (the σ-algebra of Lebesgue measurable sets in $\mathbf{R}$) would yield a contradiction.

2. Let E be a Borel set. Then $g^{-1}(E)\in\mathscr{B}$ and $T^{-1}(g^{-1}(E))\in\mathscr{A}$, i.e. $f^{-1}(E)\in\mathscr{A}$. Thus f is $\mathscr{A}$-measurable.

If $g = \chi_B$ ($B\in\mathscr{B}$) then $f = \chi_A$ where $A = T^{-1}(B)$. By our definition of ν,

$$\nu(B) = \mu(A).$$

If either side is finite we have

$$\int g\,d\nu = \int f\,d\mu.$$

The proof now follows the familiar pattern using linearity and the Monotone Convergence Theorem.

3. It follows from the identities

$$T^{-1}(B_1\setminus B_2) = T^{-1}(B_1)\setminus T^{-1}(B_2),$$
$$T^{-1}(\bigcup B_n) = \bigcup(T^{-1}(B_n))$$

that $\mathscr{B}$ is a σ-algebra of subsets of Y, and $\mathscr{B}$ is clearly the *largest* σ-algebra for which T is $(\mathscr{A},\mathscr{B})$-measurable. For any Borel set E,

$$f^{-1}(E) = T^{-1}(g^{-1}(E))\in\mathscr{A} \quad \text{if and only if} \quad g^{-1}(E)\in\mathscr{B}.$$

In this case we may omit the condition in Theorem 1 that g be $\mathscr{B}$-measurable and the reference to f being $\mathscr{A}$-measurable, because the existence of either integral will imply the $\mathscr{A}$-measurability of f and the $\mathscr{B}$-measurability of g.

4. Let $\mathscr{A}$ be the σ-algebra of all subsets of $\mathbf{R}$ and $\mathscr{B}$ the σ-algebra of Lebesgue measurable subsets of $\mathbf{R}$. Then the identity transformation on $\mathbf{R}$ is $(\mathscr{A},\mathscr{B})$-measurable but not $(\mathscr{B},\mathscr{A})$-measurable.

§15.5

1. (i) Any step function on $\mathbf{R}^k$ is the limit of a sequence of continuous functions (of compact support). Conversely, any continuous function on $\mathbf{R}^k$ is the limit of a sequence of step functions. Thus M coincides with the class of Baire functions.

(ii) Any continuous function on $\mathbf{R}^k$ is the limit of a sequence of continuous functions of compact support. Here again M is the class of Baire functions.

§15.6

1. One half is easy: given f, we take $V = \{x\in X: f(x) < \frac{1}{2}\}$ and

$$W = \{x\in X: f(x) > \tfrac{1}{2}\}.$$

The other part of the proof is almost the same as the proof of Theorem 1 except that there is now no mention of *compactness*, as Lemma 2 is replaced by the condition of normality in the following form: given a closed set A and an open set U containing A, there exists an open set V such that

$$A\subset V\subset \overline{V}\subset U.$$

This is immediate if we take $B = U^c$ and find disjoint open sets V and W so that

$$A \subset V \subset W^c \subset U.$$

2. The closed subsets of a compact Hausdorff space are precisely the same as the compact subsets (cf. Appendix to Volume 1, Ex. 23). Lemma 2 now gives the normality if we take $U = B^c$ and $W = \bar{V}^c$.

3. The verification of the topological space axioms for $(X^*, \mathscr{T}^*)$ is straightforward, using the following facts: any compact subset of a Hausdorff space is closed; the union of any two compact subsets of X is compact; the intersection of any (non-empty) collection of compact subsets of X is compact. In verifying the Hausdorff property for X^*, a point x of X may be separated from ω by choosing an open subset G_x containing x whose closure is compact.

The classical example of stereographic projection helps to motivate this construction. A sphere S in $\mathbf{R}^3$ is compact. Let V be a point of S. Then stereographic projection with vertex V gives a homeomorphism of $S \setminus \{V\}$ onto the plane $\mathbf{R}^2$ (Fig. S. 10). An open subset of $S \setminus \{V\}$ corresponds to an open subset of $\mathbf{R}^2$; also an open subset of S containing V has closed complement in the compact set S and corresponds to the complement of a compact subset of $\mathbf{R}^2$. The locally compact space $S \setminus \{V\}$ is 'compactified' by replacing the missing point V. If we carry over the topological ideas from S to $\mathbf{R}^2$ in this way and make V correspond to a fictitious 'point at infinity' ω, we produce a compact space $\mathbf{R}^2 \cup \{\omega\}$ homeomorphic to S.

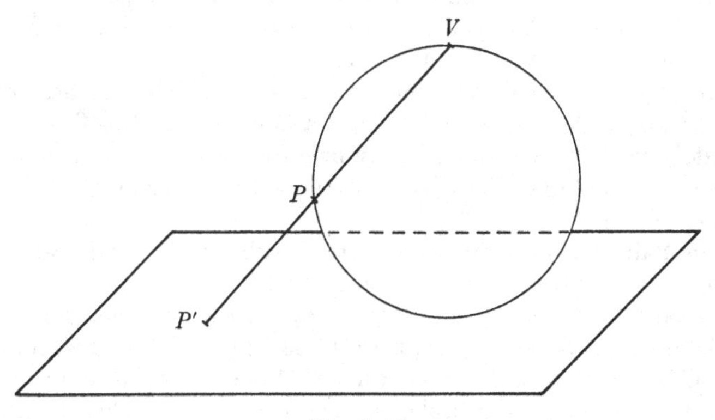

Fig. S. 10

4. Let X^* be the one-point compactification of the locally compact Hausdorff space X as constructed in Ex. 3. As X^* is normal (Ex. 2) we may apply the general form of Urysohn's Lemma (Ex. 1) to the disjoint closed subsets $A \cup \{\omega\}$ and B of X^*.

If we regard the map in Fig. 20, p. 177 as part of the plane $\mathbf{R}^2$, then the sea A must be regarded as 'infinite'. But we may also regard the map as part of the (compact) spherical surface of the world, in which case A and B are both compact!

5. Let $G_n = \{x \in X : f_n(x) > \frac{1}{2}\}$. Then G_n is open and $\bigcap G_n = A$, whence A is a G_δ. Let $F_n = \{x \in X : f_n(x) \geqslant \frac{1}{2}\}$. Then F_n is closed and $\bigcap F_n = A$, whence A is closed.

6. It is clear that $\{\phi_n\}$ is a decreasing sequence of functions in $C_c(X)$ and that $\phi_n(x) \to 1$ if $x \in Q$. Suppose that $x \notin Q$, then $x \notin G_N$ for some N (depending on x) and $\phi_N(x) = 0$.

7. Let $A = U^c$, $B = K$ and let ψ be the continuous function of compact support provided by Theorem 1. Then

$$V = \{x \in X : \psi(x) > \tfrac{1}{2}\} \quad \text{and} \quad Q = \{x \in X : \psi(x) \geqslant \tfrac{1}{2}\}$$

satisfy $K \subset V \subset Q \subset U$. By our definition on p. 180 Q is a Baire set. Also V is open and Q is a closed subset of the support of ψ and therefore compact. As

$$V = \bigcup_n \{x \in X : \psi(x) \geqslant \tfrac{1}{2} + 1/n\} \quad \text{and} \quad Q = \bigcap_n \{x \in X : \psi(x) > \tfrac{1}{2} - 1/n\},$$

V is a σ-compact Baire set and Q is a G_δ.

8. Let K be a compact subset of the separable Hausdorff space X. To any point p of K^c there exist disjoint open subsets G_p, H_p such that $p \in G_p$ and $K \subset H_p$ (Lemma 1). As X is separable, we also have $p \in G_{N(p)} \subset G_p$ for some integer $N(p)$. It follows that $K = \bigcap H_{N(p)}$ and so K is a G_δ.

9. Note that the discrete space X is locally compact as we may take $G_x = \bar{G}_x = \{x\}$ in the definition given on p. 56. The compact subsets of X are the finite subsets. An open subset of X^* is either a subset of X or a set $A \cup \{\omega\}$, where $A \subset X$ and $X \setminus A$ is finite.

(i) An element of $C_c(X)$ is a function $f: X \to \mathbf{R}$ which vanishes outside a finite subset of X. An element of $C(X^*)$ is a function $g: X^* \to \mathbf{R}$ which takes only finitely many values outside any neighbourhood $(g(\omega) - \epsilon, g(\omega) + \epsilon)$ of $g(\omega)$. [An element f of $C_c(X)$ may be extended to an element of $C(X^*)$ by taking $f(\omega) = 0$.]

(ii) The Baire sets in X are the countable subsets of X and their complements in X. Any subset of X is a Borel set in X.

(iii) Let C be a subset of X. Then C is open in X^* and hence is a G_δ; for C to be compact in X^* we also require C to be finite. Let C be a G_δ containing ω. Then C is the intersection of a sequence of open sets containing ω and so the complement of C is countable. In this case C is also a closed, and therefore compact, subset of X^*. In the light of Theorem 2 the Baire sets in X^* consist of all countable subsets of X together with their complements in X^*. Any subset of X^* is a Borel set.

(iv) The set $\{\omega\}$ is compact but is not a Baire set because its complement X is uncountable.

10. Let f be measurable with respect to $\mathscr{S}$; then

$$N(f) = \bigcup_{n=1}^{\infty} (N(f) \cap A_{-n}) \in \mathscr{S}.$$

If we also assume that $X \in \mathscr{S}$, then $X \setminus N(f) \in \mathscr{S}$, i.e. $\{x \in X : f(x) = 0\} \in \mathscr{S}$, and it follows at once that $A_c \in \mathscr{S}$ for all c in **R**.

The adaption of Theorem 3 (and also of Theorem 12.1.1 on which it depends) is straightforward.

11. To show that ν is regular use the proof of Theorem 15.1.3 as in the paragraph following Theorem 4. The uniqueness of ν may be deduced from Ex. 13.2.5. For the last part, adapt the solution to Ex. 13.2.5 by giving X the discrete topology (as in Ex. 9).

12. (i) This follows from the last part of Ex. 11.4.7.

(ii) Here we use Theorem 2 and Proposition 2.

(iii) It is clear that

$$d(x,y) = d(y,x), \quad d(x,y) \geqslant 0 \quad \text{and} \quad d(x,z) \leqslant d(x,y) + d(y,z).$$

We define the metric δ on the quotient space $\tilde{X}$ as follows:

$$\delta(\tilde{x}, \tilde{y}) = d(x,y) \quad (x, y \in X)$$

and verify that this definition does not depend on the representatives x, y chosen for the equivalence classes $\tilde{x}, \tilde{y}$. By this construction

$$\delta(\tilde{x}, \tilde{y}) = 0 \Leftrightarrow \tilde{x} = \tilde{y}.$$

(iv) For each x_0 in X, the mapping $x \to d(x_0, x)$ is continuous and hence the mapping $\phi: x \to \tilde{x}$ of X onto the metric space $\tilde{X}$ is continuous. Thus $\tilde{K} = \phi(K)$ is a compact subset of $\tilde{X}$ and as such is a G_δ (cf. Proposition 2 and the simple form of Urysohn's Lemma for a metric space, Theorem 10.1.2). Now $\tilde{K} = \bigcap H_n$, where $\{H_n\}$ is a sequence of open subsets of $\tilde{X}$. As ϕ is continuous, $G_n = \phi^{-1}(H_n)$ is an open subset of X for each n. It only remains to prove that $K = \bigcap G_n$.

If $x \in Q_n$ and $d(x,y) = 0$ then $y \in Q_n$: thus Q_n is the 'complete' inverse image $\phi^{-1}(\tilde{Q}_n)$. The inverse images of subsets of $\tilde{X}$ form a σ-algebra of subsets of X which contains all the sets Q_n and hence contains K. Thus, finally,

$$K = \phi^{-1}(\tilde{K}) = \phi^{-1}(\bigcap H_n) = \bigcap G_n,$$

as required.

§16.1

1. $\mu_F^+(I)$, $\mu_F^-(I)$ and $|\mu_F|(I)$ are the positive, negative and total variations of F on the interval I as defined in Exx. 3.5.1, 3.

2. See the solution to Ex. 11.4.13. According to Proposition 11.4.2 (iii) and the remarks after the proof, μ is a real measure. Now $\mathscr{R}$ is an *algebra* of subsets of A_1 and μ is unbounded, therefore μ cannot be expressed as the difference of two positive measures on $\mathscr{R}$.

3. As $\quad S \cap T = \varnothing, \quad \lambda_S(E) = \mu(E \cap S) + \nu(E \cap S) = \mu(E)$

and $\quad \lambda_T(E) = \mu(E \cap T) + \nu(E \cap T) = \nu(E).$

4. $T = \bigcup_m \left(\bigcap_{n > m} A_n \right).$

5. Without loss of generality $a_n \neq 0$ for all n. Suppose that Σa_n is convergent but not absolutely convergent. Then Σa_n^+ and Σa_n^- are divergent. Find $N(1)$ so that

$$\sum_1^{N(1)} a_n^+ > 1 + a_1^-,$$

then $N(2) > N(1)$ so that

$$\sum_1^{N(2)} a_n^+ > 2 + a_1^- + a_2^-,$$

and so on. The series

$$a_1^+ + \ldots + a_{N(1)}^+ - a_1^- + a_{N(1)+1}^+ + \ldots + a_{N(2)}^+ - a_2^- + \ldots$$

is a 'rearrangement' $\Sigma a_{\tau(n)}$ which diverges.

6. In the light of Ex. 5, absolute convergence is essential.

7. An atom of μ is a subset of X which contains just one of the points x_n with $p_n \neq 0$. Cf. Ex. 11.4.18.

8. Apply Theorem 2 and Ex. 11.4.19.

9. It is clear that M is a linear space over **R**. We verify only the triangle inequality for the given norm as the other axioms are obviously satisfied. For any A and B in $\mathscr{A}$,

$$\begin{aligned}(\mu+\nu)(A) - (\mu+\nu)(B) &= \mu(A) - \mu(B) + \nu(A) - \nu(B) \\ &\leqslant |\mu|(X) + |\nu|(X)\end{aligned}$$

and so, by equation (3) on p. 187,

$$|\mu+\nu|(X) \leqslant |\mu|(X) + |\nu|(X).$$

Suppose that $\{\mu_n\}$ is a sequence of real measures on $\mathscr{A}$ and $\|\mu_m - \mu_n\| \to 0$ as $m, n \to \infty$. For any A in $\mathscr{A}$

$$|\mu_m(A) - \mu_n(A)| \leqslant |\mu_m - \mu_n|(X) \to 0$$

and so the Cauchy sequence $\{\mu_n(A)\}$ converges to a real number $\mu(A)$, say. It follows at once that μ is additive. We shall prove that μ is σ-additive and that $\|\mu - \mu_n\| \to 0$.

Given any $\epsilon > 0$, we may find k such that

$$\|\mu_m - \mu_n\| < \epsilon \quad \text{for} \quad m, n \geqslant k.$$

In particular this implies that

$$|\mu(A) - \mu_n(A)| \leqslant \epsilon \quad \text{for} \quad n \geqslant k \quad \text{and } A \text{ in } \mathscr{A}.$$

Now let $A_n \uparrow A$ $(A_n \in \mathscr{A})$. As μ_k is σ-additive there exists N such that

$$|\mu_k(A) - \mu_k(A_n)| < \epsilon \quad \text{for} \quad n \geqslant N.$$

Thus

$$|\mu(A)-\mu(A_n)| \leqslant |\mu(A)-\mu_k(A)|+|\mu_k(A)-\mu_k(A_n)|+|\mu_k(A_n)-\mu(A_n)|$$
$$< 3\epsilon \quad \text{for} \quad n \geqslant N.$$

In other words $\mu(A_n) \to \mu(A)$ and so μ is σ-additive.

Referring once again to equation (3) on p. 187, we deduce from

$$|\mu_m-\mu_n|\,(X) < \epsilon \quad \text{for} \quad m, n \geqslant k$$

that

$$|\mu-\mu_n|\,(X) \leqslant \epsilon \quad \text{for} \quad n \geqslant k.$$

In other words $\|\mu-\mu_n\| \to 0$.

§ 16.2

1. As m varies $\sum_{n=1}^{m} |\lambda(A_n)|$ may be taken arbitrarily close to $\sum_{n=1}^{\infty} |\lambda(A_n)|$, so the suprema are the same.

No: $A_0 = \left(E \setminus \bigcup_{n=1}^{\infty} A_n\right) \in \mathscr{S}$, and we may as well consider the sequence $A_0, A_1, A_2, \ldots$.

2. Given a set E in $\mathscr{S}$; consider the restrictions of μ and ν to $\mathscr{S} \cap E$ and apply the Hahn Decomposition Theorem 16.1.2. In this way we may express E as the disjoint union of four sets $E_1, \ldots, E_4$ in $\mathscr{S}$, where the restriction of σ to $\mathscr{S} \cap E_j$ for $j = 1, \ldots, 4$ is of the form $\pm\mu \pm i\nu$. It readily follows that $|\lambda|\,(E_j) = |\sigma|\,(E_j)$ for $j = 1, \ldots, 4$ and hence $|\lambda|\,(E) = |\sigma|\,(E)$.

3. The proof that $M_{\mathbf{C}}$ is a complex Banach space is almost as for Ex. 16.1.9 but we use equation (1) on p. 195 in place of equation (3) on p. 187 (both in proving the triangle inequality and in showing that $\|\lambda-\lambda_n\| \to 0$ at the end).

$\|\ \|'$ is *not* a norm as it fails to satisfy the axiom

$$\|c\lambda\|' = |c|\,\|\lambda\|' \quad (c \in \mathbf{C})$$

e.g. when λ is real and $c = 1+i$.

§ 16.3

2. Suppose that μ and ν are mutually singular. Then there exist disjoint sets S, T in $\mathscr{A}$ such that μ, ν are concentrated on S, T, respectively. Without loss of generality $S \cup T = X$. Any positive element of $L(\mathscr{A})$ is of the form

$$\psi = a_1\chi_{A_1} + \ldots + a_m\chi_{A_m}$$

with disjoint A_i in $\mathscr{A}$ and $a_i \geqslant 0$. Let $B_i = A_i \cap T$, $C_i = A_i \cap S$. Then

$$\theta = a_1\chi_{B_1} + \ldots + a_m\chi_{B_m}, \quad \phi = a_1\chi_{C_1} + \ldots + a_m\chi_{C_m}$$

satisfy $\theta+\phi = \psi$ and $F(\theta) = G(\phi) = 0$, so that F and G are mutually singular.

Conversely, suppose that F and G are mutually singular. To each integer $n \geq 1$ there exist positive elements θ_n, ϕ_n of $L(\mathscr{A})$ such that $\theta_n+\phi_n = 1$, $F(\theta_n) < \frac{1}{2}^n$ and $G(\phi_n) < \frac{1}{2}^n$. Let

$$S = \{x\in X\colon \Sigma\phi_n(x) \text{ is divergent}\}$$

and let $T = X \setminus S$. First of all S and T belong to $\mathscr{A}$ because

$$S = \bigcap_{k=1}^{\infty} \left(\bigcup_{m=1}^{\infty} \left\{x\in X\colon \sum_{n=1}^{m} \phi_n(x) \geq k\right\}\right).$$

As $\Sigma F(\theta_n) < 1$ and $\Sigma G(\phi_n) < 1$, it follows that S is ν-null and T is μ-null. In other words, μ is concentrated on S and ν is concentrated on T.

3. Let μ be a real measure on a σ-algebra $\mathscr{A}$ of subsets of X. According to Theorem 16.1.1 $\mu = \mu^+-\mu^-$, where μ^+, μ^- are the positive measures on $\mathscr{A}$ defined by

$$\mu^+(E) = \sup\{\mu(A)\colon A\in\mathscr{A}, A\subset E\}, \quad \mu^-(E) = \sup\{-\mu(A)\colon A\in\mathscr{A}, A\subset E\}.$$

Let F be the linear functional on $L(\mathscr{A})$ defined by

$$F(\chi_A) = \mu(A) \quad (A\in\mathscr{A})$$

and let F^+, F^- be similarly associated with μ^+, μ^-. Referring to the formulae (1) and (2) on p. 200 we see that F^+, F^- are the positive and negative parts of F. As such they are mutually singular (Ex. 10.2.4), whence μ^+, μ^- are mutually singular by Ex. 2.

4. Let us spell out the construction on p. 202. We begin with a relatively bounded linear functional F on L satisfying the Daniell condition, and express F as $P-N$, where P and N are the positive and negative parts of F. Now P, N and $T = P+N$ are elementary integrals on L and so extend by the Daniell construction to Daniell integrals P^1, N^1, T^1 on spaces $L^1(P)$, $L^1(N)$, $L^1(T)$, respectively. We have seen in Ex. 1 that

$$L^1(T) = L^1(P) \cap L^1(N).$$

Let us agree to define the linear functional F^1 on $L^1(T)$ by the equation

$$F^1(f) = P^1(f) - N^1(f)$$

for all f in $L^1(T)$. We are now asked to prove that the restrictions of P^1 and N^1 to $L^1(T)$ are the positive and negative parts of F^1.

For the moment denote the positive and negative parts of F^1 by G^1, H^1, respectively and let G, H be their restrictions to L. Using the 'best possible' property of Ex. 10.2.2 (both for $L^1(T)$ and L) we deduce that $P = G$ and $N = H$. For any f in $L^{\text{inc}}(T)$, there is an increasing sequence $\{\phi_n\}$ of elementary functions such that $\phi_n \uparrow f$ almost everywhere with respect to T. Thus

$$P(\phi_n)\uparrow P^1(f) \quad\text{and}\quad G(\phi_n)\uparrow G^1(f)$$

(the latter by the Daniell condition for G^1). It follows by linearity that $P^1(f) = G^1(f)$ and $N^1(f) = H^1(f)$ for all f in $L^1(T)$.

5. (i) Let $$\zeta_1 = \frac{\zeta\psi_1}{\psi_1+\psi_2}, \quad \zeta_2 = \frac{\zeta\psi_2}{\psi_1+\psi_2}.$$

Express each function ζ, ψ_1 and ψ_2 in terms of disjoint sets and then use Proposition 11.1.1 to express them simultaneously in terms of disjoint sets. It is then clear that ζ_1 and ζ_2 belong to $L_{\mathbf{C}}$.

(ii) As on p. 74 we only need to verify the truncation property

$$\zeta \in L_{\mathbf{C}} \Rightarrow \operatorname{tr}_{\psi} \zeta \in L_{\mathbf{C}}$$

for any positive function ψ in $L_{\mathbf{R}}$. This follows at once if we express ζ and ψ simultaneously in terms of disjoint sets.

§17.1

1. $L^p(\mu)$ is the space of all sequences $\{x_n\}$ of real numbers for which $\Sigma |x_n|^p$ is convergent. This space is usually denoted by l^p (Ex. 7.7.1). As $\varnothing$ is the only μ-null set, $L^p(\mu)$ and $\mathscr{L}^p(\mu)$ in this case may safely be identified.

2. If $m < n$, then $\|f_m - f_n\| = 2(1-m/n)$ and so $\{f_n\}$ does not satisfy the Cauchy condition.

3. $\|f_n\| = 2^{-k/p} \to 0$ as $n \to \infty$. For any x in $[0,1]$, $\{f_n(x)\}$ alternates between 0 and 1.

4. See The Projection Theorem 7.3.3 and Ex. 7.3.7.

5. See Ex. 7.4.2.

§17.2

1. $\chi_E(1+g+\ldots+g^{n-1}) \in L^1(\lambda)$ and is bounded, so belongs to $L^2(\lambda)$.

2. Condition (i) is obvious as $\lambda = \nu_1 - \nu_1'$.
As $\nu_2 \perp \mu$ there exist disjoint sets A, B in $\mathscr{A}$ such that μ, ν_2 are concentrated on A, B, respectively. Similarly μ, ν_2' are concentrated on disjoint sets A', B' in $\mathscr{A}$. We may take

$$S = A \cap A' \quad \text{and} \quad T = B \cup B'.$$

Finally, for any E in $\mathscr{A}$, $\lambda(E) = \lambda(E \cap T) = 0$ because $\mu(E \cap T) = 0$.

5. The only set A for which $\mu(A) = 0$ is the empty set: thus $\nu \ll \mu$.
Suppose that such a function h exists. As $h \in L^1(\mu)$, h must vanish outside a countable subset S of X and then

$$\nu(A) = \int_{A \cap S} h \, d\mu = \nu(A \cap S) = 0$$

for all A in $\mathscr{A}$.

6. As for Ex. 5.

7. Let $g = d\mu/d\lambda$ to within a λ-null function. Then

$$\int_E \chi_A d\mu = \int_E \chi_A g d\lambda \quad \text{for all } A, E \text{ in } \mathscr{A}$$

because each side equals $\mu(E \cap A)$. By linearity we may replace χ_A in this equation by an arbitrary $\mathscr{A}$-simple function ϕ. If we express h as the limit (everywhere) of an increasing sequence of such functions ϕ, the Monotone Convergence Theorem gives the required result.

8. Take $\lambda = \nu$ in Ex. 7. In this case $\mu(A) = 0 \Leftrightarrow \nu(A) = 0$.

9. (i) Let $\sigma = |\mu| + |\nu|$. Then $|\lambda(A)| \leqslant \sigma(A)$ for all A in $\mathscr{A}$ and so there exists a complex valued function h in $L_{\mathbf{C}}^1(\sigma)$ such that

$$\lambda(A) = \int_A h d\sigma \quad (A \in \mathscr{A}). \tag{1}$$

As $|h| \in L_{\mathbf{C}}^1(\sigma)$, we may define

$$\tau(A) = \int_A |h| d\sigma \quad (A \in \mathscr{A}). \tag{2}$$

Clearly $\tau \ll \sigma$. But we also deduce from (1) and (2) that $|\mu| \leqslant \tau$ and $|\nu| \leqslant \tau$, whence $\sigma \ll \tau$. According to Exx. 7 and 8 (which extend at once to complex measures) we then have

$$\lambda(A) = \int_A \zeta d\tau,$$

where $\zeta = h/|h|$ and $|\zeta| = 1$ almost everywhere with respect to τ (or σ).

(ii) and (iii) follow by applying Proposition 16.2.1 and the uniqueness part of the Radon–Nikodym Theorem.

10. We may as well discard any μ_n for which $\mu_n(X) = 0$ and then define

$$\mu(A) = \sum_{n=1}^{\infty} \frac{\mu_n(A)}{2^n \mu_n(X)} \quad \text{for all } A \text{ in } \mathscr{A}.$$

The σ-additivity of μ follows from a standard theorem about double series (Ex. 14.2.1).

11. $$\mu(E) = 0 \Rightarrow E = \varnothing \Rightarrow \nu(E) = 0,$$

but $\nu(E) \geqslant 2$ for any $E \neq \varnothing$.

§17.3

1. Let μ, ν be finite measures on the σ-algebra $\mathscr{A}$ of subsets of X. By the standard construction of Theorem 11.4.2 we obtain Daniell integrals P, Q whose corresponding measures are the completions $\tilde{\mu}$, $\tilde{\nu}$ of μ, ν, respectively (Theorem 13.4.2).

Now suppose that $\nu \ll \mu$. Then $\tilde{\mu}(E) = 0 \Rightarrow \mu(B) = 0$ (for some $B \in \mathscr{A}$, $B \supset E$) $\Rightarrow \nu(B) = 0 \Rightarrow \tilde{\nu}(E) = 0$. Thus $\tilde{\nu} \ll \tilde{\mu}$. According to Stone's Theorem 8.5.1 we may apply Theorem 2 and deduce the existence of a function h in $L^1(P) = L^1(\tilde{\mu})$ such that

$$\int \phi \, d\tilde{\nu} = \int \phi h \, d\tilde{\mu} \tag{1}$$

for all ϕ in $L(\mathscr{A})$. We may express h as the limit (everywhere) of an increasing sequence of $\tilde{\mathscr{A}}$-simple functions and so (by the definition of $\tilde{\mathscr{A}}$ on p. 117) we may replace h in (1) by an element of $L^1(\mu)$ which equals h outside a $\tilde{\mu}$-null set. The required result now follows by taking $\phi = \chi_E$ in (1).

2. Let f be a positive element of $L^1(\nu)$. There exists an increasing sequence $\{\phi_n\}$ of $\mathscr{A}$-simple functions converging to f everywhere: apply the Monotone Convergence Theorem. For an arbitrary f in $L^1(\nu)$ use the decomposition $f = f^+ - f^-$.

Let fh be a positive element of $L^1(\mu)$. Consider first the case where $\nu(X)$ is finite. Let $f_0 = (fh)\,1/h$, i.e. with our usual convention about vanishing denominators,

$$\begin{aligned} f_0(x) &= f(x) \quad \text{if} \quad h(x) \neq 0, \\ &= 0 \qquad\; \text{if} \quad h(x) = 0. \end{aligned}$$

As $1 \in L^1(\nu)$ it follows from the first part that $h \in L^1(\mu)$ and so h is μ-measurable. Thus f_0 is μ-measurable and there is an increasing sequence $\{\phi_n\}$ of $\mathscr{A}$-simple functions converging to f_0 everywhere. By the Monotone Convergence Theorem we have

$$\int f_0 \, d\nu = \int f_0 h \, d\mu = \int fh \, d\mu.$$

In particular $f = \chi_X$ gives $f_0 = \chi_S$, where

$$S = \{x \in X : h(x) \neq 0\}$$

and we deduce that $\nu(S) = \nu(X)$. Thus $f = f_0$ almost everywhere with respect to ν and

$$\int f \, d\nu = \int fh \, d\mu.$$

The extension to the σ-finite case follows the standard pattern.

3. In part (iii) of the proof we used the stronger Stone condition to deduce that

$$f_n = n(\min\{\phi, c\} - \min\{\phi, c - 1/n\})$$

belongs to L. Assuming that 1 is Q-measurable, we at least know that $f_n \in L^1(Q)$, so we may use part (ii) and apply the Monotone Convergence Theorem as before to deduce equation (6).

Also in part (iii) we need the P-measurability of 1 to deduce the μ-measurability of the elementary set E_n^{ϵ}.

In part (iv) we produced a sequence $\{\phi_n\}$ of elements of L suitably truncated by K. By the weaker Stone condition for Q, $\phi_n \in L^1(Q)$ and we may use part (ii) again in the application of the Dominated Convergence Theorem.

4. See the solution to Ex. 6.4.5.

5. In part (iv) where we introduce $R = P+Q$, we have implicitly used the stronger Stone condition for L as there was no need to check the weaker Stone condition for R! In the more general case we refer to Ex. 13.3.5.

§17.4

1. The given proofs apply almost verbatim.

2. Write $\operatorname{ess\,sup} g = K$. To each $n \geqslant 1$ there is a function f_n in $\mathcal{J}$ such that $f_n < K+1/n$. In other words $g(x) < K+1/n$ for all x outside a null set S_n. Thus $g(x) \leqslant K$ for all x outside the null set $\bigcup S_n$. This shows that K is an essential upper bound. The rest is clear.

3. Using Ex. 2, $|g| \leqslant \|g\|_\infty$ almost everywhere.

4. The only null set is $\varnothing$. $f \in L^1(\mu)$ if and only if $f(b) = 0$; thus $L^1(\mu)$ and $L^1(\mu)^*$ are both one-dimensional.

All $f: X \to \mathbf{R}$ belong to $L^\infty(\mu)$ and $\|f\|_\infty = \max\{|f(a)|, |f(b)|\}$. Thus $L^\infty(\mu)$ is like $\mathbf{R}^2$ with $\|x\| = \max\{|x_1|, |x_2|\}$. It is easy to verify that $L^\infty(\mu)^*$ is like $\mathbf{R}^2$ with $\|x\| = |x_1|+|x_2|$.

5. We cannot prove that the set E is null.

6. The inequality $\|F_h\| \leqslant \|h\|_\infty$ follows from Hölder's Inequality as in the proof of Theorem 2.

Let us assume the finite subset property. As before write $M = \|F_h\|$. We are given that

$$\left|\int fh\,d\mu\right| \leqslant M\,\|f\|$$

for all f in $L^1(\mu)$. Let $\qquad E = \{x \in X: |h(x)| > M\}.$

If $\mu(E) < \infty$, then $\chi_E \in L^1(\mu)$, and the argument of Theorem 1 shows that $\mu(E) = 0$. On the other hand, we cannot have $\mu(E) = \infty$; otherwise we could find $A \subset E$ such that $0 < \mu(A) < \infty$ and the same argument would give $\mu(A) = 0$.

Suppose that there is a set E such that $\mu(E) = \infty$ and $\mu(A) = 0$ for all measurable subsets A of E of finite measure. Then $\mu(A) = 0$ for all measurable subsets A of E of σ-finite measure. Take $h = \chi_E$. In this case $\|h\|_\infty = 1$ and $\int fh\,d\mu = 0$ for all f in $L^1(\mu)$ (because f vanishes outside a subset of σ-finite measure: cf. p. 148). Thus $\|h\|_\infty = 1$ and $\|F_h\| = 0$.

Clearly the measure μ of Ex. 4 does not satisfy the finite subset property.

7. As usual, by Hölder's Inequality $\|F_h\| \leqslant \|h\|_1$. To deduce the reverse inequality take $f = \operatorname{sgn} h$.

8. If $F(f) = \int fh$ then the Daniell condition follows from the Monotone Convergence Theorem applied to h^+ and h^-. Conversely, if the Daniell condition holds we may apply the argument of Theorem 1.

9. (i) The functions of L^q which vanish outside E form another L^q space·

(ii) The function h_E of (i) is unique to within a null function.

(iii) Consider $F(|h_E|^{q-1}\operatorname{sgn} h_E)$ as in the proof of Theorem 2.

(iv) For each $n \geqslant 1$ find a σ-finite measurable subset A_n such that $\|h_{A_n}\| > M - 1/n$ and let $E_n = A_1 \cup \ldots \cup A_n$. Then $S = \bigcup E_n$ is σ-finite and, using (ii), $\|h_{E_n}\| \uparrow \|h_S\| = M$.

(v) By (ii) $h_E = h_S$ almost everywhere on S. But $\|h_E\| = \|h_S\|$ and so $h_E = 0$ almost everywhere on $E \setminus S$. As h_E, h_S both vanish outside E, we have $h_E = h_S$ almost everywhere on X.

(vi) To any f in L^p, there is a σ-finite measurable subset N outside which f is zero. Take $E = N \cup S$ in (v) and deduce that

$$F(f) = \int fh_E = \int fh.$$

(vii) Use Hölder's Inequality and (iii).

Exercises 10–15 *are virtually the same as Exercises* 7.2.5, 7.3.4–6, 7.4.7, 10 *in Volume* 1.

The Hahn–Banach Theorem

A. (i)
$$F(x_2-x_1) \leqslant \|x_2-x_1\| = \|(x_2+y)-(x_1+y)\|$$
$$\Rightarrow \; F(x_2)-F(x_1) \leqslant \|x_2+y\| + \|x_1+y\|$$
$$\Rightarrow \; -\|x_1+y\| - F(x_1) \leqslant \|x_2+y\| - F(x_2)$$
$$\Rightarrow \; \sup_{x \in M}\{-\|x+y\| - F(x)\} \leqslant \inf_{x \in M}\{\|x+y\| - F(x)\}.$$

We may take for r any real number between these bounds.

G is a linear functional extending F to N. Suppose that $a \neq 0$; the constructed inequalities are equivalent to

$$-\|a^{-1}x+y\| \leqslant F(a^{-1}x)+r \leqslant \|a^{-1}x+y\|,$$

i.e.
$$|G(a^{-1}x+y)| \leqslant \|a^{-1}x+y\|$$

or
$$|G(x+ay)| \leqslant \|x+ay\|.$$

The last inequality also holds for $a = 0$ and so $\|G\| \leqslant 1$. As G extends F we must have $\|G\| \geqslant 1$.

(ii) To any linearly ordered family $\{G_\alpha\}$ of elements of $\mathscr{C}$ we may find a supremum in $\mathscr{C}$, viz. the linear functional G whose domain of definition is $\bigcup_\alpha M_{G_\alpha}$ and where $G(x) = G_\alpha(x)$ for any α such that $x \in M_{G_\alpha}$. Using the linear ordering we check that this definition of $G(x)$ is unambiguous and that G is indeed an element of $\mathscr{C}$. The conditions of Zorn's Lemma are therefore satisfied and we conclude that there is a maximal element T of $\mathscr{C}$.

(iii) and (iv) are now immediate.

B. Define $F(\lambda b) = \lambda \|b\|$ $(\lambda \in \mathbf{R})$ and extend F by the Hahn–Banach Theorem.

C. It is clear that ϕ_x is linear, and then that ϕ is linear. From

$$|\phi_x(F)| \leqslant \|F\|\,\|x\| \quad \text{we get} \quad \|\phi_x\| \leqslant \|x\|.$$

For any b in B consider the linear functional T of Ex. B: then

$$\phi_b(T) = T(b) = \|b\| \quad \text{and} \quad \|T\| = 1,$$

whence $\|\phi_b\| \geqslant \|b\|$.

D. Let F be an element of $(\mathscr{L}^p)^{**}$, i.e. let F be a bounded linear functional on $(\mathscr{L}^p)^* = \theta\mathscr{L}^q$ (Theorem 2). Then $F \circ \theta$ is a bounded linear functional on $\mathscr{L}^q$ and so (Theorem 2) there is an element $\tilde{f}$ of $\mathscr{L}^p$ such that

$$F(\theta(\tilde{h})) = \int fh \quad \text{for all } \tilde{h} \text{ in } \mathscr{L}^q.$$

Now $$\phi_{\tilde{f}}(\theta(\tilde{h})) = (\theta(\tilde{h}))\,(\tilde{f}) = \int fh \quad \text{for all } \tilde{h} \text{ in } \mathscr{L}^q.$$

Thus $$\phi_{\tilde{f}} \circ \theta = F \circ \theta \quad \text{and} \quad \phi_{\tilde{f}} = F.$$

E. $\|\{s_n\}\| = \sup |s_n|$. As $|\lim s_n| \leqslant \sup |s_n|$, $\|F\| \leqslant 1$, and the sequence with all $s_n = 1$ shows that $\|F\| \geqslant 1$. Thus $\|F\| = 1$ and we may extend F to a bounded linear functional T on l^∞ with $\|T\| = 1$. Let ϕ_n be the sequence with $\phi_n(i) = 0$ for $i \leqslant n$ and $\phi_n(i) = 1$ for $i > n$. Then

$$\phi_n \downarrow 0 \quad \text{and} \quad T(\phi_n) = 1 \quad \text{for all } n.$$

We deduce from Ex. 8 that not all linear functions F on l^∞ are given by

$$F(\{s_n\}) = \Sigma a_n s_n,$$

where the series Σa_n is absolutely convergent.

F. Suppose that X is the union of disjoint sets X_n, where $0 < \mu(X_n) < \infty$ for $n = 1, 2, \ldots$. Let M be the linear space consisting of all f in $L^\infty(\mu)$ which are constant on each X_n and for which the sequence $\{f(X_n)\}$ is convergent, and define

$$F(f) = \lim_{n \to \infty} f(X_n).$$

We may extend F to an element T of $L^\infty(\mu)^*$. Let

$$\begin{aligned} \phi_n(x) &= 0 \quad \text{if} \quad x \in \bigcup_{i=1}^{n} X_i, \\ &= 1 \quad \text{otherwise.} \end{aligned}$$

Then $\phi_n \downarrow 0$ and $T(\phi_n) = 1$ for all n. As the Daniell condition is not satisfied, the natural mapping θ of $L^1(\mu)$ into $L^\infty(\mu)$ (Ex. 7) is certainly not *onto*.

G. Two continuous functions are equal *almost everywhere* on $[-1,1]$ if and only if they are equal *everywhere* on $[-1,1]$. As C is a linear subspace of $L^\infty(\mu)$ we may regard C as a linear subspace of $\mathscr{L}^\infty(\mu)$.

Extend F to T by the Hahn–Banach Theorem. Now choose for ϕ_n a function in C satisfying $\phi_n(0) = 1$ and $\phi_n(x) \downarrow 0$ otherwise (e.g.

$$\phi_n(x) = (1-|x|)^n$$

for $|x| \leqslant 1$). Then $\phi_n \downarrow 0$ (regarded as a sequence in $\mathscr{L}^\infty(\mu)$) and $T(\phi_n) = 1$.

The Daniell construction provides an *unbounded* linear functional on $\mathscr{L}^\infty$.

REFERENCES

[1] Asplund, E. and Bungart, L. *A First Course in Integration.* New York, Holt, Reinhart and Winston, 1966.

[2] Baire, R. Sur les fonctions des variables réelles. *Ann. Mat. Pura Appl.* (3) **3** (1899), 1–122.

[3] Berberian, S. K. *Measure and Integration.* New York, Macmillan, 1965.

[4] Bourbaki, N. *Intégration,* Ch. II, Espaces de Riesz. *Act. Sci. et Ind.* 1175, Hermann, 1965.

[5] Carathéodory, C. *Vorlesungen über Reelle Funktionen.* New York, Chelsea, 3rd Ed. 1968 (First Ed., Leipzig, 1918).

[6] Clarkson, J. A. Uniformly convex spaces. *Trans. Amer. Math. Soc.* **40** (1936), 396–414.

[7] Daniell, P. J. A general form of integral. *Ann. Math.* (2) **19** (1917/18).

[8] Goffman, C. *Real Functions,* Volume 8 of the Complementary Series in Mathematics, Boston, Prindle, Weber and Schmidt, 1953.

[9] Hahn, H. and Rosenthal, A. *Set Functions.* University of New Mexico Press, Albuquerque, 1948.

[10] Halmos, P. R. *Lectures on Boolean Algebras.* Princeton, Van Nostrand, 1963.

[11] Halmos, P. R. *Measure Theory.* Princeton, Van Nostrand, 1950.

[12] Kingman, J. F. C. and Taylor, S. J. *Introduction to Measure and Probability.* Cambridge University Press, 1966.

[13] Lang, S. *Introduction to Algebraic Geometry.* New York, Interscience, 1958.

[14] Loomis, L. H. and Sternberg, S. *Advanced Calculus.* New York, Addison-Wesley, 1968.

[15] Nathanson, I. P. *Theory of Functions of a Real Variable,* Volume 2, Translated by L. F. Boron. New York, Frederick Ungar, 1961.

[16] Riesz, F. Sur les opérations fonctionelles linéaires. *Compt. Rend. Acad. Sci. Paris.* **149** (1909), 974–7.

[17] Riesz, F. and Sz.-Nagy, B. *Functional Analysis,* translated by L. F. Boron. New York, Frederick Ungar, 1955.

[18] Royden, H. L. *Real Analysis.* New York, Macmillan, 1968.

[19] Rudin, W. *Real and Complex Analysis.* New York, McGraw-Hill, 1966.

[20] Shilov, G. E. and Gurevich, B. L. *Integral, Measure and Derivative: A Unified Approach.* Translated by R. A. Silverman. Englewood Cliffs, Prentice-Hall, 1966.

[21] Stone, M. H. Notes on Integration I–IV. *Proc. Natl. Acad. Sci. U.S.* **34** (1948), 336–42, 447–55, 483–90; **35** (1949), 50–8.

[22] Zaanen, A. C. *An Introduction to the Theory of Integration.* New York, Interscience, 1958.

INDEX